Synthetic Chemistry of Stable Nitroxides

Leonid B. Volodarsky, Ph.D.
Professor of Chemistry
Laboratory of Nitrogen Compounds
Institute of Organic Chemistry
Novosibirsk, Russia

Vladimir A. Reznikov, Ph.D.
Laboratory of the Chemistry of Stable Nitroxides
International Tomographic Center
Novosibirsk, Russia

Victor I. Ovcharenko, Ph.D.
Laboratory of Polyspin Coordination Compounds
International Tomographic Center
Novosibirsk, Russia

CRC Press
Taylor & Francis Group
Boca Raton London New York

CRC Press is an imprint of the
Taylor & Francis Group, an **informa** business

PREFACE

Stable nitroxide radicals (nitroxides) have attracted the attention of researchers for three decades due to their paramagnetism, which is a unique property in organic chemistry. They have a characteristic EPR spectrum that consistently changes depending on the environment. Due to this, the highly sensitive and informative EPR method was extended to molecular biology, biophysics, polymer chemistry, analytical chemistry, and other fields. On the other hand, the development of nitroxide chemistry in combination with coordination chemistry leads to compounds with new electrophysical properties. Synthesis of stable nitroxides with various functional groups depends first of all on their utility in scientific and applied research using EPR spectroscopy. The development of synthetic nitroxide chemistry in turn leads to the emergence of new directions of their application. A typical example is creation of pH-sensitive spin labels and probes. Intensive studies are underway on the application of nitroxides in NMR and EPR tomography, for oxygen determination, for investigating conformational transfer processes, protein coagulation, etc. These problems require the constantly growing synthetic provision. It is also worthwhile to note the growing use of oxoammonium salts derived from piperidine and pyrrolidine nitroxides as oxidizers in different systems, including catalytic ones.

The aim of this book is to illustrate the possibilities and limitations of organic chemistry in synthesis of nitroxides. The book is a manual for the application of known reactions and methods in the routine work of a researcher to aid in choosing a desired structure of radical and synthetic procedure, rather than being a summary of all known facts in synthetic chemistry of nitroxides. One can expect that the book will not only promote expansion of the use of nitroxides, but also lead to new fields of their application. This volume on the whole illustrates the ability of synthetic organic chemistry to meet the requirements of specialists in various fields who are interested in using nitroxides.

On behalf of authors of the book, I would like to thank Dr. V. V. Martin for participating in the preparation of Chapter 1, Introduction to Nitroxide Chemistry, and for adding helpful comments to the outline of the manuscript at the early stage of its preparation.

The authors also thank Mrs. L. G. Smolina and Mrs. N. N. Voronova for their help in preparing the manuscript.

Leonid B. Volodarsky
Novosibirsk, Russia

THE AUTHORS

Leonid B. Volodarsky, Ph.D., is the Head of the Laboratory of Nitrogen Compounds at the Novosibirsk Institute of Organic Chemistry, Siberian Division of the Russian Academy of Sciences, and a Professor of Chemistry at Novosibirsk State University. Dr. Volodarsky received his B.S. and M.S. in Chemistry from the Leningrad Technological Institute in 1954. During the next five years he was employed in industry. In 1959 he joined the Novosibirsk Institute of Organic Chemistry, Siberian Division of the Academy of Sciences of the U.S.S.R., as a postgraduate student, and in 1963 he received his Ph.D. from this institute. Dr. Volodarsky prepared a thesis on organic hydroxylamine derivatives and received a Doctor of Chemical Science degree (a senior doctoral degree in the U.S.S.R.) from the same institute in 1972. At this time he joined Novosibirsk State University as an Associate Professor. He became a Full Professor in 1976 and Head of the laboratory in 1980.

His research interests are in the areas of organic hydroxylamine derivatives, heterocyclic N-oxides and nitrones, and stable nitroxide radicals. His associates are physical chemists, biophysicists, pharmocologists, and inorganic, analytic, and organic chemists, including both researchers and practioners. His publications include more than 280 articles (25 in Western Europe and the U.S.). He is an editor and a co-author of the two-volume book *Imidazoline Nitroxides*, CRC Press (1988). He holds 70 Russian patents and a few foreign patents. He has presented lectures at the ECHEM Conference on Stable Free Radicals, Finland, August 1981; at the FECHEM Conference on Heterocycles in Bioorganic Chemistry, the Netherlands, May 1982; at the International Conference on Nitroxide Radicals, Novosibirsk, Russia, September 1989; and at the 13th International Congress of Heterocyclic Chemistry, Oregon, U.S., August 1991. He was the Chairman of the Conference on Nitroxide Radicals in Novosibirsk and Vice Chairman of the All-Union Conference on Nitrogen Heterocycles, Novosibirsk, Russia, September 1987.

Vladimir A. Reznikov, Ph.D., was born in 1955. In 1977 he graduated from Novosibirsk State University and was appointed as researcher in the Novosibirsk Institute of Organic Chemistry. He earned his Ph.D. in Chemistry from the same institute in 1982 with a thesis on the chemistry of stable nitroxides of 3-imidazolines. Dr. Reznikov prepared a thesis on synthetic chemistry of imidazoline, imidazolidine, and pyrrolidine nitroxides and received a Doctor of Chemical Science degree (a senior doctoral degree in Russia) from the same institute in 1992. At present he is a Head of Laboratory of the Department of Chemistry of Stable Nitroxides at the International Tomographic Center in Novosibirsk. His main research interests are in the field of heterocyclic chemistry and chemistry of stable nitroxides. His associates are biochemists and biophysicists, and physical, inorganic, and analytic chemists.

Victor I. Ovcharenko, Ph.D., was born in 1952. In 1974 he graduated from Novosibirsk State University and was appointed as a researcher in the Institute of Inorganic Chemistry. He earned his Ph.D. in Chemistry from the same institute in 1979 with a thesis on the chemistry of transition metal complexes with nitroxides. Dr. Ovcharenko prepared his thesis on synthetic chemistry of the transition metal complexes with 3-imidazoline nitroxides, magneto-structural correlations intrinsic for these compounds, and molecular design of 2- and 3-dimensional heterospin compounds, which are able to magnetic phase transition in ferromagnetic state. He received a Doctor of Chemical Science degree (a senior doctoral degree in Russia) from the Institute of Inorganic Chemistry in 1992. In 1992 he became a Professor of General and Inorganic Chemistry at the University of Novosibirsk. At present he is a Head of Laboratory of Polyspin Coordination Compounds at the International Tomographic Center in Novosibirsk. His main research interests are the coordination chemistry of free radicals and magnetic materials.

TABLE OF CONTENTS

Chapter 1

INTRODUCTION TO NITROXIDE CHEMISTRY

I. THE SUBJECT OF NITROXIDE CHEMISTRY

One of the basic tendencies in the development of organic chemistry is the growing interest of researchers in particles with unsaturated valence, the so-called free radicals. These former putative reactive intermediates containing an unpaired electron have now become important objects of study.[1] Progress in modern techniques of free radical generation and investigation clarified their significant impact on many chemical transformations, including major processes of industrial chemistry and energetics.[2] Free radicals also proved to play an important role in living nature, particularly in breathing, photosynthesis, and carcinogenesis.[3-5]

Free radical chemistry might now be considered as a developed field of chemistry studying the generation of free radical particles, their structure, and physical (in particular, spectral) properties as well as free radical conversions in chemical reactions and biological systems. Progress in this area during recent years resembles, according to one of its founders, W. A. Walters, that of a branched chain reaction.[6]

Perhaps one of the most meaningful events in the history of this area is the discovery of miscellaneous long-lived radicals, which transformed free radicals from reactive intermediates only postulated or registered under special conditions into true chemical compounds capable of being isolated in a pure form.[7,8] Some very interesting and valuable compounds among other stable radicals[9] are nitroxide radicals (nitroxides) — derivatives of nitrogen oxide with a disubstituted nitrogen atom formally containing a one-valent oxygen atom as the third substituent.[10-12] The presence of the unpaired electron provides for paramagnetic properties, unique for organic matter,[13] to these compounds, and above all, the possibility of detection by electron paramagnetic (spin) resonance spectroscopy (EPR or ESR), based on an ability of the compounds containing unpaired electrons to change the spin state upon absorbency of microwave energy in the magnetic field.[14] High sensitivity of EPR as well as chemical stability and availability of nitroxides allow them to be used as reporter molecules, the so-called spin labels and spin probes in a wide variety of scientific and technical fields, from molecular biology and medicine to criminology and oil production.[15-17] Recent advances in EPR development, including new methods of computer-assisted signal processing and spectra simulation, dramatically increase the informativeness of the spin labeling method compared to pioneering works.[18] Today, spin labeling is the major consumer of nitroxides of various structures determined by the requirements of the particular application problem. Due to the paramagnetic nature of nitroxides, they are able to influence the relaxation time in nuclear magnetic resonance (NMR) experiments,[19] and consequently are applied as contrast-enhancing agents in medical NMR tomography.[20,21] Another characteristic feature of nitroxides derived from the presence of an unpaired electron in the molecule is a tendency of transition to a more oxidized state upon release of the electron (one-electron oxidation) or to the less oxidized level upon one-electron reduction or interaction with the free radical.[22] The ability of nitroxide radicals to trap short-lived free radical intermediates and act as mediators in redox reactions allows for their use as antioxidants,[11] stabilizers of monomers and polymers,[23,24] reagents for the chemical or electrochemical oxidation-reduction,[25,26] and working matter for electric energy accumulators.[27]

The outstanding stability of nitroxides, exceeding those for any other types of free radicals, allows for the use of everyday preparative methods of organic chemistry for their syntheses,[28] instead of the special generation techniques of free radical chemistry.[29] Thus, it is possible to locate nitroxide chemistry as a field situated inbetween "traditional" organic and free radical chemistry, in which the object of study is nitroxides and their transformations. Conception of nitroxide chemistry consists of the experimental methods and theoretical notions and includes as major parts

1. Synthesis, isolation, and purification of individual nitroxides
2. Determination of their structure
3. Investigation of physical, including spectral, properties and of their relationship with the structure
4. Study of nitroxide reactivity
5. Application of nitroxide radicals in various fields

Here, we want to review the synthetic chemistry of nitroxides not only as a combination of methods aimed at the synthesis of nitroxides with desired structure, but as a major part of the organic chemistry of nitroxide which in turn can be determined as the study of a nitroxide structure-property relationship by means of general organic chemistry techniques. The subject of the synthetic chemistry of nitroxides covers areas including methods of nitroxide group design and syntheses of the nitroxide radicals of different structures as well as reactivity of either the nitroxide group itself or the paramagnetic molecule as a whole. Synthetic chemistry of nitroxides definitely plays a fundamental role in nitroxide applications as well, providing ground for making the use of nitroxides for solving different applied problems possible. Syntheses of new structures and corresponding new properties give rise to new applications areas, and conversely, applications of nitroxides demand the synthesis of new structures. This structure or application circle is definitely a powerful source of the progress in this area. At the same time, synthetic chemistry of nitroxides could be thought of as being a tool kit for making a particular nitroxide for a particular application, narrowing the subject somewhat. We would like to discuss here the synthetic chemistry of nitroxides apart from the particular application problems, paying more attention to the general questions of the syntheses and properties of this interesting class of organic nitrogen compounds. Investigation of the chemical properties of nitroxides may be of general interest because of their contribution to understanding the effect of the paramagnetism of a molecule on various chemical transformations, including those which do not involve the radical center itself.[30]

II. STABILITY OF NITROXIDE RADICALS AND THE SCOPE OF THIS BOOK

Under the term "stable nitroxide radicals" we include those compounds from this class which " . . . can be obtained in pure form, stored and handled in the laboratory with no more precaution than that normally observed when working with conventional organic compounds."[31] Certainly, "stable" here has the same meaning as the terms "long-lived" or "persistent" for specifying free radicals capable of existing for some time.[32] "Stability" therefore is a kinetic attribute, indicating the "living ability" of the radical and characterizing the magnitude of its reactivity, which is determined by the activation barrier for possible degradation reactions. The ability to handle the radical without special precautions at room temperature means that activation energy and the rate constant of degradation, respectively, must satisfy the following conditions:

$$\Delta G^{\neq} > 20\text{-}25 \text{ kcal/mol and } k_{25°} < 10^{-4} \text{ sec}^{-1}$$

The stability of the nitroxide group (nitroxide radical center) itself originates from the strong three-electron nitrogen-oxygen bond.[33] The peculiarities of this bond, particularly high delocalization energy (about 32 kcal/mol) contribute to a low energy content, e.g., greater thermodynamic stability of the radical center.[33,34] This is also the reason for the absence of dimerization for most of the known nitroxides, because the formation of a weak O–O bond in a possible dimer cannot compensate for the loss of delocalization energy for two nitroxide groups, thus not leading to a net energy gain.[34] The stability of molecules with nitroxide moiety depends mostly upon the surroundings of the radical center, which determine the possible degradation reaction pathways. Considering the effect of structure on the stability, it should be particularly emphasized that factors lowering the energy level, e.g., influencing the thermodynamic stability, are of less importance compared to kinetic factors enhancing the activating barrier for degradation reactions. This is because the nitroxide group is stable by itself due to its electron structure without additional stabilizing factors, such as the participation of a neighboring group. Furthermore, influence of the adjacent groups while increasing the thermodynamic stability of the molecule can even decrease its persistency because they bring new degradation reactions to life. As an example, conjugation of the radical center with the benzene ring in nitroxide **2** (Scheme 1) leads to the electron density delocalization and lowering of the energy state, compared to compound **1,** thus increasing thermodynamic stability. At the same time, it delivers a higher unpaired electron density to the ring carbon atoms setting up new reaction centers. This delocalization seems to be the main cause of the significant difference in the stability of nitroxides **1** and **2**; the first is indefinitely stable in pure form while the second is able to persist only for a limited time in the solution and rapidly disproportionates yielding products **3** and **4**.[35]

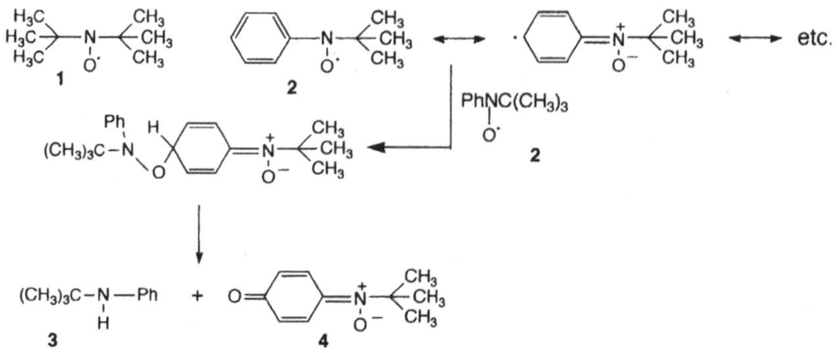

SCHEME 1.

Similar reactions contribute to the relatively low stability of aryl-substituted nitroxides. As a rule, most of the nitroxides having carbon-carbon or carbon-heteroatom multiple bonds in the α-position to a radical center display relatively weak stability; however, a number of compounds belonging to this type are relatively stable. Insufficiency of the spin density delocalization to the carbon centers makes some of the acyl nitroxides **5** (Scheme 2) stable enough for isolation,[36] although these compounds in general are more reactive than bis(*tert*-alkyl) nitroxides such as **1**, due to the higher spin density on the nitroxide oxygen atom and stronger oxidation properties of the nitroxide group.[37] Significant stability of the nitronyl nitroxides **6** can be attributed to the dissipation of a spin density through the two N–O centers.[38,39]

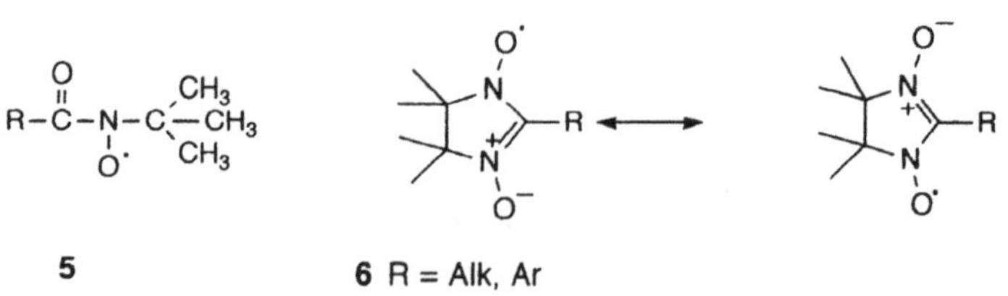

SCHEME 2.

Compounds **7** to **14** listed below (Scheme 3) are examples of nitroxides containing multiple bonds attached to a radical center. All these radicals are able to persist for some time in solutions; however, they have not been isolated as individual compounds. For more examples of similar nitroxides see reviews in References 31 and 40 to 42.

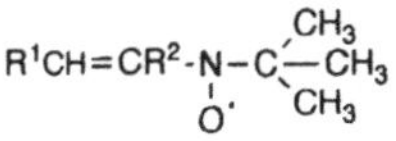

$$R^1CH=CR^2-N-C-CH_3 \quad (CH_3, CH_3, \dot{O})$$

7[43] R^1, $R^2 = CO_2CH_3$; $R^1 = PhSO_2$, $R^2 = H$, Ph

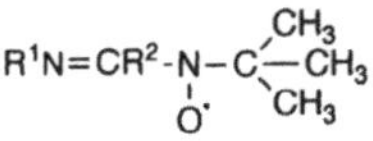

$$R^1N=CR^2-N-C-CH_3 \quad (CH_3, CH_3, \dot{O})$$

8[44] R^1, $R^2 = H$, Alk, Ar

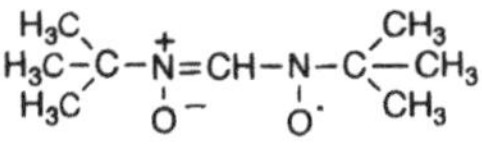

$$H_3C-\overset{H_3C}{\underset{H_3C}{C}}-\overset{+}{N}=CH-N-C-CH_3$$

9[45]

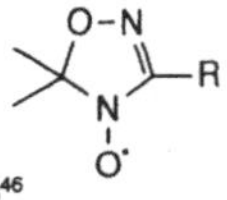

$$\text{(structure)}$$

10[46]

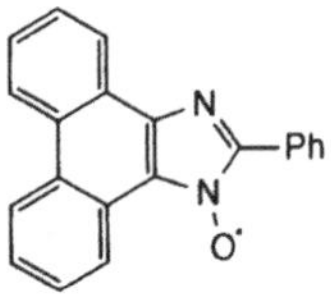

11[47]

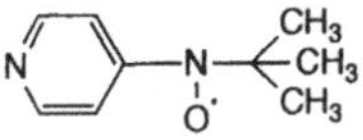

$$N-\text{(pyridine)}-N-C(CH_3)_3$$

12[48]

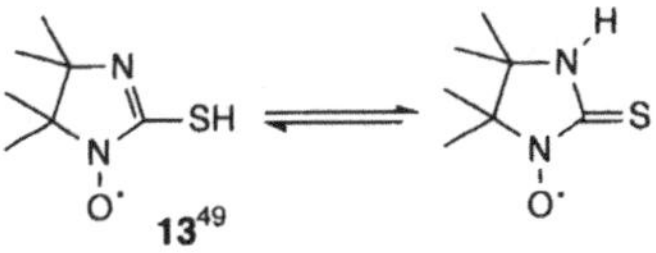

$$\text{(structures)} \quad \text{SH} \rightleftharpoons \text{S}$$

13[49]

$$\text{(indolinone structure)}=O$$

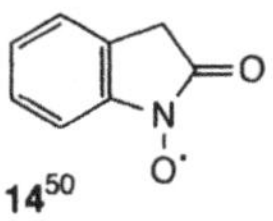

14[50]

SCHEME 3.

Low stability is also typical for the radicals with nitroxide nitrogen-heteroatom attachment. Some examples of the radicals of this type **15** to **18** are listed below (Scheme 4; see reviews in References 31 and 40 to 42 for more examples).

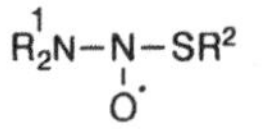

$$R^1_2N-N-SR^2 \quad (\dot{O})$$

15[51] R^1, $R^2 = $ Alk, Ar

$$X-N-C-CH_3 \quad (CH_3, CH_3, \dot{O})$$

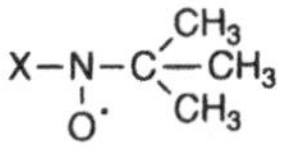

16[52] $X = SO_2Ph$, $PO(OCH_3)_2$, $NAlk_2$, OAlk, SAlk

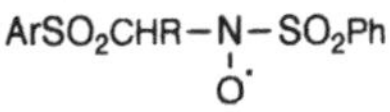

$$ArSO_2CHR-N-SO_2Ph \quad (\dot{O})$$

17[53] $R = H$, Et

$$Ar-N-BH_3^- \ Na^+ \quad (\dot{O})$$

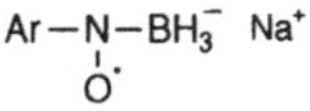

18[54].

SCHEME 4.

The stability of nitroxides containing two sp^3-carbon atoms as nitrogen substituents highly depends upon the presence of the hydrogen atoms at these carbons. The occurrence of even one of the β-hydrogen atoms makes the disproportionation of nitroxide **19** to nitrone **20** and hydroxylamine **21** possible (Scheme 5),[55] which

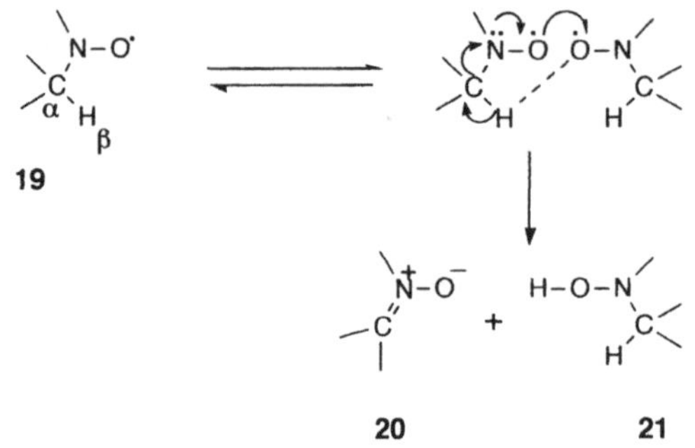

SCHEME 5.

is a leading degradation reaction for these types of nitroxides and a major cause of their low persistence. At the same time, some nitroxides containing β-hydrogens are relatively stable and even isolatable. The absence of disproportionation of bicyclic nitroxides like compound **22** (Scheme 6) is related to the necessity of violation of Bredt's rule for the formation of possible nitrone **23**.[56]

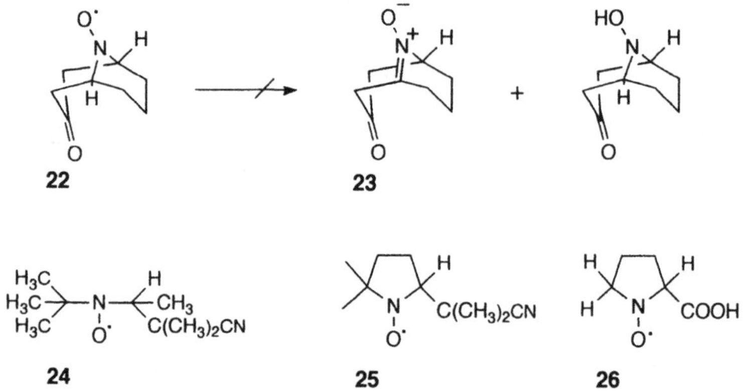

SCHEME 6.

The cause of stability of nitroxide **24**, which has been isolated in individual form, appears to be the steric hindrance in the transition state for the disproportionation reaction.[57] The cyclic analog of nitroxide **24** — radical **25** — is less stable and persists for a limited time in solution.[57] Surprisingly, prolinoxide **26** contains three β-hydrogen atoms but, nevertheless, is able to exist in both the crystal state and in a solution of phosphate buffer.[58] In this case, disproportionation may be retarded by the negatively charged carboxylate group,[41] because the transition state for the abstraction of a hydrogen atom must have an anionic character.

However, as a general case, dialkyl nitroxides with β-hydrogens are unstable and cannot be isolated. The absence of the possibility of the mentioned disproportionation reaction is one of the major reasons for the outstanding stability of bis(*tert*-alkyl) nitroxides. Stability of bis(trifluoromethyl) nitroxide **27** (Scheme 7), flanked

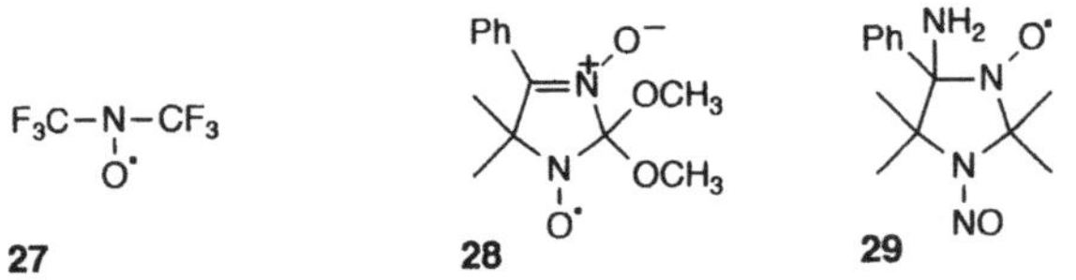

SCHEME 7.

with relatively small substituents,[59] as well as stability of nitroxides like compounds **28** and **29** with alkoxy- or amino groups at the α-carbon atoms is also due to the inability to disproportionate to a nitrone and hydroxylamine.[60,61] It can be concluded that stability of the nitroxide radicals is contributed by the following factors

1. Electron structure of the nitroxide group; this factor provides for thermodynamic stability of the radical center as well as prevents dimerization, which is common for other types of free radicals. The same factor is the cause of nitroxides being relatively weak oxidants or reductants, consequently, making them inert to a majority of ordinary chemical compounds.
2. Steric hindrance of the radical center by substituents at the nitrogen atom; this factor prevents nitroxides from participation in bimolecular reactions.
3. Absence of effective mechanism of degradation for most bis(*tert*-alkyl) nitroxides.

In ranking the significance of the above-mentioned factors, the dominant contribution comes from the electronic structure, while steric hindrance is of less importance. Thus, contrary to stable nitroxide **30**, aminyl radical **31** (Scheme 8) is very unstable despite four blocking methyl groups.[62] Intermediate in the oxidation of correspondent hydrazine derivative **32** diazen (*N*-nitren) **33**,[63] which can be considered as the structural analog of the nitroxide, rapidly undergoes further reactions giving a dimerization-type product — tetrazen **34**.[64] Similarly, easy dimerization is the characteristic feature of the nitroxide's thio analogs **35**, exhibiting the free radical character only upon thermo- or photolysis.[65]

From the point of view of synthetic chemistry, those radicals that can be used in common laboratory conditions are of the most importance. This requirement is satisfied by the majority of compounds with two *tert*-alkyl substituents at the nitrogen atom, including those having a nitroxide group involved in the heterocyclic system. Therefore, bis(*tert*-alkyl) nitroxides will be the focus of attention in this book, while nitroxides with other substituents at the radical center (heteroatoms, multiple bonds) will be mentioned less often.

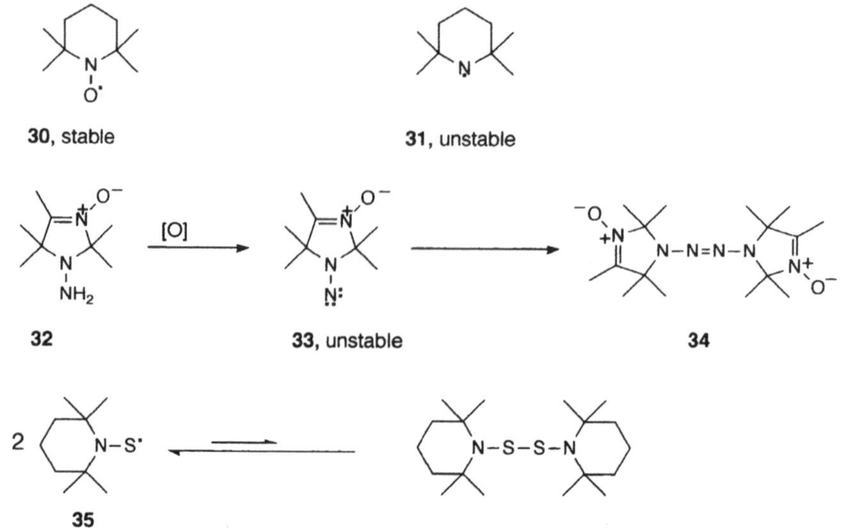

30, stable   **31, unstable**

32   **33, unstable**   **34**

35

SCHEME 8.

III. BRIEF HISTORICAL REVIEW

The first nitroxide, to be isolated and characterized was the inorganic compound, Fremy's salt **36**, synthesized in 1845.[10] Of course, the free radical nature of this compound was not discovered, because the concept of the free radical itself did not yet exist in chemistry. In 1901 almost simultaneously with Gomberg's paper on some unusual properties of hexaphenyl ethane,[66] which is considered to be the beginning of free radical chemistry,[67] papers by Piloty and Schwerin describing the synthesis of the compound named porphyrexide appeared.[68,69] This heterocyclic compound was considered by the authors as the derivative of a four-valent nitrogen **37** (Scheme 9), but indeed was the first organic nitroxide radical **38**, as well as the

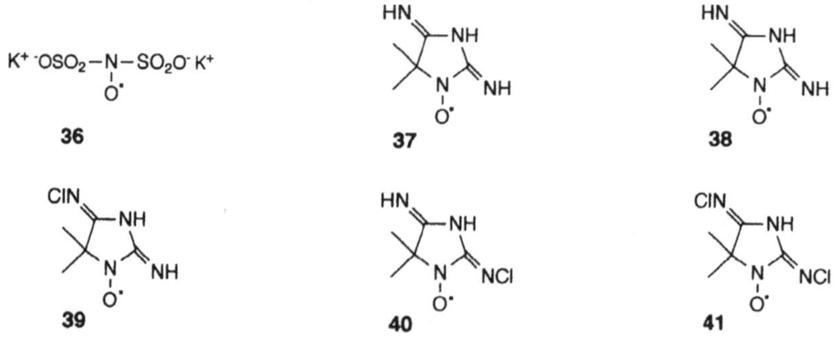

$$K^+ \ ^-OSO_2-N-SO_2O^- \ K^+$$

36   **37**   **38**

39   **40**   **41**

SCHEME 9.

first organic free radical ever isolated in an individual state. Chlorination of porphyrexide gave mono **39, 40** and bis **41** chloro derivatives, which were stable free
radicals as well.[70]

The first to point out the free radical nature of porphyrexide was Wieland who
in 1914 discovered a new class of organic nitroxide radicals with two aromatic
substituents on the nitrogen atom **42**.[71] He also demonstrated the relationship between these compounds and Gomberg's triphenyl methyl.[72] For the synthesis of
diaryl nitroxides **42** he suggested the reaction of nitroso arenes with aryl magnesium
halide, following oxidation of the product, changing diaryl hydroxylamine **43** into
target nitroxide (Scheme 10). Wieland, and later Meyer synthesized a number of
diaryl nitroxides **42** and proved their stability,[73,74] in particular the absence of
dimerization either in a solid state or in solution.

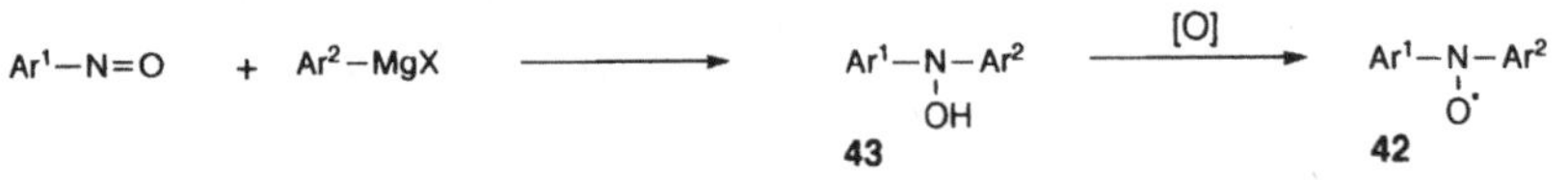

SCHEME 10.

After that, syntheses of nitroxides having only one aromatic substituent at the
radical nitrogen were reported. While studying the reaction of phenyl hydroxylamine
with acetone in 1926, Banfield and Kenyon isolated a stable free radical and
attributed the structure **44** (Scheme 11) to it.[75] The true structure **45** of this compound
was established afterward.[76] Phenyl decalin nitroxide **46** was obtained by Hückel
in 1938 from the oxidation of disubstituted hydroxylamine **47** formed in the reaction
of phenyl magnesium bromide with nitroso decalin **48**, i.e., via the reaction sequence
proposed by Wieland. Unfortunately, attempts to apply aliphatic Grignard reagents
to this reaction, particularly *tert*-butyl magnesium chloride, which could lead to
nitroxides with no aromatic substituents, were unsuccessful.[77]

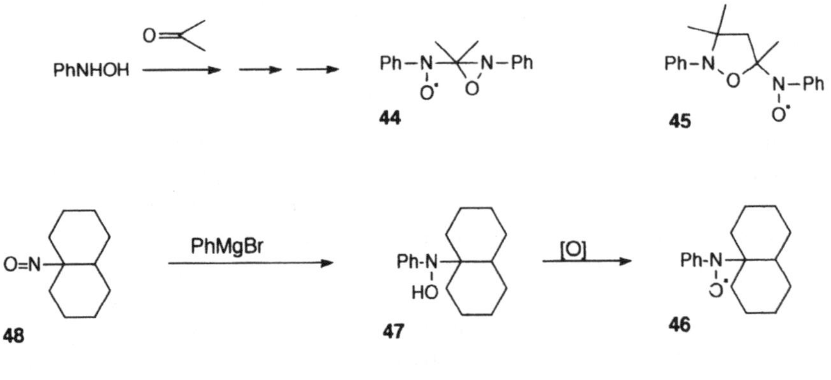

SCHEME 11.

The first example of a synthesis of a nitroxide containing two tert-alkyl substituents at nitroxide nitrogen was reported in 1959, about 20 years later. Probably, such a delay was occasioned in some way by the predominant notion regarding the necessity of the unpaired electron to be delocalized through the conjugated system for radical stability, although it has been predicted that dialkyl nitroxides containing no β-hydrogens might be stable.[78] It seems reasonable that bis(*tert*-alkyl)-substituted nitroxides could have been discovered earlier, because researchers already had various types of nitroxide precursors in hand. Triacetoneamine **49**, for instance, has been known for almost a century,[79] and a vast variety of its derivatives were obtained and studied.[80,81] Nevertheless, oxidation of these sterically hindered amines that could produce nitroxides had not been performed. Very close to the synthesis of nitroxide radicals was the group of chemists headed by Todd (Bonnett, Brown, Clark, Sutherland, Lamchen et al.), who developed in the late 1950s a method of syntheses of alkyl nitrones **50** via oxidation of hydroxylamine derivatives **51** derived from less substituted nitrones **52** and organometallic compounds (Scheme 12).[82,83] This reaction sequence later became a foundation of one of the most powerful methods of nitroxide syntheses.[31,84]

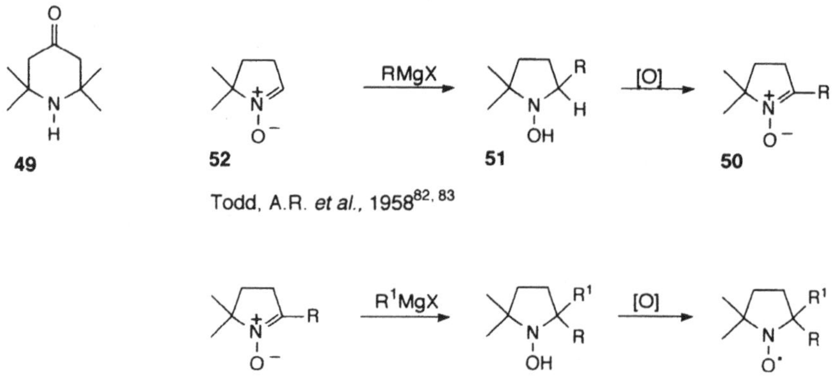

Todd, A.R. *et al.*, 1958[82, 83]

Keana, J.F.W. *et al.*, 1976[31, 84]

SCHEME 12.

Note also the works of Haszeldine, who synthesized bis(trifluoromethyl) hydroxylamine **53** in the reaction sequence which was postulated to pass through bis(trifluoromethyl) nitroxide **27** (Scheme 13).[85] About a decade later, oxidation of hydroxylamine **53** was performed almost simultaneously by two other research groups and led to the discovery of this interesting stable nitroxide **27** which is gaseous under normal conditions.[86,87]

Certainly, one of the reasons that dialkyl-substituted nitroxides were not discovered and studied earlier was the absence of demand for such compounds. After the classic works of Wieland and Hückel properties of already known nitroxides

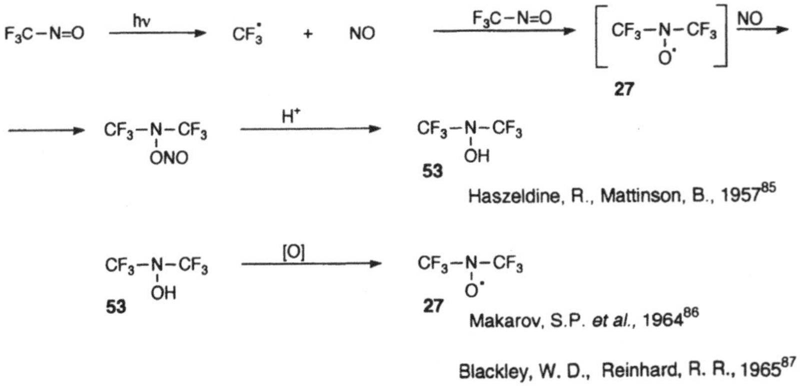

$$F_3C-N=O \xrightarrow{h\nu} CF_3^{\cdot} + NO \xrightarrow{F_3C-N=O} \left[\underset{O^{\cdot}}{CF_3-N-CF_3} \right] \xrightarrow{NO}$$

27

$$\xrightarrow{} \underset{ONO}{CF_3-N-CF_3} \xrightarrow{H^+} \underset{\underset{\textbf{53}}{OH}}{CF_3-N-CF_3}$$

Haszeldine, R., Mattinson, B., 1957[85]

$$\underset{\underset{\textbf{53}}{OH}}{CF_3-N-CF_3} \xrightarrow{[O]} \underset{\underset{\textbf{27}}{O^{\cdot}}}{CF_3-N-CF_3}$$

Makarov, S.P. *et al.*, 1964[86]

Blackley, W. D., Reinhard, R. R., 1965[87]

SCHEME 13.

were not under further investigation. Supposedly, this class of compounds was seen by most of the researchers as a chemical rarity inducing more curiosity than scientific interest. On the other hand, it reflected the general decrease of interest in free radicals due to rapid growth of the ionic theory,[88] successfully interpreting a large number of chemical reactions. The renaissance of free radical chemistry was chiefly due to the discovery of EPR by Zavoisky in 1944, which made the detection of free radicals and study of their structure possible. Nevertheless, until the 1960s the progress in free radical chemistry proceeded slowly. Therefore, the synthesis and isolation of the first bis(*tert*-alkyl) nitroxide **30** from 2,2,6,6-tetramethylpiperidine **54** performed by Lebedev and Kazarnovsky (Scheme 14) was not properly rated even by themselves who published the report on this very important event in nitroxide chemistry in a province journal.[89] Later, the same authors synthesized nitroxide **55** from triacetonamine **49**, although their attempts to isolate radical **55** in an individual state failed.[90] Nevertheless, the method suggested by Lebedev and Kazarnovsky — the oxidation of the sterically hindered amines with hydrogen peroxide in the presence of a catalyst — turns out to be general and widely applicable for the synthesis of a variety of nitroxides.[91]

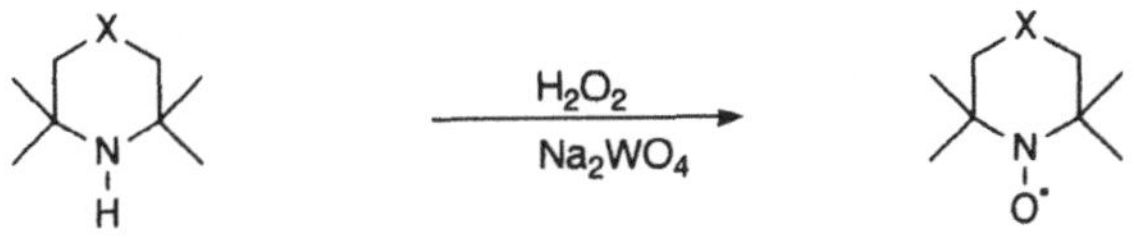

$$\underset{\substack{\textbf{54, 49}}}{\overset{X}{\underset{\underset{H}{N}}{\bigcirc}}} \xrightarrow[\text{Na}_2\text{WO}_4]{\text{H}_2\text{O}_2} \underset{\substack{\textbf{30, 55}}}{\overset{X}{\underset{\underset{O^{\cdot}}{N}}{\bigcirc}}}$$

54,30. X = CH$_2$

49,55. X = CO

SCHEME 14.

Actually, the chemistry of nitroxides has developed rapidly since the mid-1960s. The causes of such an acceleration were the creation of a comprehensive method for nitroxide syntheses based on the reaction discovered by Lebedev and Kazarnovsky. Also important was the idea of a nitroxide group being a common organic functional group with its own reactivity, so the starting compounds for the syntheses of the nitroxide radicals may be other nitroxides. Note, however, that this notion is a straightforward consequence of the pioneering chlorination of porphyrexide in 1903 by Piloty and Vogel.[70] Nevertheless, the introduction by Rozantsev and Neiman of nitroxide reactions without direct involvement of the nitroxide center ("radical reactions involving no free valence") into synthetic chemistry substantially affected the development of this area.[92,93] The increase of interest toward nitroxide chemistry was connected with the rapid growth of free radical chemistry and was influenced by the discovery of the ability of nitroxide to retard free radical reactions and thus display the properties of polymer stabilizers,[94] antioxidants,[10] as well as biological activity.[95] The most powerful impetus for development of nitroxide chemistry came from physical chemistry and biochemistry, specifically the successful application in biological studies by Hamilton and McConnell of the spin labeling method based on EPR spectroscopy.[96] Later spin labeling was developed by different scientific groups in the U.S. (Berliner,[97] Chignell,[98] Gaffney,[99] Swartz,[100] Hyde,[101] Griffith, et al.[102]), Russia (Likhtenstein,[103] Wasserman[104]), and in other countries. This caused increasing demand for nitroxides with various functional groups, hence, since the mid-1960s intensive works have been carried out by the research groups headed by Rozantsev (Russia),[105] Rassat (France),[106] and later by Sosnovsky,[107] Keana (U.S.),[31] and Hideg et al. (Hungary),[108] who supplied the spin-labeling applications with a variety of nitroxides. Remarkable results from the new methods of the nitroxide group creation and reactivity have been published by Aurich (Germany).[40] Studies by Kevan,[109] Bowman,[110] and Freed (U.S.),[111] Luckhurst (U.K.),[112] and Buchachenko,[8] Lebedev et al. (Russia),[113] contributed significantly to spectroscopy of nitroxides. Note the works on the interaction of the nitroxide radical center with other types of paramagnetic metals, which were made by G. and S. Eaton (U.S.)[114] who are also developing a new method of mapping the unpaired electron density in the sample; namely EPR imaging.[115] Of exclusive importance are the studies on nitroxide radical center reactivity, performed by Ingold's group from Canada.[116] Syntheses and reactivity of aromatic nitroxides and acyl nitroxides are the areas of intensive research by the groups of Forrester and Perkins.[117,118] The scientific group headed by the first author of this book collected interesting data on the chemistry of nitroxide derivatives of heterocyclic compounds as well as developed general methods for the syntheses of nitroxides with amino or alkoxy groups flanking the radical center.[119,120]

The synthetic methods brought to life as the result of the nitroxide chemistry development led to a significant amount of various nitroxide types and derivatives, some of which are commercially available. The development of nitroxide chemistry continues today. Of great importance is the development of EPR itself, increasing the sensitivity, the use of different microwave bands,[121] methods of signal detection,[122] as well as of the related spectroscopic methods: electron nuclear double resonance (ENDOR),[123] spin echo,[124] EPR,[115] and proton-electron double imaging (PEDRI)[125] etc., which extends the limits of the spin labeling. At the same time nitroxides are finding applications in other fields. Along with biochemistry and molecular biology nitroxide spin markers are used in analytical chemistry,[126] polymer chemistry,[127] oil production,[17] and in areas such as polymer stabilization[24] or NMR imaging,[20,21,128] where their application is not directly related with the ability to give EPR spectra. The study of the recently discovered influence of the nitroxide radical on the reactivity of the side functional groups[30] seems to be an interesting research area for understanding the effect of spin density on reactivity.

Continuous interest of researchers in this class of compounds as well as expansion of nitroxide application areas indicate that the history of nitroxides is still on the way to its summit.

IV. CLASSIFICATION AND NOMENCLATURE OF NITROXIDES

As mentioned previously, the scope of this book will be limited mostly to compounds containing carbon substituents at the nitroxide nitrogen. Classification of such nitroxides is usually based on the character of these substituents. Depending upon whether the nitroxide group has two tert-alkyl groups not connected to each other or being a part of a cyclic system, it is ordinary to speak about open chain or cyclic nitroxides. In turn, cyclic nitroxides, being heterocyclic derivatives, may be conveniently classified on a parent heterocycle basis.

When speaking about nitroxide nomenclature, it is necessary to have in mind two different aspects related to the names: (1) root name, the name of a "family" of the compounds which can be used as a key word for the computerized search in databases and (2) systematic names of the particular compounds. Unfortunately, there is no consensus among chemists on either of these issues.

Initially, this class of compounds was named "iminoxyl"[11] radicals; then the terms "aminoxyl", "aminyloxide,"[12] "nitroxyl"[10] radicals as well as "iminoxyls", "nitroxyls", "nitroxides",[34] and "aminyloxides"[40] were applied. Later alkylidenenitroxide radicals, $R_2C=N-O\cdot$ derived from oximes were named as iminoxyl radicals.[28] In the current situation it seems reasonable to establish the most widely used term for the parent name. A study of the literature data made by Rassat in 1981,[130] and repeated in 1991 by Volodarsky, Reznikov, and Ovarchenko using the CAS ON-LINE data file shows that the terms "nitroxide" and "nitroxyl" appear more frequently than others. In the papers published before 1981 nitroxide has been used in 1192 papers compared to 344 papers containing nitroxyl, 36 with both of these terms, and 223 using iminoxyl.[113] In 1991 the numbers were the following: 2531, nitroxide; 994, nitroxyl; and 95, iminoxyl. In connection to these, the term aminyloxide proposed by the International Union of Pure and Applied Chemistry (IUPAC) (Rule C-81.1) has been almost neglected by the chemists as the revised term aminoxyl, adopted by IUPAC in 1985 (used only 26 and 42 times, respectively). Rozantsev's proposal on using the term nitroxide for conjugated radicals like **2** and nitroxyls for bis(*tert*-alkyl) conjugated compounds such as **1** to emphasize the difference between these two types has not gained support.[131] In 1985 IUPAC accepted nitroxides as the name of the class of compound, but nitroxyl is still being used for this purpose.[34] Another reason for the frequency of nitroxyl is the literal translation from the Soviet literature where the term nitroxyl radicals ("nitroksil'nye radikaly"), which emphasized the free radical character of these compounds (yl), is used officially. Thus, 476 of the 994 works concerning nitroxyl have been written in Russian.

To construct the systematic name for particular compounds the rules of radico-functional (Chapter C-02 of the IUPAC rules) or substitutive (Chapter C-01) nomenclatures are applied.[132] The former considers the N–O group as aminoxyl (formed according to Rule C-81.2 as an amine having oxygen as the substituent), where the "amine" part gets its name according to IUPAC rules applying to a secondary amine with the appropriate substituents at nitrogen. Alternatively, nitroxide can also be used as the functional name. Compound **56** is named, therefore, as bis(*p*-methoxyphenyl) aminoxyl and compound **27** is named bis(trifluoromethyl) nitroxide (Scheme 15). Generally this kind of nomenclature is more convenient for the open-chain nitroxides.

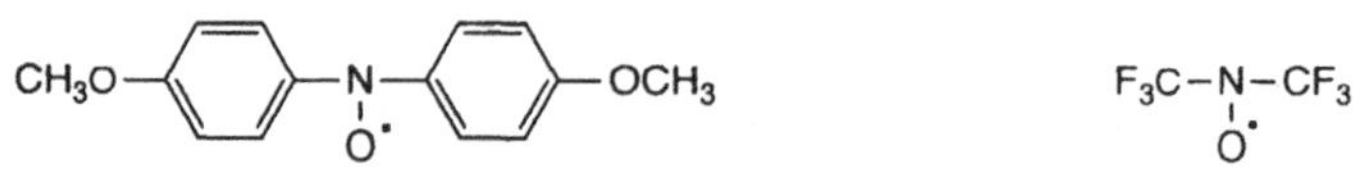

56   **27**

SCHEME 15.

For the cyclic nitroxides the same procedure combined with the rules of heterocycle nomenclature are applied. Nitroxide function is considered to include the nitrogen atom as being a part of the heterocycle (to be the amine part of aminoxyl) and the O substituent marked as "oxyl". Consequently, oxyl is accounted to be a main function and a name suffix. Thus, radicals **57, 58,** and **59** can be named as 2,2,5,5-tetramethylpyrrolidine-1-oxyl, 2,2,4,4-tetramethyl-5-phenyloxazolidine-3-oxyl, 2,2,6,6-tetramethyl-4-hydroxypiperidine-1-oxyl, respectively. However, oxyl is frequently used as the prefix, placed among other substituents according to alphabetical order, therefore, radical **58** may be named as 2,2,4,4-tetramethyl-1-oxyl-5-phenyloxazolidine (Scheme 16).

Concurrently, the suffix "yloxy" is used for marking the oxyl substituent at the nitrogen atom used, particularly in *Chemical Abstracts*. Accordingly, compound **58** is to be 2,2,4,4-tetramethyl-5-phenyloxazolidine-3-yloxy, although in the original literature oxyl is preferred.

In conclusion, it should be noted that long and complex systematic names for nitroxides brought to life a number of trivial names for several structural types

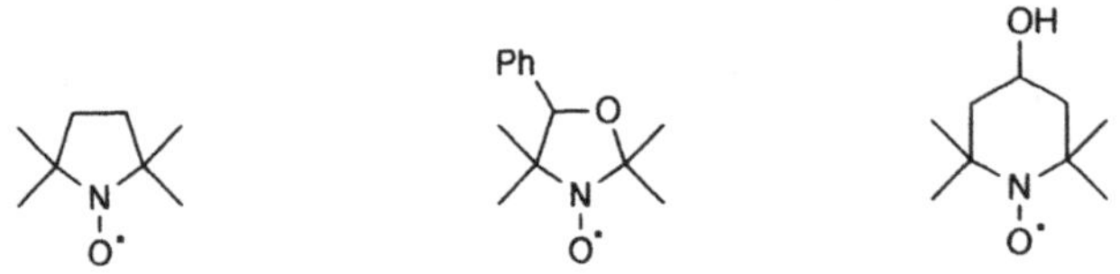

57   **58**   **59**

SCHEME 16.

("doxyls", "proxyls", "azetoxyls", etc.[31,41]) as well as the abbreviations for particular compounds (TEMPO, TEMPON, etc.[34]). Despite the obvious writing convenience, usage of the particular trivial name and/or abbreviations must be specified to avoid misleading the reader.

V. PURPOSES AND STRUCTURE OF THIS BOOK

At the present time a number of books and reviews on nitroxide chemistry and applications have been published. The questions of synthetic organic chemistry are discussed in the books by Rozantsev,[10,28] Forrester, Hay, and Thomson,[9] Berliner,[11] Volodarsky,[119,120] Zhdanov,[133] and Aurich,[34] in a continuing series edited by Berliner[134] and Zhdanov,[135-137] and in review articles by Griffith and Waggoner,[102] Rozantsev and Sholle,[91] Aurich,[40] Keana,[31,41,42] and Gaffney.[138] However, about 10 years have passed since the publication of most of the volumes mentioned, which is too much time for this area with a history consisting of 25 years. The later publications are more additions than summarizing of matter.[139,140] Furthermore, most of the publications are devoted to the synthesis of nitroxides for a single application area — spin labeling in molecular biology. This influenced the presentation of material, which becomes connected with the application area.

This book is an attempt to introduce the notion of synthetic organic chemistry outside the context of particular applications. Application topics as well as physical properties of the nitroxides will be mentioned only briefly while considering the peculiarities of different structures. Most of the attention will be paid to particular nitroxide types, to point out structural peculiarities and the synthetic potential. However, the authors did not intend to prepare a complete survey of all existing literature or create the encyclopedia of nitroxides, therefore, the literature coverage is of a more representative than comprehensive character. The main purpose of the book — introduction of the general properties of nitroxides as well as main methods and techniques of the synthesis of the compounds — should be helpful for application in specific areas.

The structure of this volume is derived from these purposes. The nitroxide chemistry survey will begin with a review of the general methods of nitroxide radical center generation. Most of the attention will be paid to those methods that are of synthetic convenience, while nitroxide generation methods for spectroscopic studies will be touched on only briefly as well as physical and spectroscopic properties of nitroxides, which are the subjects of a number of specially aimed books and reviews (see for example References 8, 11, and 18 and references cited therein). These questions will be discussed on a qualitative level without employing the mathematic apparatus of quantum mechanics, chemical kinetics, and spectroscopy. Looking over the reactivity of compounds containing a nitroxide group, reactions of the latter as well as its effect on the reactivity of other functional groups will be reviewed. In the beginning of Chapter 3 a brief survey of the main application areas of nitroxides will be made in connection with the structural requirements. Determination of possible types of nitroxides as well as a particular functionalization serves to design the possible synthetic schemes and choose the starting compounds. The authors are not considering the details of applications, in particular, biological, because it has been the topic of most books and reviews already published (see for example References 96, 102, 103, and 135 to 142 and references cited therein). Instead of this, the comparative consideration of the structural features, methods of syntheses, and reactivity of the particular types of compounds will be made. Nitroxide classification here is based on skeleton structure — the acyclic compounds, and then various heterocyclic derivatives will be considered. In Chapter 4 a review of the rapidly growing application of nitroxides to coordination chemistry will be presented.

The authors are addressing this volume to those researchers using nitroxides in everyday work — chemists, biochemists, analytical chemists, and spectroscopists. We hope it will be helpful to graduate and undergraduate students in chemistry as well as for organic chemists exploring the vast area of the organic derivatives of the nitrogen.

ACKNOWLEDGMENT

This chapter was written with the substantial participation of Dr. Vladimir V. Martin. His present address is the Department of Chemistry, University of Oregon, Eugene, OR (U.S.).

REFERENCES

1. **Kochi, J. K.,** *Free Radicals, Vol. 1.,* Wiley Interscience, New York, 1973.
2. **Asinger, F.,** *Parafins. Chemistry and Technology,* Pergamon Press, Oxford, 1968.
3. *Free Radicals in Biology and Medicine,* 2nd ed., Halliwell, B. and Gutteridge, J. M. C., Eds., Clarendon Press, Oxford, 1989.
4. **Loach, P. A. and Hals, B. J.,** Free radicals in photosynthesis, in *Free Radicals in Biology,* Vol. 1, Pryor, W. A., Ed., Academic Press, New York, 1976, chap. 5.
5. **Floyd, R. A., Ed.,** *Free Radicals in Cancer,* Marcel Dekker, New York, 1981.
6. **Nonhebel, D. C., Tedder, J. M., and Walton, J. C.,** *Radicals,* Cambridge University Press, Cambridge, 1981, xi.
7. **Buchachenko, A. L.,** *Stable Radicals,* Consultant Bureau, New York, 1965.

8. **Buchachenko, A. L. and Wasserman, A. M.**, *Stable Radicals. Electron Structure, Reactivity and Applications*, Khimia, Moscow, 1973 (in Russian).
9. **Forrester, A. R., Hay, J. M., and Thomson, R. H.**, *Organic Chemistry of Stable Free Radicals*, Academic Press, New York, 1968.
10. **Rozantsev, E. G.**, *Free Nitroxyl Radicals*, Plenum Press, New York, 1970.
11. **Berliner, L. J., Ed.**, *Spin Labelling. Theory and Applications*, Vols. *I and II*, Academic Press, New York, 1976, 1979.
12. **Aurich, H. G.**, Nitroxides, in *The Chemistry of Functional Groups*, Patai, S., Ed., John Wiley & Sons, Chichester, England, 1982.
13. **Abragam, A.**, *The Principles of Nuclear Magnetism*, Clarendon Press, Oxford, 1961.
14. **Pool, C. P.**, *Electron Spin Resonance*, 2nd ed., Wiley Interscience, New York, 1983.
15. **Hoff, A., Ed.**, *Advanced EPR: Application in Biology and Biochemistry*, Elsevier, Amsterdam, 1989.
16. **Piette, L. H. and Hsia, J. C.**, Spin labeling in biomedicine, in *Spin Labeling II*, Berliner, L. J., Ed., Academic Press, New York, 1979, 115.
17. **Bukin, I. I.**, Nitroxides in oil-field development, in *International Conference on Nitroxide Radicals*, (Abstr.), Novosibirsk, Russia, 69P, 1988.
18. **Kevan, L. and Bowman, M. K., Eds.**, *Modern Pulsed and Continuous-Wave Electron Spin Resonance*, John Wiley & Sons, Chichester, England 1990.
19. **Bennett, H. F., Brown, R. D., Koenig, S. H., and Swartz, H. M.**, Effects of nitroxides on the magnetic field and temperature dependence of $1/T_1$ of solvent water protons, *Magn. Reson. Med.*, 4, 93, 1987.
20. **Brasch, R. C., London, D. A., Wesbey, G. E., Tozer, T. N., Nitecki, D. E., Williams, R. D., Doemeny, J., Tuck, L. D., and Lallemand, D. P.**, Work in progress: nuclear magnetic resonance study of a paramagnetic nitroxide contrast agent for enhancement of renal structures in experimental animals, *Radiology*, 147, 773, 1983.
21. **Rosen, G. M.**, Method for enhancing nuclear magnetic resonance imaging using a nitroxide, *PCT Int. Appl.*, 1991.
22. **Pokhodenko, V. D., Beloded, A. A., and Koshechko, V. G.**, *Oxidation-Reduction Reactions of Free Radicals*, Naukova Dumka, Kiev, 1977 (in Russian).
23. **Rozantsev, E. G., Goldfein, M. D., and Trubnikov., A. V.**, Stable radicals and kinetics in radical polymerization of vinyl monomers, *Usp. Khim.*, 55, 1881, 1986 (in Russian).
24. **Rabek, J. F.**, *Photostabilization of Polymers. Principles and Applications;* Elsevier Applied Science, New York, 1990, chap. 6.
25. **Yamaguchi, M., Miyazawa, T., Takata, T., and Endo, T.**, Application of redox systems based on nitroxides to organic synthesis, *Pure Appl. Chem.*, 62, 217, 1990.
26. **Semmelhack, M. F., Chou, C. S., and Cortes, D. A.**, Nitroxyl-mediated electrooxidation of alcohols to aldehydes and ketones, *J. Am. Chem. Soc.*, 105, 4492, 1983.
27. **Abakumov, G. A., Muraev, V. A., Razuvaev, G. A., Tikhonov, V. D., Chechnet, Yu. V., and Nechuev, A. I.**, Electrochemical aspects of single-electron transfer in organic reactions, *Dokl. Akad. Nauk S.S.S.R.*, 230, 589, 1976 (in Russian).
28. **Rozantsev, E. G. and Sholle V. D.**, *Organic Chemistry of Free Radicals*, Khimia, Moscow, 1979 (in Russian).
29. **Lazar, M., Rychly, J., Klimo, V., Pelikan, P., and Valko, L.**, *Free Radicals in Chemistry and Biology*, CRC Press, Boca Raton, FL, 1989, chap. 2.
30. **Grigor'ev, I. A., Shchukin, G. I., and Volodarsky, L. B.**, Studies of the effect of the radical center on chemical reactivity, in *Imidazoline Nitroxides, Vol. 1. Synthesis and Properties*, Volodarsky, L. B., Ed., CRC Press, Boca Raton, FL, 1988, chap. 5.
31. **Keana, J. F. W.**, Newer aspects of the synthesis and chemistry of nitroxide spin labels, *Chem. Rev.*, 78, 37, 1978.
32. **Griller, D. and Ingold, K. U.**, Persistent carbon-centered radical, *Acc. Chem. Res.*, 9, 13, 1976.

33. **Mahoney, L. R., Mendenhall, G. D., and Ingold, K. U.,** Calometric and equilibrium studies on some stable nitroxide and iminoxy radicals. Approximate OH bond dissociation energies in hydroxylamines and oximes, *J. Am. Chem. Soc.,* 95, 8610, 1973.
34. **Aurich, H. G.,** Nitroxides, in *Nitrones, Nitronates and Nitroxides,* Patai, S. and Rappoport, Z., Eds., John Wiley & Sons, Chichester, England, 1989.
35. **Calder, A. and Forrester, A. R.,** Nitroxide radicals. Part VI. Stability of meta- and para-alkyl substituted phenyl-*t*-butylnitroxides, *J. Chem. Soc., (C),* 1459, 1969.
36. **Alewood, P. F., Hussain, S. A., Jenkins, T. C., Perkins, M. J., Sharma, A. H., Siew, N. P. Y., and Ward, P.,** Acyl nitroxides. Part I. Synthesis and isolation, *J. Chem. Soc., Perkin Trans. 1,* 1066, 1978.
37. **Hussain, S. A., Jenkins, T. C., and Perkins, M. J.,** Oxidations with acylnitroxyls, *Tetrahedron Lett.,* 3199, 1977.
38. **Ullman, E. F., Call, L., and Osiecki, J. H.,** Stable free radicals. VIII. New imino, amido and carbamoyl nitroxides, *J. Org. Chem.,* 35, 3623, 1970.
39. **Ullman, E. F., Osiecki, J. H., Boocock, D. G. B., and Darcy, R.,** Studies of stable free radicals. X. Nitronyl nitroxides monoradicals and biradicals as possible small molecule spin labels, *J. Am. Chem. Soc.,* 94, 7049, 1972.
40. **Aurich, H. G. and Weiss, W.,** Formation and reactions of aminyloxides, *Top. Curr. Chem.,* 59, 65, 1975.
41. **Keana, J. F. W.,** New aspects of nitroxide chemistry, in *Spin Labelling II,* Berliner, L. J., Ed., Academic Press. New York, 1979, 115.
42. **Keana, J. F. W.,** Nitroxide spin labels, in *Spin Labelling in Pharmacology,* Holtzman, J. L., Ed., Academic Press. Orlando, FL, 1984, 1.
43. **Aurich, H. G., Hahn, K., and Stork, K.,** Aminyloxide (nitroxide). XXX. Vinylaminyloxides. Spin density distribution and reactions, *Chem. Ber.,* 112, 2776, 1979.
44. **Aurich, H. G., Duggal, S. K., Hohlein, P., and Klingelhofer, H. G.,** Aminyloxides (nitroxides). Amidinyl *N*-oxides and *N,N'*-dioxides as secondary radical in the oxidation coupling of amines and hydroxylamines with nitrones, *Chem. Ber.,* 114, 2240, 1981.
45. **Aurich, H. G. and Hohlein, P.,** Aminyloxide. XIX. Bildung von Amidinyl-*N*-Oxiden und Amidinyl-*N,N*-Dioxiden durch Oxidative Kupplung, *Tetrahedron Lett.,* 279, 1974.
46. **Aurich, H. G. and Stork, K.,** Bildung von Aminyloxiden bei Reactionen von Nitriloxiden mit Hydroxylaminen, *Chem. Ber.,* 108, 2764, 1975.
47. **Colonna, M., Greci, L., and Poloni, M.,** Stable nitroxide radicals. Reaction between 2-cyano-benzoquinoline and 4-cyanobenzoquinoline *N*-oxides and the Grignard reagent, *J. Heterocycl. Chem.,* 17, 1473, 1980.
48. **Cazianis, C. T. and Eaton, G. R.,** Spin-labelled ligands, *Can. J. Chem.,* 52, 2454, 1974.
49. **Darcy, R.,** Tautomerism in a thioamide-nitroxide: solvent effects in terms of an ESR parameter for a 2-thiocarbonylimidazole-1-oxyl, *J. Chem. Soc. Perkin Trans. 2,* 1089, 1981.
50. **Balaban, A. T. and Pascaru, I.,** Factors effecting stability and equilibria of free radicals. VI. Oxidation of cyclic hydroxamic acids to nitroxides, *J. Magn. Reson.,* 7, 241, 1972.
51. **Waters, W. A.,** A new group of nitroxide free radicals formed from aliphatic nitrosamines and thiols, *J. Chem. Soc. Chem. Commun.,* 741, 1978.
52. **Aurich, H. G. and Czepluch, H.,** Aminyloxide (nitroxide). XXXI. A study of the spin-density distribution of various aminyl oxide types using ^{17}O-labeled radicals, *Tetrahedron,* 36, 3543, 1980.
53. **Rawson, G. and Engbert, J. B. F. N.,** Mannich-type condensation products of sulfinic acids with aldehydes and hydroxylamines or hydroxamic acids, *Tetrahedron,* 26, 5356, 1970.
54. **Crozet, M. P. and Tordo, P.,** Boronitroxides. II. On the reaction of sodium borohydride with nitroso compounds in alcohols, *Inorg. Chim. Acta,* 53, L57, 1981.
55. **Bowman, D. F., Gillan, I., and Ingold, K. U.,** Kinetic applications of electron paramagnetic resonance. III. Self reactons of dialkyl nitroxide radicals, *J. Am. Chem. Soc.,* 93, 6555, 1971.
56. **Briere, R. and Rassat, A.,** Synthesis et etude cinetique de la decomposition du *t*-butyl isopropyl nitroxide. Effet isotopique, *Tetrahedron,* 32, 289, 1976.

57. **Iwamura, M. and Inamoto, N.,** Novel formation of nitroxide radicals by radical addition to nitrones, *Bull. Chem. Soc. Jpn.,* 40, 703, 1967.

58. **Lin, J. S., Tom, T. C., and Olcott, H. S.,** Proline nitroxides, *J. Agric. Food Chem.,* 22, 526, 1974.

59. **Clidewell, C., Rankin, D. W. H., Robiette, A. G., Sheldrick, G. M., and Williamson, S. M.,** Molecular structure of bis(trifluoromethyl)nitroxyl: electron diffraction study, *J. Chem. Soc. (A),* 478, 1971.

60. **Grigor'ev, I. A., Volodarsky, L. B., Starichenko, V. F., Shchukin, G. I., and Kirilyuk, I. A.,** Route to stable nitroxides with alkoxy groups at α-carbon — the derivatives of 2- and 3-imidazolines, *Tetrahedron Lett.,* 26, 5058, 1985.

61. **Grigor'ev, I. A., Starichenko, V. F., Kirilyuk, I. A., and Volodarskii, L. B.,** Synthesis of stable nitroxyl radicals with amino group at the carbon atom alpha to the radical center by oxidative amination of 4*H*-imidazole *N*-oxides, *Izv. Akad. Nauk. S.S.S.R., Ser Khim.,* 661, 1989 (in Russian).

62. **Sholle, V. D., Rozantsev, E. G., Prokof'ev, A. I., and Solodovnikov, S. P.,** E.P.R. study of the 2,2,6,6-tetramethyl-4-oxo-1-piperidinoxy free radical, *Izv. Akad. Nauk S.S.S.R., Ser. Khim.,* 2628, 1967 (in Russian).

63. **Lemal, D. M.,** Aminonitrenes (1,1-diazenes), in *Nitrenes,* Lwowsky, W., Ed., Wiley Interscience, New York, 1970, 345.

64. **Martin, V. V. and Volodarsky, L. B.,** Synthesis and some reactions of sterically-hindered 3-imidazoline-3-oxides, *Khim. Heterocycl. Soedin.,* 103, 1979 (in Russian).

65. **Bennett, J. E., Sieper, H., and Tavs., P.,** 2,2,6,6-Tetramethylpiperidyl-1-thiyl. A stable new radical, *Tetrahedron,* 23, 1697, 1967.

66. **Gomberg, M.,** Ueber das triphenylmethyl, *Chem. Ber.,* 33, 2726, 1901.

67. **Ihde, A. J.,** The history of free radicals and M. Gomberg's contributions, *Pure Appl. Chem.,* 15, 1, 1967.

68. **Piloty, O. and Schwerin, B. G.,** Ueber die Existenz von deriten des vierwerthigen Stickstoffs, (I. Mitthellung), *Chem. Ber.,* 34, 1870, 1901.

69. **Piloty, O. and Schwerin, B. G.,** Ueber die Existenz von deriten des vierwerthigen Stickstoffs, (II. Mitthellung), *Chem. Ber.,* 34, 2354, 1901.

70. **Piloty, O. and Vogel, W.,** Ueber die constitution des porphyrexids, einer analogous des isatins, *Chem. Ber.,* 36, 1283, 1903.

71. **Wieland, H. and Offenbacher, M.,** Diphenylstickstoffoxyd, ein neues organisches radical mit verwertigem stickstoff, *Chem. Ber.,* 47, 2111, 1914.

72. **Wieland, H. and Roth, K.,** Weitere untersuchengen uber derivate des vierwertigem stickstoff, *Chem. Ber.,* 53, 210, 1920.

73. **Meyer, K. H. and Gottlieb-Billroth, H.,** Uber die einwirkung der salpetersaure auf phenolather, *Chem. Ber.,* 52, 1476, 1919.

74. **Meyer, K. H. and Reppe, W.,** Uber die reduktionsstufen von arylderivaten der Salpetersaure, *Chem. Ber.,* 54, 327, 1921.

75. **Banfield, F. H. and Kenyon, J.,** The constitution of the condensation product of β-phenylhydroxylamine and acetone, *J. Chem. Soc.,* 1612, 1926.

76. **Baldry, P. J., Forrester, A. R., and Thomson, R. H.,** Nitroxide radicals. Part XVIII. Further spectroscopic investigation of the Banfield and Kenyon nitroxide, *J. Chem. Soc. Perkin Trans. 2,* 76, 1976.

77. **Hückel, W. and Liegel, W.,** Ein neues radikal mit vierwertigem stickstoff. Phenyl-9-*trans*-dekalyl-stickstoffoxyd, *Chem. Ber.,* 71, 1442, 1938.

78. **Johnson, D. H., Rogers, M. A. T., and Trappe, G.,** Aliphatic hydroxylamines. Part II. Autoxidation, *J. Chem. Soc.,* 1093, 1956.

79. **Heintz, W.,** Ammoniakderivate des Acetons, *Ann. Chem.,* 174, 133, 1874.

80. **Yakhontov, L. N. and Krasnokutskaya, D. M.,** Advances in the chemistry of α,α'-disubstituted pyridines, *Russ. Chem. Rev.,* 50, 565, 1981 (in English).

81. **Lutz, W. B., Lazarus, S., and Meltzer, R. I.,** New derivatives of 2,2,6,6-tetramethyl piperidine, *J. Org. Chem.,* 27, 1695, 1962.

82. **Bonnett, R., Brown, R. F. C., Clark, V. M., Sutherland, I. O., and Todd, A.,** Experiments towards the synthesis of corrins. Part II. Preparation and reactions of D^1-pyrroline-1-oxides, *J. Chem. Soc.,* 2094, 1959.

83. **Brown, R. F. C., Clark, V. M., Lamchen, M., and Todd, A.,** Experiments towards the synthesis of corrins. Part VI. The dimerization of D^1-pyrrolidine-1-oxides to 2-(1'-hydroxypyrrolidine-2'-yl)-D^1-pyrroline-1-oxides, *J. Chem. Soc.,* 2116, 1959.

84. **Keana, J. F. W., Lee, T. D., and Bernard, E. M.,** Side chain substituted 2,2,5,5-tetramethylpyrrolidine-*N*-oxyl (proxyl) nitroxides. A new series of lipid spin labels showing improved properties for the study of biological membranes, *J. Am. Chem. Soc.,* 98, 3052, 1976.

85. **Haszeldine, R. and Mattinson, B.,** Perfluoroalkyl derivatives of nitrogen. Part VI. *N,N*-bis Trifluoromethylhydroxylamine, the structure of trifluoromethane dimer and the direction of free-radical addition to a nitrosogroup, *J. Chem. Soc.,* 1741, 1957.

86. **Makarov, S. P., Yakubovich, A. Ya., Dubov, S. S., and Medvedev, A. N.,** Synthesis of the stable free radical hexaflorodimethyl oxide of nitrogen structure and properties, *Proc. All-Union Conf., Fluoroorganic Compounds,* Novosibirsk, 1964, 22 (in Russian).

87. **Blackley, W. D. and Reinhard, R. R.,** A new stable radical, bis(trifluoromethyl) nitroxide, *J. Am. Chem. Soc.,* 87, 802, 1965.

88. **Lazar, M., Rychly, J., Klimo, V., Pelikan, P., and Valko, L.,** *Free Radicals in Chemistry and Biology,* CRC Press, Boca Raton, FL, 1989, chap. 1.

89. **Lebedev, O. L. and Kazarnovsky, S. N.,** Catalytic oxidation of aliphatic amines with hydrogen peroxide, *Treatises on Chemistry and Chemical Technology,* Gorky, 3, 649, 1959 (in Russian).

90. **Lebedev, O. L., Khidekel, M. L., and Razuvaev, G. A.,** Isotopic analysis of nitrogen by electron paramagnetic resonance method, *Dokl. Akad. Nauk S.S.S.R.,* 140, 1327, 1961 (in Russian).

91. **Rozantsev, E. G. and Sholle, V. D.,** Synthesis and reactions of stable nitroxyl radicals, *Synthesis,* 190, 1971.

92. **Neiman, M. B., Rozantsev, E. G., and Mamedova, Yu. G.,** Free radical reactions involving no unpaired electrons, *Nature,* 196, 472, 1962.

93. **Rozantsev, E. G. and Neiman, M. B.,** Organic radical reactions involving no free valence, *Tetrahedron,* 20, 131, 1964.

94. **Emanuel, N. M. and Buchachenko, A. L.,** *Chemical Physics of Aging and Stabilization of Polymers,* Nauka, Moscow, 1982 (in Russian).

95. **Konovalova, N. P., Bogdanov, G. N., Miller, V. B., Neiman, M. B., Rosantsev, E. G., and Emanuel, N. M.,** Antitumor activity of stable free radicals, *Dokl. Akad. Nauk S.S.S.R.,* 157, 707, 1964.

96. **Hamilton, C. L. and McConnell, H. M.,** Spin labels, in *Structural Chemistry and Molecular Biology,* Rich, A. and Davidson, N., Eds., W. H. Freeman, San Francisco, 1968, 115.

97. **Berliner, L. J.,** The spin-label approach to labeling membrane protein sulfhydryl groups, *Ann. N.Y. Acad. Sci.,* 414, 153, 1983.

98. **Chignell, C. F.,** Spin labeling in pharmacology, in *Spin Labeling II;* Berliner, L. J., Ed., Academic Press, New York, 1979, chap. 5.

99. **Gaffney, B. J., Elbrecht, C. H., and Scibilia, I. P.,** Enhanced sensitivity to slow motions using nitroxide ^{15}N spin labels, *J. Magn. Res.,* 44, 436, 1981.

100. **Swartz, H. M.,** The use of nitroxides in viable biological systems: an opportunity and challenger for chemists and biochemists, *Pure Appl. Chem.,* 62, 235, 1990.

101. **Hyde, J. S. and Subczynski, W. K.,** Spin-label oximetry in *Spin Labeling VIII,* Berliner, L. J. and Reuben, J., Eds., Plenum Press, New York, 1989, chap. 8.

102. **Griffith, O. H. and Waggoner, A. S.,** Nitroxide free radicals: spin labels for probing biomolecular structure, *Acc. Chem. Res.,* 2, 17, 1969.

103. **Likhtenstein, G. I., Kulikov, A. V., Kotelnikov, A. I., and Levonenko, L. A.,** Methods of physical labels — a combined approach to the study of microstructure and dynamics in biological systems, *J. Biochem. Biophys. Methods,* 12, 1, 1986.

104. **Wasserman, A. M., Alexandrova, T. A., and Buchachenko, A. L.**, The study of rotational mobility of stable nitroxyl radicals in polyvinylacetate, *Eur. Polym. J.*, 13, 1976, 691.
105. **Rozantsev, E. G. and Sholle, V. D.**, Synthesis and reactions of stable nitroxyl radicals, *Synthesis*, 401, 1971.
106. **Rassat, A.**, Magnetic properties of nitroxide multiradicals, *Pure Appl. Chem.*, 62, 223, 1990.
107. **Sosnovsky, G. A. and Konieczny, M.**, Preparation of spin-labeled phosphates using imidazole as transfer agent, *Z. Naturforsch., Teil B*, 32, 1977, 82.
108. **Hideg, K. and Hankovszky, H. O.**, The chemistry of spin-labeled amino acids, peptides. Some newer mono- and bifunctionalized nitroxide free radicals, in *Paramagnetic Models of Biologically Active Compounds*, Zhdanov, R. I. and Rozantsev, E. G., Eds., CRC Press, Boca Raton, FL, 1984.
109. **Kevan, L.**, Developments in electron spin-echo-modulation analysis, in *Modern Pulsed and Continuous-Wave Electron Spin Resonance*, Kevan, L. and Bowman, M. K., Eds., Wiley Interscience, Chichester, England, 1990, chap. 5.
110. **Bowman, M. K.**, Fourier transform electron spin resonance, in *Modern Pulsed and Continuous-Wave Electron Spin Resonance*, Kevan, L. and Bowman, M. K., Eds., Wiley Interscience, Chichester, 1990, chap. 1.
111. **Freed, J. H.**, Theory of slow tumbling ESR spectra for nitroxides, in *Spin Labeling: Theory and Applications;* Berliner, L. J., Ed., Academic Press, New York, 1976, chap. 3.
112. **Luckhurst, G. R.**, Biradicals as spin probes, in *Spin Labeling: Theory and Applications;* Berliner, L. J., Ed., Academic Press, New York, 1976, chap. 4.
113. **Lebedev, Ya. S.**, High-frequency continuous-wave electron spin resonance, in *Modern Pulsed and Continuous-Wave Electron Spin Resonance*, Kevan, L. and Bowman, M. K., Eds., Wiley Interscience, Chichester, England, 1990, chap. 6.
114. **Eaton, S. S. and Eaton, G. R.**, Interaction of spin labels with transition metals, *Coord. Chem. Rev.*, 26, 207, 1978.
115. **Eaton, S. S. and Eaton, G. R.**, Electron spin resonance imaging, in *Modern Pulsed and Continuous-Wave Electron Spin Resonance*, Kevan, L. and Bowman, M. K., Eds., Wiley Interscience, Chichester, England, 1990, chap. 9.
116. **Bowry, V., Lusztyk, J., and Ingold, K. U.**, Calibration of very fast alkyl radical "clock" rearrangements using nitroxides, *Pure Appl. Chem.*, 62, 213, 1990.
117. **Forrester, A. R., Hepburn, S. P., and McConnachie, G.**, Nitroxide radicals. Part XVI. Unpaired electron distribution in para-substituted aryl *t*-butyl nitroxides and 2-naphthyl phenyl nitroxides, *J. Chem. Soc. Perkin Trans.*, 1, 2213, 1974.
118. **Perkins, M. J., Berti, C., Brooks, D. J., Grierson, L., Grimes, J. A.-M., Jenkins, T. C., and Smith, S. L.**, Acyl nitroxides: reactions and reactivity, *Pure Appl. Chem.*, 62, 195, 1990.
119. **Volodarskii, L. B., Grigor'ev, I. A., and Sagdeev, R. Z.**, Stable imidazoline nitroxides, in *Biological Magnetic Resonance*, Vol. 2, Berliner, L. J. and Reuben, J., Eds., Plenum Press, New York, chap. 4.
120. **Volodarsky, L. B., Ed.**, *Imidazoline Nitroxides. Vol. 1. Synthesis and Properties*, CRC Press, Boca Raton, FL, 1988.
121. **Allgeier, J., Disselhorst, J. A. J. M., Weber, R. T., Wenckebach, W. Th., and Schmidt, J.**, High-frequency pulsed electron spin resonance, in *Modern Pulsed and Continuous-Wave Electron Spin Resonance*, Kevan, L. and Bowman, M. K., Eds., Wiley Interscience, Chichester, England, 1990, chap. 6.
122. **Schweiger, A.**, New trends in pulsed electron spin-resonance methodology in *Modern Pulsed and Continuous-Wave Electron Spin Resonance*, Kevan, L. and Bowman, M. K., Eds., Wiley Interscience, Chichester, England, 1990, chap. 2.
123. **Kurrek, H., Kirste, B., and Lubitz, W.**, *Electron Nuclear Double Resonance Spectroscopy of Radicals in Solutions*, VCH Publishers, New York, 1988.
124. **Salikhov, K. M., Semenov, A. G., and Tsvetkov, Yu. D.**, *Electron Spin Echoes and Their Application*, Nauka, Novosibirsk, Russia, 1976.

125. **Lurie, D. J., Nicholson, I., Foster, M. A., and Mallard, J. R.**, Free radical imaged *in vivo* in the rat by using proton-electron double-resonance imaging, *Philos. Trans. R. Soc., London, Ser. A,* 333, 453, 1990.

126. **Zolotov, Yu. A., Petrukhin, O. M., Nagy, V. Yu., and Volodarskii, L. B.**, Stable free-radical complexing reagents in applications of electron spin resonance to the determination of metals, *Anal. Chem. Acta,* 115, 1, 1980.

127. **Buchachenko, A. L., Kovarskii, A. L., and Wasserman, A. M.**, Study of polymers by the paramagnetic probe method, in *Advances in Polymer Science,* Rogovin, Z. A., Ed., John Wiley & Sons, New York, 1976, 273.

128. **Keana, J. F. W., Lex, L., Mann, J. S., May, J. M., Park, J. H., Pou, S., Prabhu, V. S., Rosen, G. M., Sweetman, B. J., and Wu, Y.**, Novel nitroxides for spin-labelling, -trapping, and magnetic resonance imaging applications, *Pure Appl. Chem.,* 62, 201, 1990.

129. **Grigor'ev, I. A., Schukin, G. I., Mamatyuk, V. I., and Volodarsky, L. B.**, Effect of the nitroxyl radical center on keto-enol equilibrium of β-ketoesters of 3-imidazoline derivatives, *Izv. Akad. Nauk S.S.S.R., Ser. Khim.,* 653, 1985 (in Russian).

130. **Rassat, A.**, Communication at the *EUCHEM Conf. Free Nitroxide Radicals,* Hamennlinna, Finland, 1981.

131. **Rozantsev, E. G.**, Some problems of nitroxyl chemistry, *Pure Appl. Chem.,* 62, 223, 1990.

132. *Nomenclature of Organic Chemistry,* International Union of Pure and Applied Chemistry, Butterworths, London, 1971.

133. **Zhdanov, R. I.**, *Paramagnetic Models of Biologically Active Compounds,* Nauka, Moscow, 1981 (in Russian).

134. **Berliner, L. J. and Reuben, J., Eds.**, *Organic Magnetic Resonance: Spin Labeling VIII,* Plenum Press, New York, 1989.

135. **Emanuel, N. M. and Zhdanov, R. I., Eds.**, *Spin Labels and Probes in Biology and Medicine. Spin Labeling Method. Problems and Outlook,* Nauka, Moscow, 1986 (in Russian).

136. **Rozantsev, E. G. and Zhdanov, R. I., Eds.**, *Spin Labels and Probes in Biology and Medicine. Nitroxyl Radicals. Chemistry, Synthesis and Applications,* Nauka, Moscow, 1987 (in Russian).

137. **Likhtenstein, G. I. and Zhdanov, R. I., Eds.**, *Spin Labels and Probes in Biology and Medicine. Biomolecules in Spin Labeling,* Nauka, Moscow, 1988 (in Russian).

138. **Gaffney, B. J.**, The chemistry of spin labels, in *Spin Labeling: Theory and Applications;* Berliner, L. J., Ed., Academic Press, New York, 1976, chap. 5.

139. **Volwerk, J. J. and Griffith, O. H.**, Electron spin resonance of biological membranes: spin-labeled lipids and proteins, *Magn. Res. Rev.,* 13, 135, 1988.

140. **Zhdanov, R. I. and Rozantsev, E. G., Eds.**, *Paramagnetic Models of Biologically Active Compounds,* CRC Press, Boca Raton, FL, 1984.

141. **Likhtenstein, G. I.**, *Spin Labeling Methods in Molecular Biology,* Wiley Interscience, New York, 1976.

142. **Kuznetsov, A. N.**, *Spin Probe Method,* Nauka, Moscow, 1976 (in Russian).

Chapter 2

GENERATION AND CHEMICAL PROPERTIES OF THE NITROXYL GROUP

I. METHODS OF GENERATING THE NITROXYL GROUP

The main preparative methods for generating the nitroxyl group are the oxidation reactions of sterically hindered *N,N*-disubstituted hydroxylamines, secondary, to a lesser degree, tertiary amines. Other reactions include the reductions of nitro and nitroso compounds and oxidative transformations of nitrones, as well as oxidation of unstable aminyl radicals generated by specific methods. Synthetic chemistry of stable nitroxides uses the two former transformations exclusively, the latter reactions often lead to unstable radicals. In particular, there are well-known reactions of nitrones with short-lived radicals leading to nitroxides containing hydrogen atoms at the α-carbon atom of the nitroxyl group. Formation of such radicals (spin adducts) may be recorded spectrometrically, for example, by ESR spectroscopy, but as a rule they may not be isolated individually, while being a useful tool to study different systems where free radical processes occur.[1]

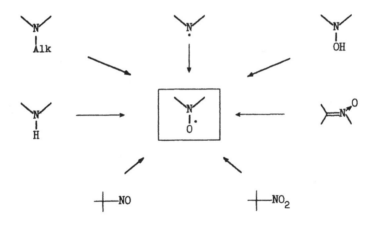

SCHEME 1.

A key stage in the synthesis of a large series of nitroxides is the interaction of nitrones with organometallic compounds (Chapters 3 and 5). The initially formed hydroxylamines are oxidized into corresponding nitroxides, often without isolating them. However, these and related reactions in this chapter are not discussed. The topic of this chapter is the second stage of the two-step process and is considered in the following sections.

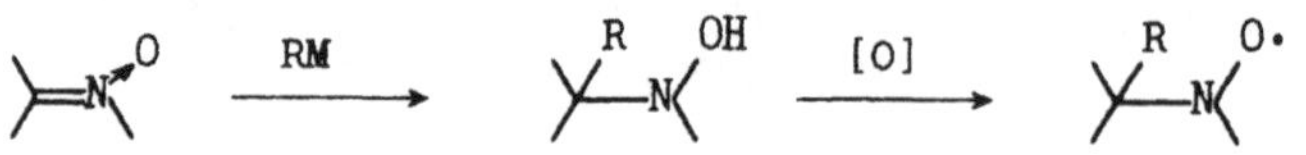

SCHEME 2.

A. OXIDATION OF HYDROXYLAMINES

The *N,N*-disubstituted hydroxylamines are oxidized by different oxidizing reagents, generally under mild conditions, so that most other functional groups remain
unaffected. In particular, this method was used to generate the first stable nitroxides:
the Fremy's salt **1** and nitroxide **2**, synthesized by Piloty and Schwerin back in
1901.[2]

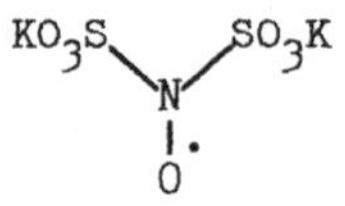

The oxidation is occasionally so easy that it may be difficult to isolate a
hydroxylamine without a corresponding radical impurity. The kinetics of the reaction
were studied for nitroxide **3**, as an example. The reaction rate has been shown to
be the first-order by oxygen and the hydroxide anion, due to the single-electron
oxidation of the hydroxylamine anion. The reaction proceeds by the following
mechanism.[3]

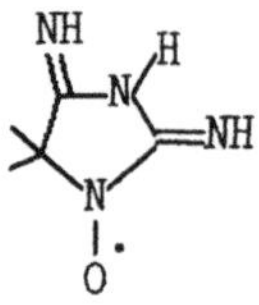

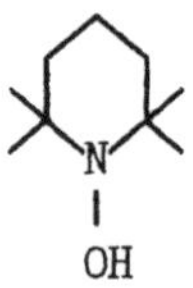

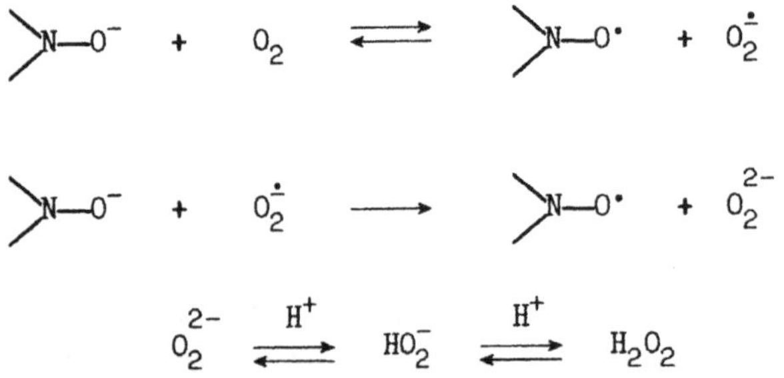

In the organic solvents different bases may be used for ionization of the hydroxylamino group, e.g., potassium *tert*-butoxide.[4]

The oxidation reaction is markedly faster in the presence of copper salts in
accordance with the equation: $V \approx K[Cu^{2+}] [OH^-]$. This is attributed to the fact
that the copper ions in this process are electron carriers from the hydroxylamine
anion to oxygen.[3]

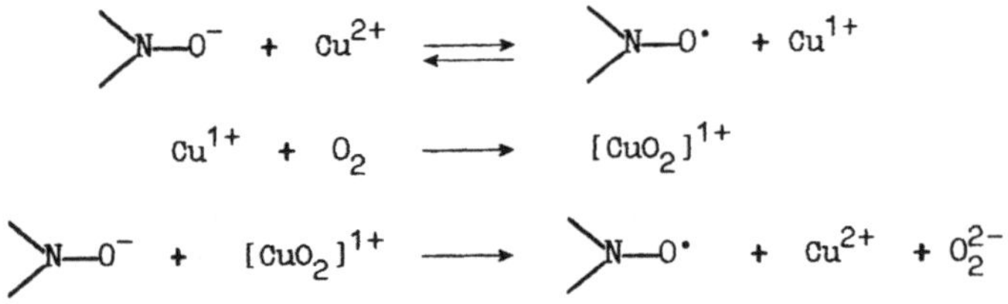

Because alkaline media accelerate the reaction, an ammonia solution of Cu^{2+} in aqueous ethanol or methanol is used. A large number of pyrrolidine nitroxides have been synthesized in these conditions (Chapter 3.II.B). It should be noted that nitroxides with hydrogen at the α-carbon of the nitroxyl group are oxidized in these conditions to nitrones, which allows the transition from less to more substituted nitrones.

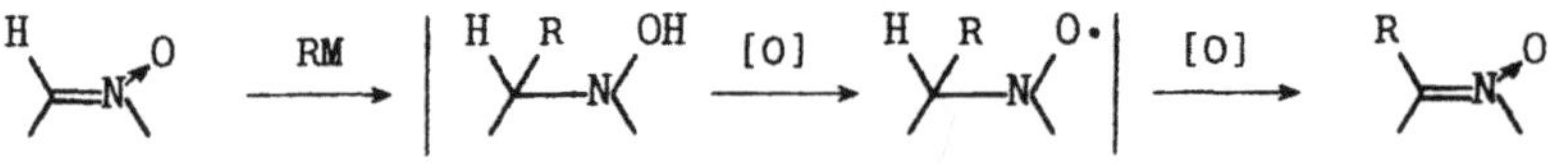

SCHEME 3.

The oxidizers most widely used for this reaction are metal oxides PbO_2, MnO_2,[5] and more rarely mercury (II) oxide.[6] To prepare diphenyl nitroxide, *N,N*-diphenyl-hydroxylamine was oxidized with silver (I) oxide.[7] Other oxidizers that may be used are nitrous acid,[8] Fremy's salt,[9] and other stable nitroxides reduced in this reaction to hydroxylamines.[5] The hydroxylamine-nitroxide equilibrium has been studied using hydroxylamino derivatives **4,5** containing ^{15}N isotope to create the relative oxidative ability scale of cyclic nitroxides and study structural effects on the oxidative ability of the nitroxyl group.[10]

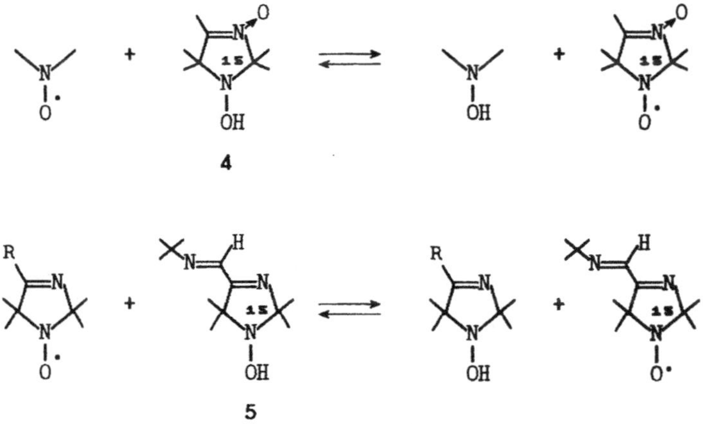

SCHEME 4.

Three main factors governing the stability of the hydroxylamino group as compared to the corresponding nitroxide have been identified: (1) size of heterocycle with nitroxyl group; (2) the nature of heterocycle when its size is the same; and (3) electronic effects of functional groups in different positions of the heterocycle.

The stability of the nitroxyl group against oxidizers increases with the number of units in heterocycle, for bicyclic compounds with bridge hydroxylamino group, and for compounds with higher electron-accepting characters of their substituents.[5]

The oxidation of hydroxylamines to nitroxides is generally an easy process, but it may be difficult for molecules containing a large number of accepting substituents. Thus, as shown in Reference 11 the hydroxylamine **6** is not oxidized to a radical by MnO_2, PbO_2, or H_2O_2 in the presence of Na_2WO_4 (compare to Section I.B): nitroxide **7** was obtained by oxidation of **6** with nitrous acid.[11]

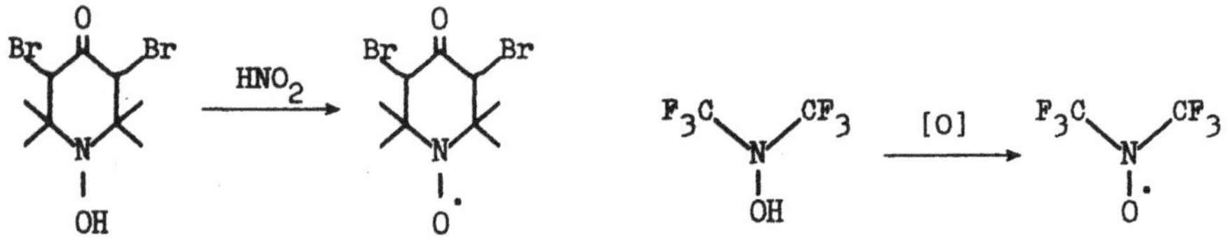

SCHEME 5.

However, *N,N*-bistrifluoromethylhydroxylamine **8** is oxidized to radical **9** by different oxidizers. The most effective oxidizers have been shown to be $Ce(SO_4)_2$ and PbO_2.[12]

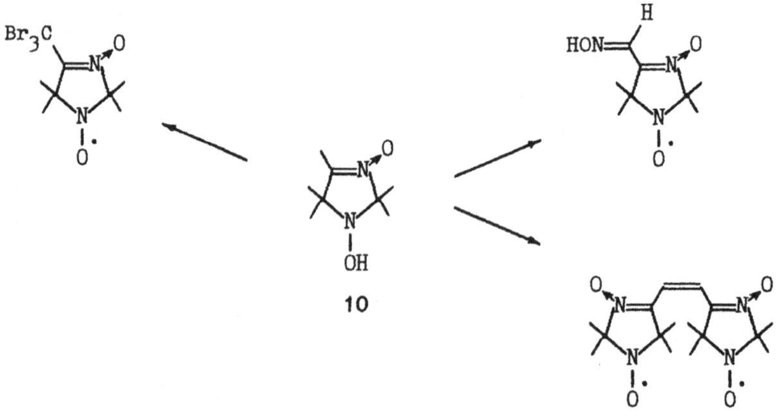

SCHEME 6.

In some cases, the hydroxylamino group is oxidized to the nitroxyl group in the process of chemical modifications of other functional groups in a molecule. Thus, nitroxides were the reaction products in the bromination or nitrozation of hydroxylamine **10**. The reaction of **10** with nitrosobenzene in the presence of NaNH$_2$ in liquid ammonia also leads to biradical **11**.[5]

The hydroxylamino group is also oxidized to the nitroxyl group by other reagents, Br$_2$, I$_2$, NBS,[13] potassium ferrocyanide, etc. The only restriction for the use of an oxidizer is the possibility of further oxidation of the nitroxyl group into the oxammonium group (II.B).

B. OXIDATION OF AMINES

The oxidation of secondary amines is an important preparative method for the synthesis of nitroxides, although it proceeds under much more rigid conditions than the oxidation of hydroxylamines. This is partly due to the fact that modern synthetic chemistry of stable nitroxides started its development from piperidine derivatives, specifically from triacetonamine (which was practically a single starting substrate) where the molecule contains a sterically hindered secondary amino group. The corresponding hydroxylamino derivatives are much less accessible (compare to Chapter 3.III.C.3). This is a consequence of the fact that the starting acyclic compounds for heterocycles, the functionally substituted hydroxylamine derivatives with a hydroxylamino group at the tertiary carbon atom, are also much less available than corresponding amino derivatives. Possibly due to this, subsequent construction of heterocyclic precursors of nitroxides often leads to compounds with a sterically hindered amino group (Chapter 3.III.D). The exceptions are 3-imidazoline and 3-imidazoline-3-oxide derivatives synthesized from the available α-hydroxylaminooximes and α-hydroxylaminoketones (Chapter 3.III.D.1.d). Due to this, methods for the oxidation of the amino group to nitroxyl group have been developed.

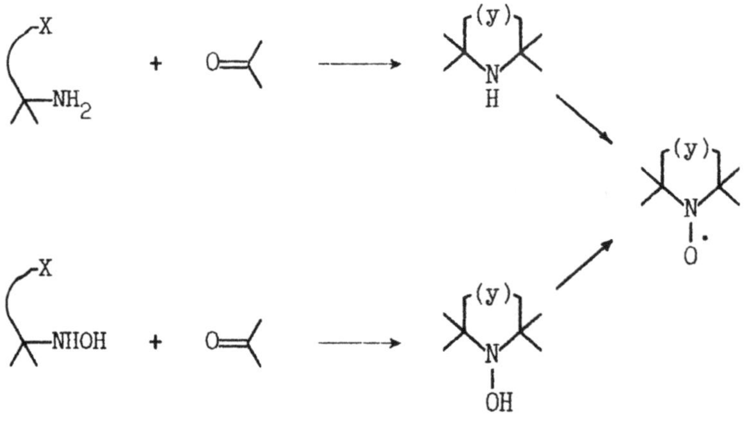

SCHEME 7.

The first stable piperidine nitroxide was registered in the oxidation reaction of 2,2,6,6-tetramethylpiperidine by sodium tungstate.[14] Later on, this commonly used method of oxidation was optimized but remained unchanged in essence. Presently, amines are oxidized using hydrogen peroxide in the presence of sodium tungstate and Trilon B in water or aqueous methanol,[15-19] or hydrogen peroxide in aqueous acetonitrile without a catalyst.[20] The reaction is carried out at room temperature or with heating.* The reaction scheme involves the intermediate formation of hydroxylamine N-oxide in which the homolytical decay ends up with a nitroxide formation.[7] The authors of the monograph in Reference 21 indicate that the reaction mechanism has not been established.

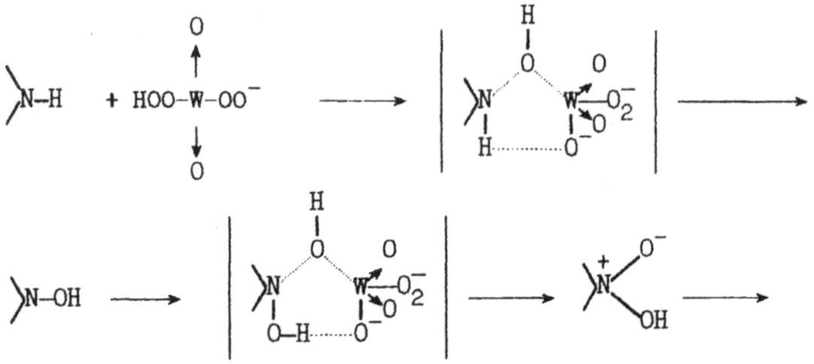

More recently, the kinetics and mechanism of the reaction for piperidine derivatives have been studied.[22] It has been shown that the reaction rate is first-order for the substrate and the catalyst (the tungstate anion). It depends in a complex way on the concentration of hydrogen peroxide and the pH of a medium. The following scheme has been suggested for the process.

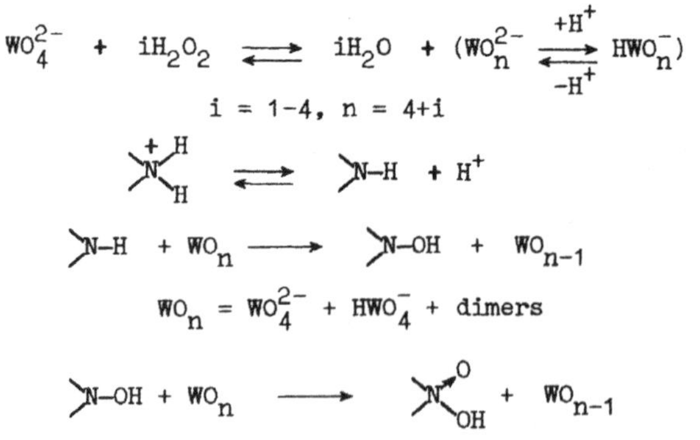

---

* The preparative procedures for oxidation of some amines into nitroxides are given in Reference 7.

$$H^+ + \quad \underset{O^-}{\overset{O}{\diagup\!\!N}} \quad \rightleftharpoons \quad \underset{OH}{\overset{O}{\diagup\!\!N}} \quad \rightleftharpoons \quad \overset{+}{\diagup\!\!N}{=}O \;+\; OH^-$$

$$H_2O_2 \quad \rightleftharpoons \quad HO_2^- \;+\; H^+$$

$$\overset{+}{\diagup\!\!N}{=}O \;+\; HO_2^- \quad \longrightarrow \quad \diagup\!\!N{-}O^\bullet \;+\; O_2^{\bar{\bullet}} \;+\; H^+$$

$$\overset{+}{\diagup\!\!N}{=}O \;+\; O_2^{\bar{\bullet}} \quad \longrightarrow \quad \diagup\!\!N{-}O^\bullet \;+\; O_2$$

Thus, it has been shown that the reaction really proceeds via the stage of the hydroxylamino derivative, while hydroxylamine *N*-oxide is converted into the nitroxyl group via formation of the oxammonium group which is reduced by hydrogen peroxide to form the resulting nitroxyl group. The oxidation of amines by peracids occurs by a similar mechanism. The low yields of nitroxides seem to be due to the fact that the intermediate oxammonium salts may be unstable and decompose with frame destruction.[22]

A large number of nitroxides were synthesized by oxidation of secondary amines by hydrogen peroxide in the presence of phosphotungstic acid.[23,24] It has been shown, however, that this oxidative system is of limited applicability and in some cases it does not give satisfactory results.[25]

The nitroxyl group may also be generated by oxidation of a tertiary amino group, but this reaction occurs under more rigid conditions. There are comparatively few examples of successful conversion of the N–CH$_3$ group to the nitroxyl group (see, e.g., References 18 and 16 to 29). To convert the *N*-benzylamino group into a nitroxyl group, it is subjected to hydrogenolysis to the secondary amino group which is further oxidized to give the corresponding nitroxide (Scheme 3.14).[30]

The comparatively rigid conditions of oxidation in the H_2O_2/Na_2WO_4 system impose certain limitations on the stability of other functional groups in a molecule. Thus, the oxidation of aldehyde **12** gives acid **13** in a moderate yield (compare to Chapter 3.III.A),[9] and oxidation of diamine **14** under this conditions leads to oxime **15**.[31]

Due to this, in order to obtain the amine **16,** the starting diamine **14** is converted to the acetamide derivative **17** in which oxidation and hydrolysis lead to the desired product.[31] Under the oxidation reaction conditions, other functional groups may be involved as well: the cyano group; the mercapto group, etc. This imposes certain limitations on this method for the generation of nitroxides, but these limitations are not absolute and may occasionally be eliminated by selecting adequate reaction conditions. Thus, 2,2,6,6-tetramethyl-1,2,5,6-tetrahydropiperidine may be oxidized into the corresponding nitroxide virtually without the C=C bond being involved.[7] A recent example of oxidation of the N–CH$_3$ group into the nitroxyl group in the presence of the nitrile group makes it possible to consider nitrile oxide **18** as a synthon for introduction of the nitroxyl group into a molecule of the substrate containing multiple bonds.[32]

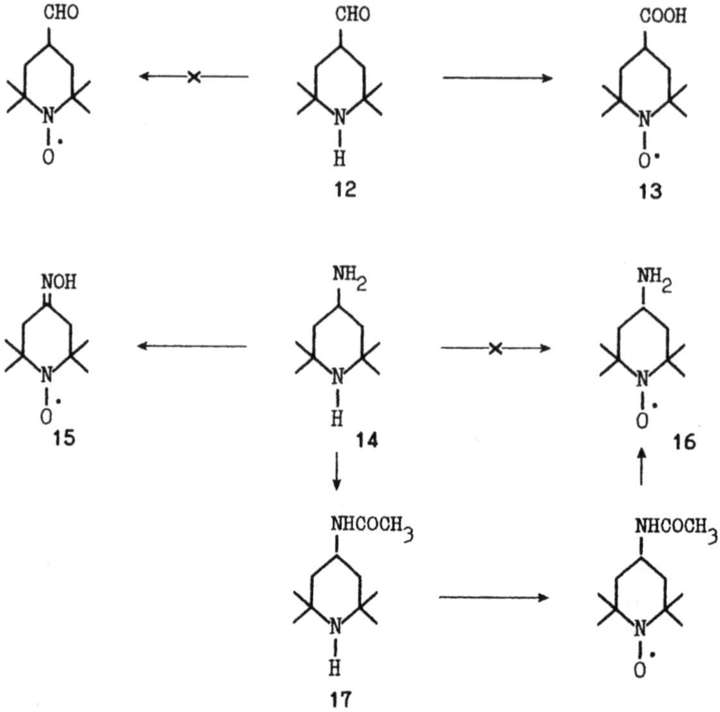

SCHEME 8.

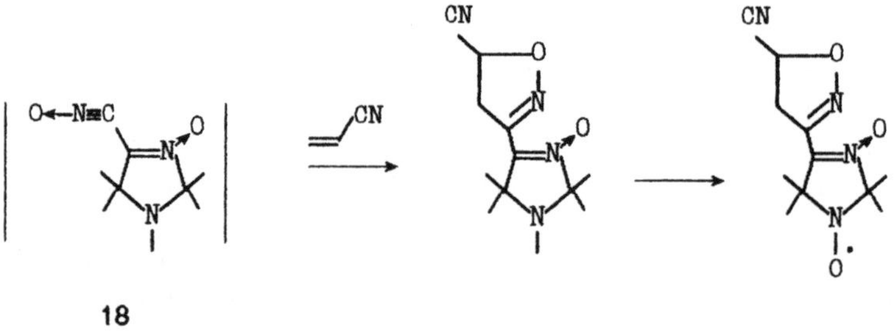

SCHEME 9.

It has been noted that the oxidation of some sterically hindered amines facilitated by H_2O_2/Na_2WO_4 generally takes place only when ultrasound irradiation is used.[33,34]

Peracids, first of all *m*-chloro- and *m*-nitroperbenzoic acids have been often used for oxidation of the secondary amino group into the nitroxyl group. In this case the reaction is conducted in an organic solvent, chloroform or methylene chloride. In these conditions, various heterocyclic nitroxides have been obtained:

imidazolidines; oxazolidines; piperidines, etc. (Chapter 3). This method is most widely used in the synthesis of oxazolidine nitroxides (doxyls). This is due to the fact that the introduction of the oxazolidine fragment into a substrate molecule containing a ketone group is one of the most effective methods for the preparation of spin probes.[35] The method is based on the condensation reaction of 2-amino-2-methylpropanol with ketones.[36]

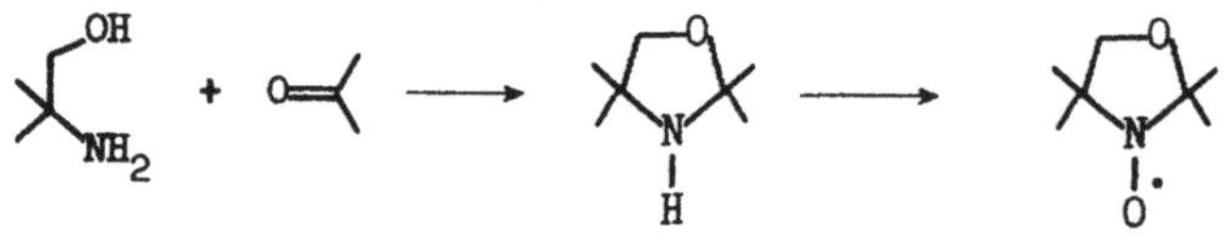

SCHEME 10.

The hydroxylamine intermediate is formed as a result of nucleophilic attack by nitrogen on oxygen of the peroxide group of the peracid.[37]

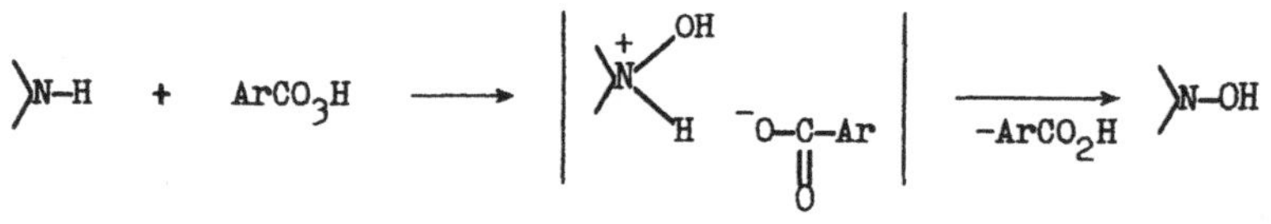

It should be noted that with increased reaction time and reagent excess, the nitroxyl group may be further oxidized into the oxammonium group.[38] This may be the reason for moderate yields of nitroxides in these conditions. Thus, radical **7** was obtained by oxidation of the corresponding amine by meta-chloroperbenzoic acid in a 42% yield.[39]

Peracids may also be used for oxidation without being isolated. Thus, the nitroxyl group may be successfully generated by using hydrogen peroxide solutions of carbamic, acetic, and trifluoroacetic acids.[40] Other oxidizers include hydroperoxides,[41-43] hydrogen peroxide in the presence of cerium salts,[44] peroxyl radicals,[45,46] silver oxide,[47] alkaline solutions of hydrogen peroxide,[48,49] lead dioxide,[50] other different peracids,[24] and ozone.[51] One more preparative method is oxidation of the amino group into nitroxyl group by hydrogen peroxide in an aqueous-hydrocarbon media in the presence of Na_2WO_4. The reaction is conducted with distilling off water azeotrope.[52]

Some recent papers have reported the possibility of oxidation of secondary amines by dimethyldioxyrane. The reaction with equimolar amounts of an oxidant leads to the corresponding hydroxylamines in high yields,[53-55] where oxidation with an excess of reagent smoothly leads to nitroxides. The reaction is also supposed to proceed via the stage of hydroxylamine *N*-oxide formation.[56]

C. OXIDATIVE TRANSFORMATIONS OF NITRONES

When speaking about the reactions of nitrones leading to nitroxides, one usually means the reactions with short-lived radicals leading to the so-called spin adducts. This group of transformations is well defined and is of great interest because various types of active radicals may be recorded by ESR due to formation of much more stable nitroxides, spin adducts.[1,57] This method of investigating active short-lived radicals was named spin traps.* The radicals are most effectively captured by aldonitrones to form nitroxides with hydrogen at the α-carbon atom of the nitroxyl group which have limited stability. Due to this, the spin adducts formed may not be generally isolated as individual compounds. It should be noted, however, that the nitroxide **19** formed in the reaction of phenyl-*N-tert*-butylnitrone with azobisisobutyronitrile under heating was isolated individually in a 3.2% yield.[59]

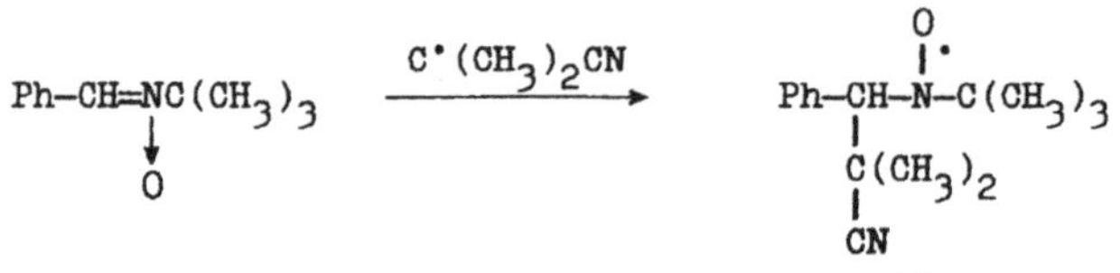

SCHEME 11.

An exception from this series of transformations is the reactions of heterocyclic nitrones with nucleophiles in the presence of oxidants. The reactions proceed via the stage of formation of cation-radicals where interaction with nucleophiles leads to nitroxides with a heteroatom at the α-carbon of the nitroxyl group according to the Scheme 12.

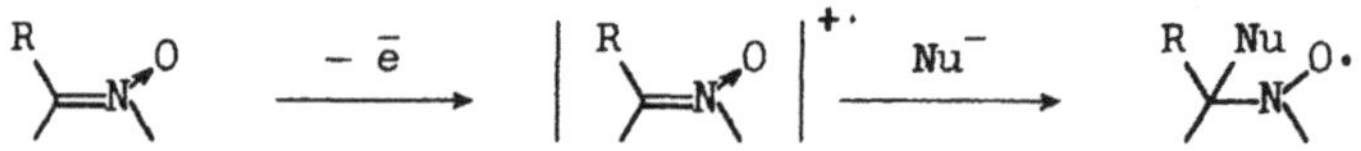

SCHEME 12.

When the nitrone group has a comparatively low oxidation potential, as, e.g., in 4*H*-imidazole *N*-oxides ($E_p \approx 1.0$ to 1.3 V relative to Ag/Ag$^+$) or alkoxynitrones ($E_p \approx 1.1$ V),[60] the reaction proceeds with PbO$_2$ or MnO$_2$ used as oxidants. Synthesis of nitroxides by oxidation of nitrones having higher E_p values ($E_p \geqslant 1.5$ V) requires

---

* Reference 58 reports new data about the use of heterocyclic nitrones 2*H*-imidazole *N*-oxides as spin traps.

stronger oxidizers, such as XeF_2. It is also possible to oxidize nitrones electro-chemically in nucleophilic media (e.g., in alcohols).[61] This approach was used to synthesize various heterocyclic nitroxides: the derivatives of 2- and 3-imidazoline-3-oxide, 2- and 3-imidazoline,[62-65] imidazolidine,[62,66,67] oxazolidine[68] and 1,2,5-oxadiazine containing one or two alkoxy groups in the α-position, the amino group, and fluorine atom.[68,69] Thus, this approach makes it possible to generate the nitroxyl group with a nontraditional environment other than tetraalkyl (Chapter 3).

In conclusion, it should be noted that the oxidation of nitrones capable of tautomeric equilibrium with the enhydroxylamino form leads to vinyl nitroxides which may be recorded by ESR as they are and, more often, as spin adducts with the starting nitrones.[57]

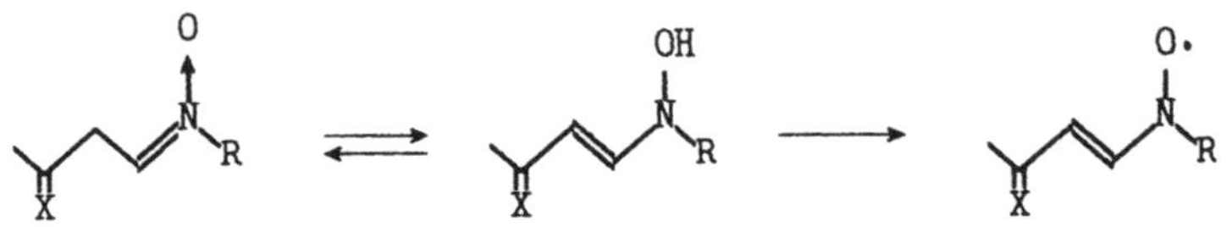

SCHEME 13.

D. METHODS OF CONVERTING THE NITROSO GROUP TO THE NITROXYL GROUP

The aliphatic and aromatic nitroso compounds, just as nitrones, are known to be effective spin traps forming spin adducts with short-lived radicals which as a rule may not be isolated individually.[57,61,71] The nitroso compounds can decompose thermally or photochemically to nitrogen oxide and alkyl radicals. The latter are captured by the starting nitroso compound to form nitroxides, including stable ones.

$$R–NO \rightarrow R^{\cdot} + NO$$
$$R^{\cdot} + R–NO \rightarrow R_2NO^{\cdot}$$

Nitroxide formation occurs most readily in the case of tertiary nitroso com-pounds. Bistrichloromethylnitroxide is formed on thermal decomposition of CCl_3NO at room temperature in darkness,[72] while diadamantylnitroxide is produced on ex-posure of nitrosoadamantane to UV irradiation.[71] This approach is widely used to generate the perfluoronitroxides.[72]

On irradiation of the intramolecular dimer of the nitroso compound **20,** one of the C–N bonds homolytically decomposes to form the C-radical which is captured by another nitroso group to yield the stable nitroxide **21.**[73]

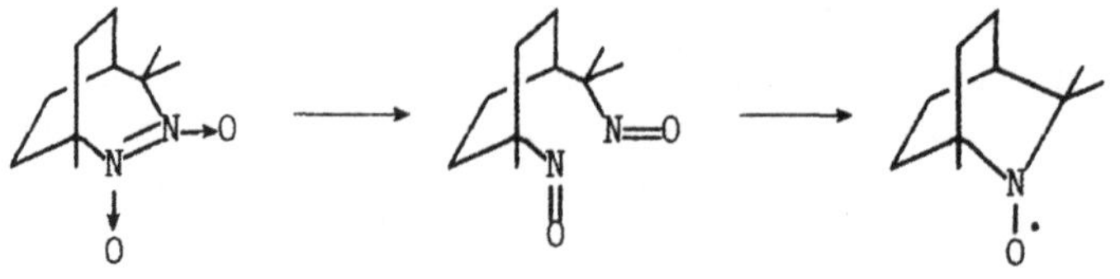

20   21

SCHEME 14.

A well-known nitroxide-forming reaction is that of nitroso compounds with olefins.[57] Formation of nitroxide **22** may be represented by one of the following sequences (Scheme 15).

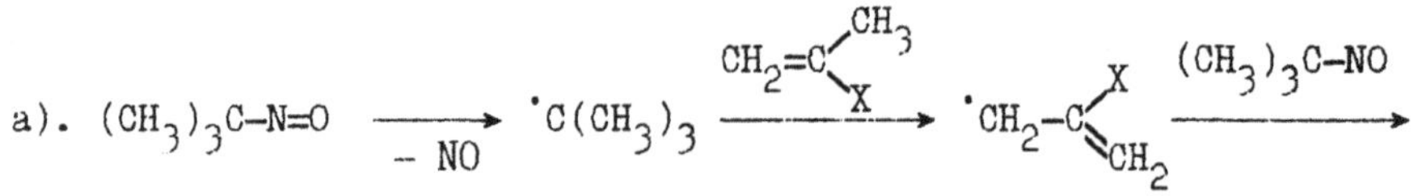

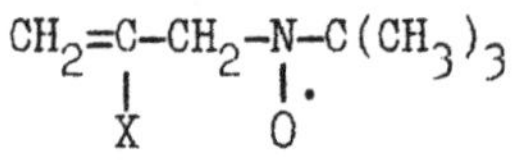

a). $(CH_3)_3C-N=O \xrightarrow{- NO} \cdot C(CH_3)_3 \xrightarrow{CH_2=C\stackrel{CH_3}{\underset{X}{}}} \cdot CH_2-C\stackrel{X}{\underset{CH_2}{}} \xrightarrow{(CH_3)_3C-NO}$

$CH_2=C-CH_2-N-C(CH_3)_3$
          | |·
          X   O·

22

b).

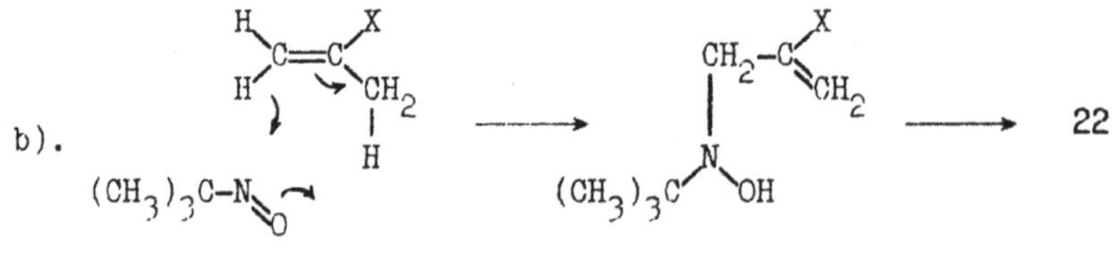

SCHEME 15.

It has been shown using partially deuterated olefins that route **b** is favored in this case.[74] The intermediate hydroxylamine is oxidized by the starting nitroso compound to nitroxide **22**.

This approach occasionally leads to stable nitroxides. Thus, caryophyllene reacts with nitroso-*tert*-butane to form nitroxide **23**.[75]

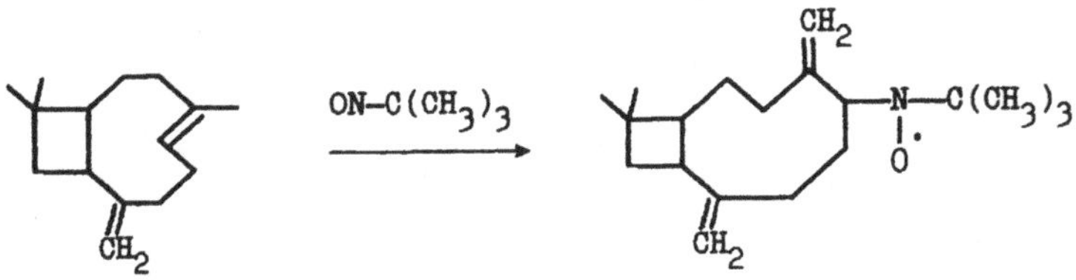

SCHEME 16.

It seems that the oxidation reaction of the hydroxylamino group by nitroso compounds is of a sufficiently general character. In particular, this accounts for the fact that the reactions of nitroso derivatives with organomagnesium compounds yield nitroxides rather than hydroxylamines (compare to Reference 76).

Another interesting example of the intramolecular spin capture is the interaction of the unsaturated nitroso compound **24** with iodine leading to the stable nitroxide **25**.[77]

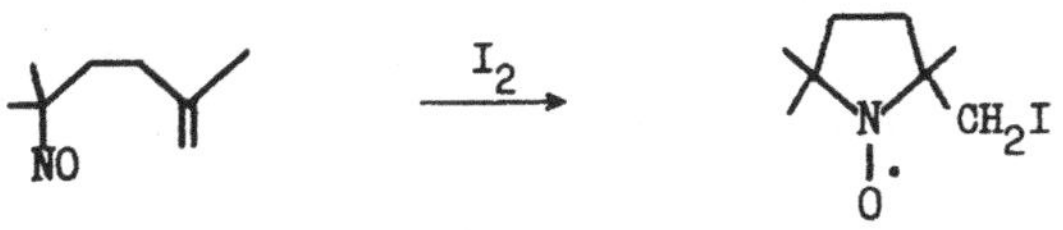

SCHEME 17.

Some nitroxide-forming transformations of nitroso compounds are described in reviews (References 57 and 78), but they are of limited interest for the synthetic chemistry of nitroxides, because the nitroxides formed in these reactions are unstable and may not be isolated individually.

E. THE NITRO GROUP AS A PRECURSOR OF THE NITROXYL GROUP

The nitroxyl group is formed by the reaction of the nitro group by different reducing agents: organomagnesium and -lithium compounds; lithium aluminohydride; or electrochemically.[79-84] The reaction of nitro-*tert*-butane with sodium metal gives di-*tert*-butylnitroxide.[79] The reaction proceeds as a one-electron reduction forming the anion-radical $(CH_3)_3CNO_2^-$ **26** which decomposes to the NO_2^- and the *tert*-butyl radical. Recombination of this radical with the anion-radical leads to the hydroxylamine *N*-oxide anion, subsequent hydrolysis of which produces a nitroxide (compare with Section I.B.).[80] The reaction mechanism seems to be similar for other reducing agents as well.

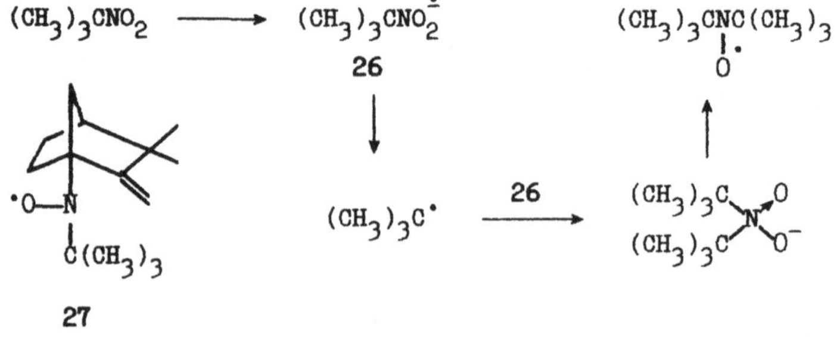

$(CH_3)_3CNO_2 \longrightarrow (CH_3)_3C\overset{\cdot}{N}O_2^{-}$ $(CH_3)_3CNC(CH_3)_3$
                                            26 $\underset{O\cdot}{|}$

27

$(CH_3)_3C\cdot \overset{26}{\longrightarrow} (CH_3)_3C\overset{(CH_3)_3C}{\underset{}{N}}\overset{O}{\underset{O^-}{}}$

SCHEME 18.

Some nonsymmetric dialkylnitroxides were obtained by this method. For example, the reaction of 1-nitrocamphene with *tert*-butylmagnesium chloride gave nitroxide **27**.[85] In the reactions of aromatic nitro compounds, alkylarylnitroxides are formed in moderate yields.[82-84] Reference 7 gives some preparative procedures for the synthesis of nitroxides by the reaction of nitro derivatives with organomagnesium compounds.

The nitro compounds can also be traps for short-lived radicals.[57] The alkoxynitroxides formed in this reaction are unstable (though they may be recorded spectrally), and easily decompose to form dialkylnitroxides or the alkoxyaminyl radicals.[86]

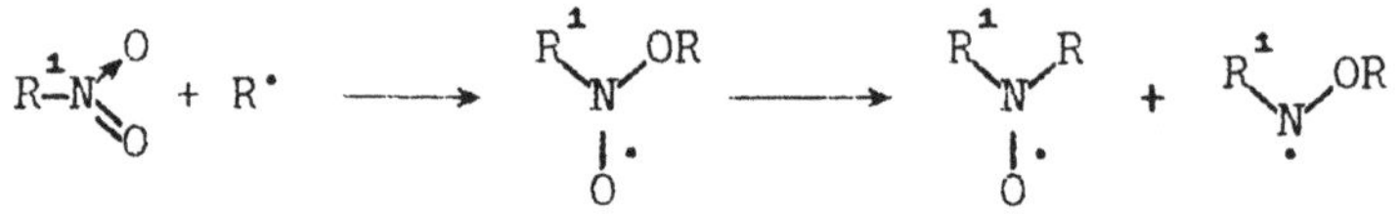

F. OTHER METHODS OF GENERATING THE NITROXYL GROUP

Nitroxides are easily formed by oxidation of aminyl radicals with oxygen. The starting aminyl radicals are unstable, as opposed to the corresponding nitroxides. To convert them into nitroxides, they are generated by one of procedures indicated in Scheme 19 in the presence of oxygen.[78]

It has been shown for the reactions of aminyl radicals with oxygen $^{17}O_2$ that the reaction takes place directly between these compounds and forms a peroxide radical where subsequent recombination with the aminyl radical leads to a nitroxide with the ^{17}O isotope.[87] Despite the seemingly general character of this approach, it is of limited applicability for the synthesis of stable nitroxides but is used to generate nitroxides in spectral studies.

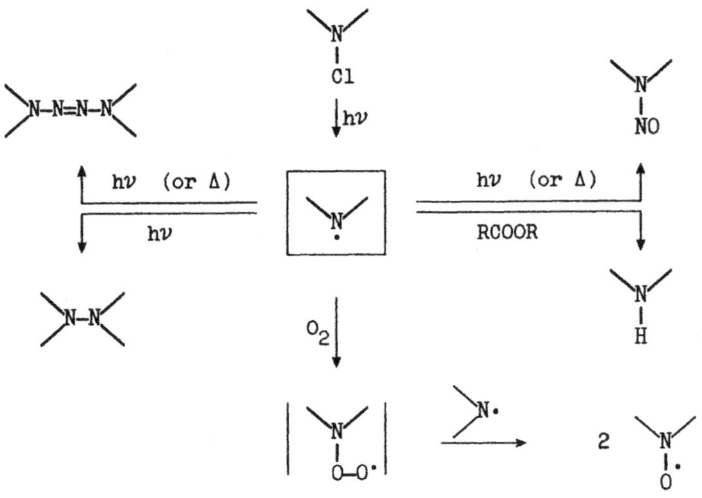

SCHEME 19.

The nitroxyl group may be generated by the one-electron reduction of the oxammonium group. While the latter is usually obtained by oxidation of the nitroxyl group, this method is applicable to obtain some diarylnitroxides.[88,89] For example, the reaction of trimethyl ether of fluoroglucine with nitric acid in acetic acid gives an oxammonium salt **28** whose reduction with KI leads to nitroxide **29**.

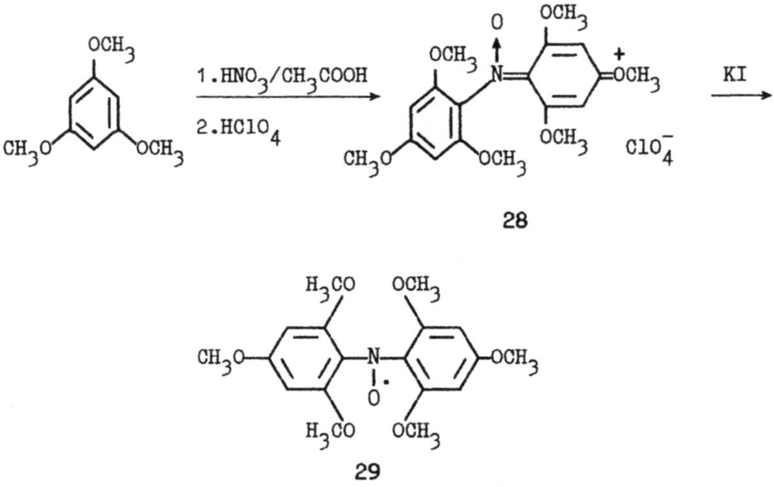

28

29

SCHEME 20.

In the reactions of *N*-substituted hydroxylamines with nitrile oxides, unstable nitroxides **30** are formed. An oxidizer in this reaction is nitrile oxide.[57] Chloroxime **31** has been shown to react with a fivefold excess of *tert*-butylhydroxylamine to give *N*-hydroxyamidoxime **32**, while with a twofold excess of *tert*-butylhydroxylamine it leads to a stable radical **33**, along with compound **32**. The reaction is supposed to occur via formation of the nitrile oxide.[90]

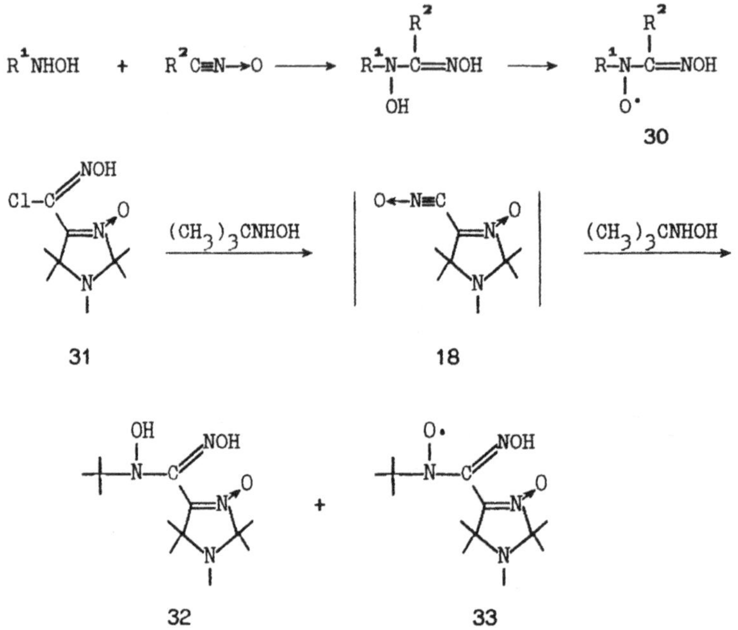

$$R^1NHOH \quad + \quad R^2C\equiv N\rightarrow O \longrightarrow R^1-N-C=NOH \longrightarrow R^1-N-C=NOH$$

with R^2 above and OH / O· below, **30**

SCHEME 21.

The *N,O*-disubstituted hydroxylamines may be occasionally transformed into nitroxides. The deprotonation of the disubstituted hydroxylamine **34** leads to anion **35** being in equilibrium with the anion **36**, owing to which its oxidation by oxygen or electrochemically affords the nitroxide **37**.[91]

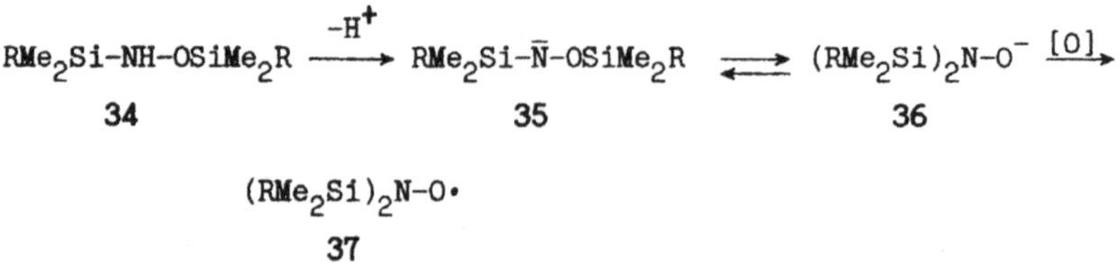

$$RMe_2Si-NH-OSiMe_2R \xrightarrow{-H^+} RMe_2Si-\bar{N}-OSiMe_2R \rightleftharpoons (RMe_2Si)_2N-O^- \xrightarrow{[O]}$$

$$\text{34} \qquad\qquad\qquad \text{35} \qquad\qquad\qquad \text{36}$$

$$(RMe_2Si)_2N-O\cdot$$

$$\text{37}$$

SCHEME 22.

Thus, the analysis of the data on the generation of nitroxides shows amines and hydroxylamines to be synthetically the most useful precursors of nitroxides. One can argue that the compounds containing a sterically hindered amino or hydroxylamino group in all cases may actually be transformed into nitroxides. This makes it possible to synthesize a wide variety of structures containing a nitroxyl group (see Chapter 3).

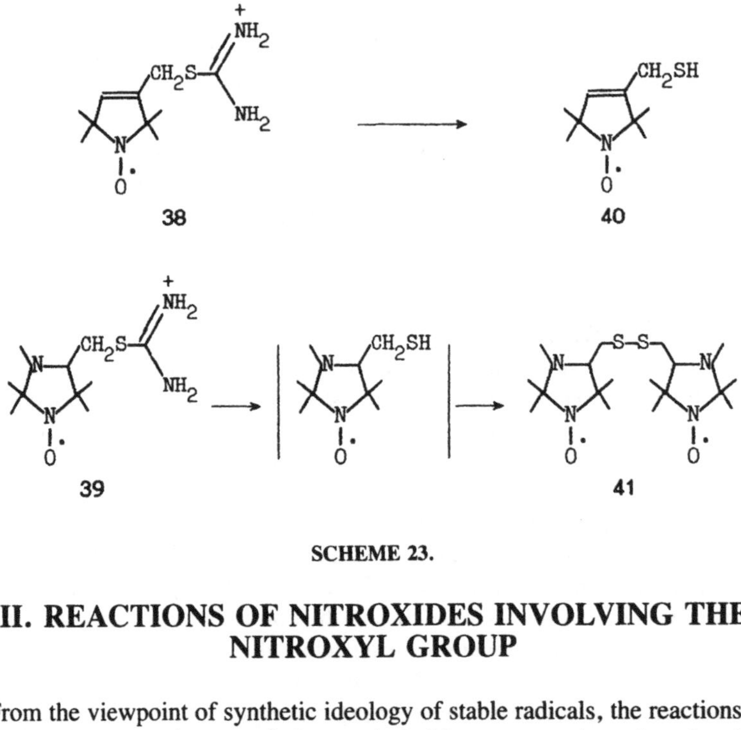

SCHEME 23.

II. REACTIONS OF NITROXIDES INVOLVING THE NITROXYL GROUP

From the viewpoint of synthetic ideology of stable radicals, the reactions at the nitroxyl group are to be regarded as undesirable processes lowering the yield of the target product or hindering its formation altogether. In this sense it is important to indicate possible conditions of such processes, to avoid them in the process of chemical modifications involving other functional groups of the molecule, or to assess the possibility for any specific structure to exist in any specific conditions. Thus, it seems unlikely that one molecule should contain both the nitroxyl group and strong reducing groups. However, the stability of the nitroxyl group is also determined by the nature of its environment and the structure of the whole frame of the molecule. Thus, hydrolysis of thiuronium salts **38,39** under the same conditions leads in the former case to a stable radical with the mercapto group **40,**[92] and in the latter, to disulfide **41,** which seems to result from the oxidation of the intermediate mercaptane by the nitroxyl group.[93]

The nitroxyl group has limited stability in acid media and in the presence of strong oxidants. Under these conditions the nitroxyl group can be disproportionate to the oxammonium and hydroxylamino group or be oxidized into the oxammonium group. When the oxammonium salt or hydroxylamine formed are unstable in the reaction conditions, the molecule of the starting nitroxide can undergo irreversible destruction. In the case of stable intermediates, in the reaction mixture diluted with water, the nitroxyl group may be regenerated as a result of the oxidation of hydroxylamine by oxammonium salt or reduction of the oxammonium group by water. In some cases this allows the nitration reaction to be performed with formal preservation of the nitroxyl group.[94-96]

The nitroxyl group may be reduced under the reaction conditions to the hydroxylamino or amino group. In the former case, the nitroxyl group is easily regenerated in mild conditions, while in the latter case subsequent oxidation can involve other functional groups in the molecule, so that the nitroxide of the desired structure will not be obtained. The aim of this section is to indicate the most typical transformations of the nitroxyl group which can proceed along with modification of other functional groups in a molecule. The reactivity of the nitroxyl group is also covered in monographs and reviews.[5,7,21,57,78,97,98]

A. REDUCTION REACTIONS

Ease of reduction of the nitroxyl group essentially depends on its environment. Increase of the electron density on oxygen and, consequently, of its nucleophilicity is determined by the contribution of the resonance structure **B**. In this connection it is interesting to note that the splitting constant values of the nitrogen atom of the nitroxyl group (α_N) and ease of nitroxyl group reduction must change oppositely. Therefore, the α_N value may be suggested for qualitative estimation of the ease of nitroxyl group reduction in nitroxides of related structure. It should be noted that the contribution of the resonance form **B** seems to decrease on introduction of electron-accepting substituents, resulting in the corresponding decrease of the α_N value. Evidently, all other conditions being equal, the more electron-accepting the substituents and the lower the α_N value, the slower the oxidation reaction of the hydroxylamino group into the corresponding nitroxide.

$$\text{A} \qquad\qquad \text{B}$$

For quantitative estimation of ease of the one-electron reduction of nitroxyl group, it was suggested that its electrochemically determined reduction potential should be used.[5] However, in biological systems, the ease of nitroxyl group reduction does not correlate with its electrochemical reductive potential and is rather determined by accessibility of the nitroxyl group or its less steric hindrance.[99]

Another approach to plot the ratio scale of reductive ability of nitroxides is the use of the above-mentioned equilibrium of nitroxides with ^{15}N-labeled hydroxylamines (Scheme 4).

The nitroxyl group is reduced by many inorganic reagents being electron donors. Evidently, not only electron but also proton donors are required for the reduction reaction to occur. The extent of nitroxyl group reduction (to hydroxylamine or amine) depends not only on the reducing agent but also on the nature of the radical.

Back in 1922 Wieland established that diarylnitroxides are reduced to hydroxylamines by zinc dust in hydrochloric acid.[100] Zinc reduces the nitroxyl group to the hydroxylamino group merely in an alcoholic solution or in the presence of ammonium chloride.[101,102] The nitroxide **42** is reduced by zinc in acetic acid at 0°C to hydroxylamine **43**, while upon heating it is reduced to diamine **44**.[103]

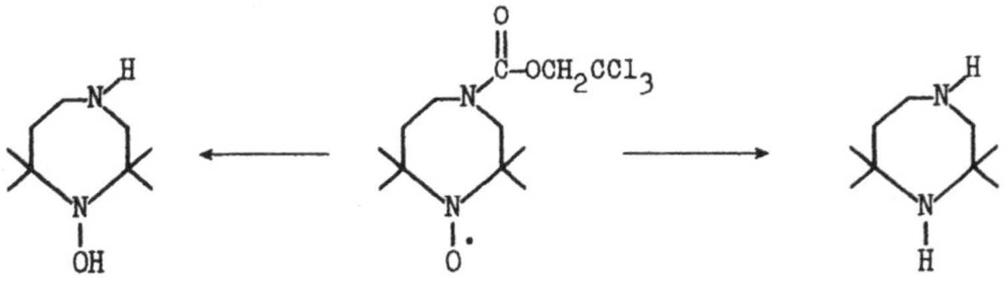

SCHEME 24.

The nitroxyl group is reduced by alkaline metals, thus, the reaction of di-*tert*-butylnitroxide with sodium metal followed by treatment with hydrogen chloride leads to di-*tert*-butylhydroxylamine hydrochloride.[79] Bis(trifluoromethyl) nitroxide **9** is easily reduced to the anion of the corresponding hydroxylamine by stainless steel, iron, tin, and lead.[104,105] A reducing agent may be a Fe(II) salt. The reaction possibly proceeds as a one-electron transfer from the iron ion to the nitroxyl group with subsequent protonation of the resulting anion. This, evidently, accounts for the fact that the reaction rate increases with decreased pH of the medium.[106] It has been remarked earlier (Section I.A) that the copper salts catalyze the oxidation reaction of the hydroxylamino group to the nitroxyl group. In view of the reversibility of all reaction stages, it is clear that the reverse process of nitroxyl group reduction is also catalyzed by copper salts, the equilibrium position is dependent on pH.[107] It has been noted that the nitroxyl group is reduced to the hydroxylamino group by methanol in the presence of Cu(II) salts, and methanol is reduced to formaldehyde.[108]

The nitroxyl group is reduced to the amino group by sodium sulfide in aprotic solvents,[109] potassium hydrosulfide in ethanol, or by metal carbonyls.[110] The reaction with sodium sulfide has an induction period and is accelerated by elementary sulfur, indicating the radical character of the process. The nitroxyl group is also reduced to the amino group by tin(II) chloride in hydrochloric acid.

The extent of reduction of the nitroxyl group by hydrogen depends on the nature of the catalysts. On a platinum Adams catalyst the reduction ends up with the formation of hydroxylamine, while using the Raney nickel often leads to amines.

The authors of Reference 111 have studied the effect of the catalyst nature on the relative ease of reduction of the nitroxyl group and other functional groups. The activity series for palladium catalysts has been established to be: $C \equiv C - C = C > C \equiv C$ $> C = C - C = C > C = C$ (α) $> {>}N{-}O > C = C$ (β) and for platinum catalysts: ${>}N{=}O$ $> C \equiv C - C = C > C \equiv C > C = C - C = C$. The biradicals are reduced slightly more easily than the monoradicals, and the process may be conducted stepwise. Nickel catalysts are similar to platinum ones, but the nitroxyl group and the $C \equiv C$ bond are reduced simultaneously.[111]

2,2,6,6-Tetramethylpiperidine-1-oxyl is reduced by copper perchlorate in orthoformic ether at room temperature to the corresponding amine.[112]

The nitroxyl group is very easily reduced to the hydroxylamino group by iodides in acid media. This reaction was suggested for quantitative determination of nitroxides.[113] It has been noted that on boiling 2,2,6,6-tetramethylpiperidine-1-oxyl with methyl iodide in benzene, the hydroiodide of the corresponding amine was formed in a high yield.[114] Quantitatively the nitroxyl group is reduced to the hydroxylamino group by formic acid,[7] sodium ascorbate, and thioalcohols.[115]

Very efficient reducers of the nitroxyl group are hydroxylamine, *N*-alkylhydroxylamines,[116,117] and hydrazine derivatives — phenylhydrazine,[118-122] methylhydrazine,[123] hydrazobenzene,[124] and unsubstituted hydrazine.[125] Because such reducers are strong nucleophiles they can simultaneously react with other functional groups in a molecule. It has been shown that in the reactions of 3-imidazoline-3-oxide nitroxides, the nitrone group of the heterocycle is deoxygenated along with reduction of the nitroxyl group.[19] Still more dramatic changes are caused by hydrazine and hydroxylamine reduction of compounds **45** and **46**.[126,127] The reaction is accompanied by cleavage of the imidazoline heterocycle with subsequent recyclization.

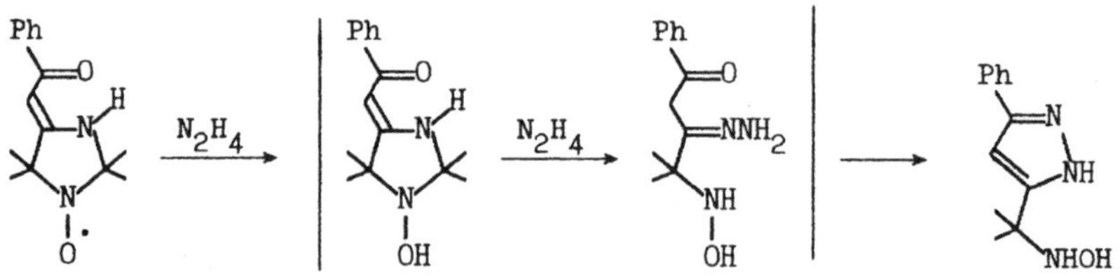

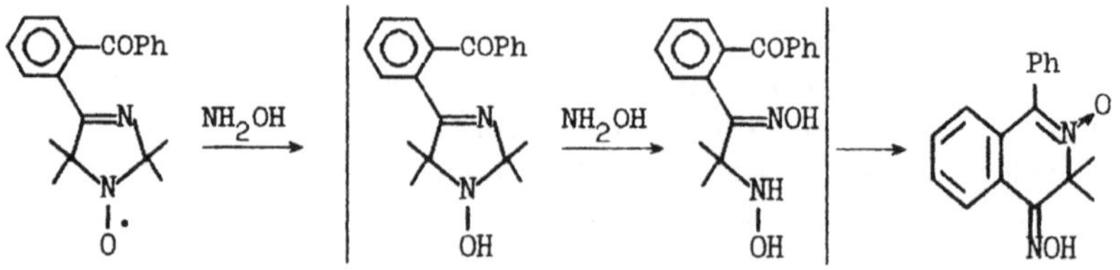

SCHEME 25.

Another interesting peculiarity of the interaction of nitroxides with substituted hydrazines (methylhydrazine and phenylhydrazine) is a formation of *O*-methyl and *O*-phenyl ethers of hydroxylamines along with the corresponding hydroxyl-amines.[118,122,123] The following reaction scheme has been suggested.[123]

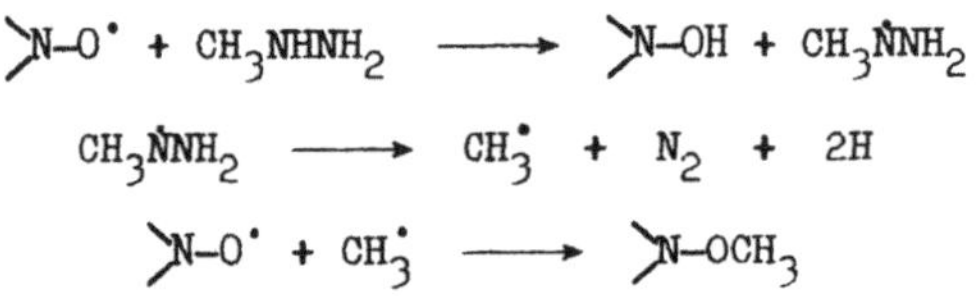

$$\text{>N-O}^{\bullet} + CH_3NHNH_2 \longrightarrow \text{>N-OH} + CH_3\overset{\bullet}{N}NH_2$$

$$CH_3\overset{\bullet}{N}NH_2 \longrightarrow CH_3^{\bullet} + N_2 + 2H$$

$$\text{>N-O}^{\bullet} + CH_3^{\bullet} \longrightarrow \text{>N-OCH}_3$$

The frame of the nitroxide molecule is often built using reactions with organometal compounds. In this case the nitroxyl group is always reduced; the structure of products and their ratio essentially depend on the structure of the starting radical. In a simple case of reactions of 3-imidazolinium nitroxides, the nitroxyl group is reduced to the hydroxylamine anion.[101] In the metallation reaction of the methyl group in the 4-position of the heterocycle in 2,2,4,5,5-pentamethyl-3-imidazoline-1-oxyl by phenyllithium, the nitroxyl group is also reduced to the anion.[128] When nitroxide **47** reacts with butyllithium, there occurs the one-electron reduction of the nitroxyl group to form the hydroxylamine anion and the butyl radical. Recombination of the latter with the starting radical leads to the butoxy derivative **48**. Also, the product of deoxygenation, amine **49**, is formed.[129]

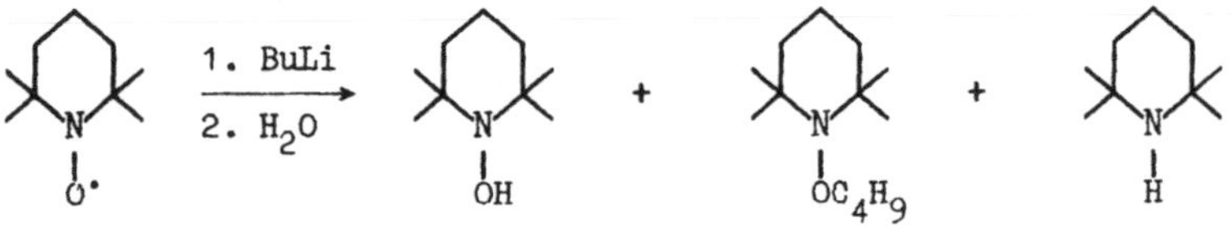

SCHEME 26.

The reaction of **47** with organomagnesium compounds proceeds in a similar manner. The ratio of reaction products, hydroxylamine and its alkoxy derivative, depends on the nature of the organomagnesium reagent. At elevated temperatures and with an excess of the reagent, the hydroxylamine anion may be reduced to the amino derivative.[130] In the reaction of 2,2,5,5-tetramethyl-4-phenyl-3-imidazoline-1-oxyl with phenyllithium, the *O*-phenyl ether was not formed, only the nitroxyl group was reduced to the hydroxylamine anion. The reaction of this nitroxide with butyllithium leads, after oxidation of reaction products with MnO_2, to approximately equivalent quantities of the starting nitroxide and the butoxy derivative.[127]

Another large group of reducers often used in the synthesis of functionally substituted nitroxides is metal hydride complexes. The occurrence of nitroxyl group reduction in this case depends on the type of nitroxide, hydride complex, and reaction conditions. Lithium aluminohydride does not reduce the nitroxyl group in pyrrolidine derivatives but does reduce it in piperidines.[131,132] In the latter case, the occurrence of the process is governed by the reaction temperature: on boiling the reaction mixture in ether, the nitroxyl group is reduced to the hydroxylamino group, while at room temperature it remains intact, permitting the reduction of other functional groups in its presence.[133] The nitroxyl group of the 3-imidazoline heterocycle is reduced by lithium aluminohydride to the hydroxylamino group. Sodium borohydride does not affect it,[101,134] nor is it reduced in the piperidine heterocycle by potassium borohydride.[135] However, in indoline nitroxides **50**, the nitroxyl group is reduced along with the keto group.[136]

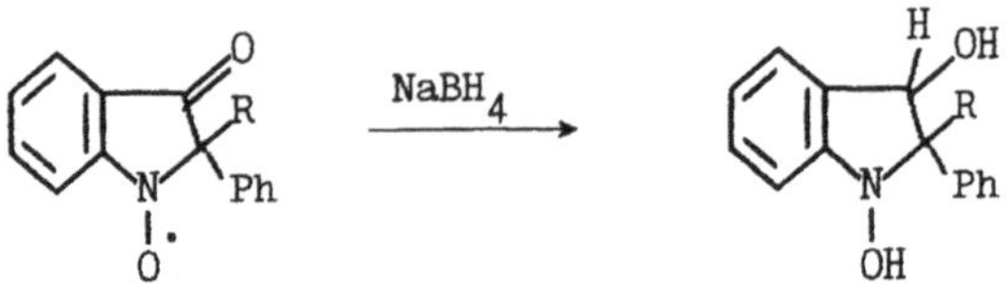

50

SCHEME 27.

Possibly, in this case, the reduction of the nitroxyl group is a secondary process ensuing from oxidation of the initially formed alcohol group by the nitroxyl group (compare to Reference 134).

As noted earlier, the electron-accepting substituents increase the oxidative ability of the nitroxyl group. In bis(trifluoromethyl) nitroxide **9,** the nitroxyl group is bonded with two strong electron-accepting groups and, therefore, is reduced very easily. Thus, nitroxide **9** eliminates the hydrogen atom from oximes to form iminoxyl radicals,[137] or is added at multiple bonds of alkenes and fluorinated olefins,[104,105] acetylenes,[138] perfluoroallenes,[139] isonitriles,[140] and ketenes,[141] or is reduced by acetylenes,[142] aldehydes,[143] alkylbenzenes,[144] alkenes, and alkanes.[145,146] It should be noted that at elevated temperatures, ordinary nitroxides can also eliminate hydrogen atoms from hydrocarbon molecules.[147]

B. OXIDATION REACTIONS

In the one-electron oxidation of the nitroxyl group, the oxammonium salts are formed. This process, which is reversible in most cases, has been studied by cyclic voltametry in acetonitrile on a platinum electrode for a wide range of heterocyclic nitroxides.[148] It has been shown that the value of oxidation potential mainly depends on the heterocycle size and on the electronic effects of substituents: increased accepting character of substituent increases the oxidation potential, following a definite tendency. The oxidation potential value only slightly depends on the solvent and is between $+0.21$ and $+1.09$ V. This value is rather high, so that the oxidation requires reasonably strong oxidants. Thus, piperidine nitroxides are not oxidized by iodine, but bromine and chlorine easily and quantitatively oxidize them into oxammonium salts.[113,149,150] The reactions of pyrrolidine nitroxides with bromine also afforded tribromides of oxammonium salts.[151,152]

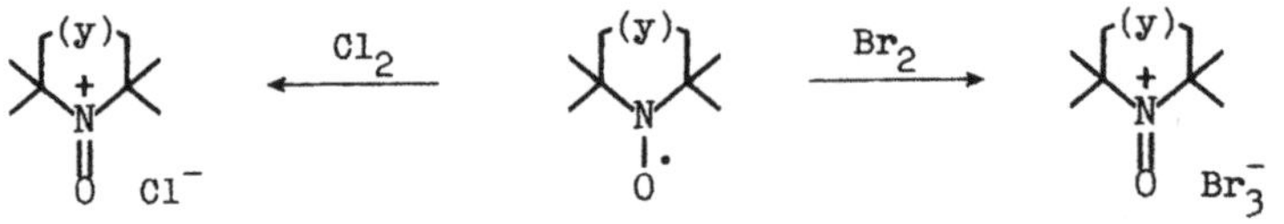

SCHEME 28.

The introduction of accepting substituents increasing the oxidation potential of the nitroxyl group hinders its transformation into the oxammonium group. Thus, 2,2,6,6-tetramethyl-4-oxo-3-chloropiperidine-1-oxyl is not oxidized by chlorine, bromine, or N-bromosuccinimide.[13]

The nitroxyl group of the piperidine nitroxide **51** is oxidized by xenon difluoride. One of the C–N bonds is cleaved to form nitroso compounds,[153] however, the nitroxyl group of the imidazolidine heterocycle is stable against this reagent.[62]

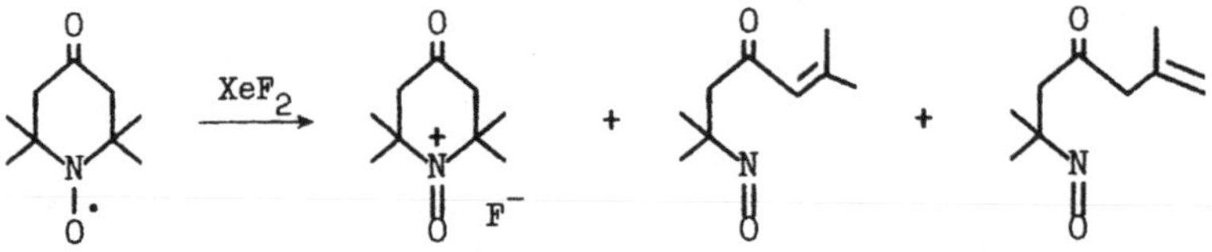

SCHEME 29.

The oxidation of the nitroxyl group also takes place in reactions with Lewis acids: antimony, tungsten, tin, and aluminium halides.[154-157]

$$2 \; \text{>N—O}^{\cdot} + 3 \; SbCl_5 \longrightarrow 2 \; \text{>N}{=}\text{O} \; [SbCl_6]^- + SbCl_3$$

The stability of oxammonium salts is governed by the structure of the heterocycle including the oxammonium group. The piperidine and pyrrolidine nitroxides form stable oxammonium salts which may be isolated individually. With an additional geminal heteroatom in the heterocycle, the oxammonium salts are predominately unstable. For example, the oxammonium salts formed in the oxidation of 3-imidazoline nitroxides by halogens or nitrous acid readily decompose with heterocycle cleavage.[158] The oxammonium salts formed in the oxidation of oxazolidine nitroxides by chlorine or nitrous acid are also unstable: the heterocycle is cleaved to produce ketones.[159,160]

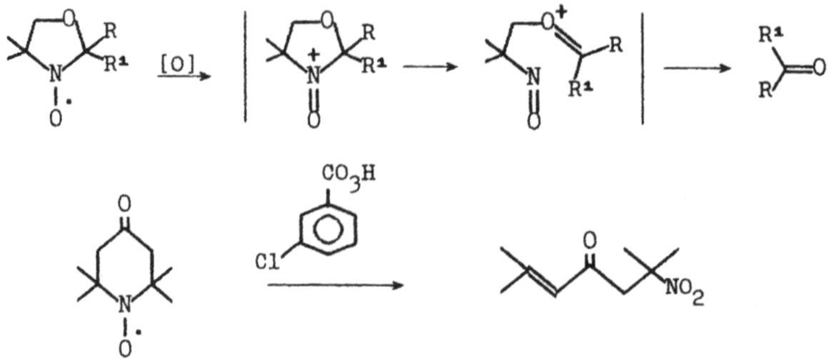

SCHEME 30.

The low yields of oxazolidine nitroxides in the reactions with *m*-chloroperbenzoic acid used to generate such compounds are attributed to the possibility of such a transformation.[38]

The oxammonium salt seems to be formed as an intermediate in the oxidation of nitroxide **51** by *m*-chloroperbenzoic acid. The reaction yields nitroketone **52** as an end product.[38] The pyrroline and pyrrolidine nitroxides are stable against *m*-chloroperbenzoic acid.

Di-*tert*-butylnitroxide is oxidized by ozone to nitro-*tert*-butane and tris(*tert*-butyl)hydroxylamine.[161] The nitroxyl group is oxidized into the oxammonium group in the reaction with triphenylmethyl perchlorate in the presence of oxygen.[162]

It is worthwhile to note an unusual reaction pathway in the oxidation of stable nitroxide **53** by MnO_2 under mild conditions. Obviously, the reaction does not proceed via the stage of the oxammonium salt but rather via the aminyl radical that is thought to be possible due to the existence of radical **53** in the spirobicyclic tautomeric form.[127]

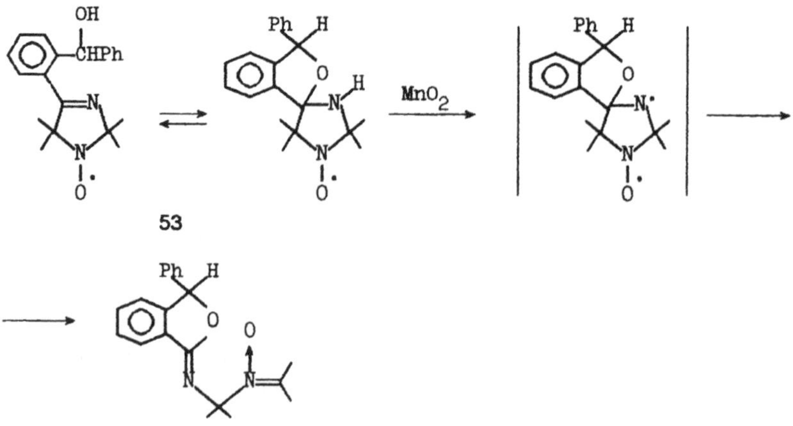

SCHEME 31.

The oxammonium salts being the products of oxidation of nitroxides, have acquired great independent value during the last few years. This is due to the fact that these compounds are strong oxidants; the oxidative processes involving such compounds proceed under mild conditions with high yields and selectivity. Because the oxammonium group is quantitatively reduced in these processes to the nitroxyl or hydroxylamino group, it may be regenerated by another oxidant added to the system, so that the process is performed in a catalytic variant. It should be noted that the auxiliary oxidant and the nitroxide introduced into the reaction generally do not react with the substrate being oxidized. This catalytic system is used, for example, for oxidation of alcohols to aldehydes and ketones, of primary amines to aldehydes or, in some cases, to nitriles, of ketones to α-diketones, and of phenols to quinones. Such transformations have been covered in reviews.[162,164] The main requirement of the oxammonium salts used in such reactions is their stability and reversibility of the redox process leading to a nitroxide or hydroxylamine. This requirement is satisfied by oxammonium salts generated from pyrrolidine nitroxides, and especially piperidine mono- and dinitroxides. The auxiliary oxidants are often Cu (II) salts, chlorine, hypochlorites,[165] and *m*-chloroperbenzoic acid. The oxammonium group is also generated by acid-catalyzed disproportionation of nitroxides.[163] For oxidation of secondary alcohols by *m*-chloroperbenzoic acid in the presence of hydroxylamine **3**, the following scheme has been suggested[166] (Scheme 32).

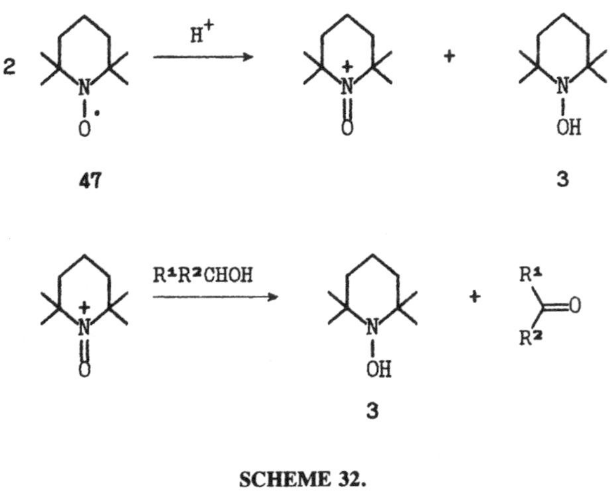

SCHEME 32.

It has been shown that the process may be performed electrochemically in a two-phase system containing the nitroxide, NaBr, and the substrate.[167] This method may be used for selective oxidation of primary alcohols in the presence of secondary alcohols. The reaction proceeds as shown in Scheme 33.

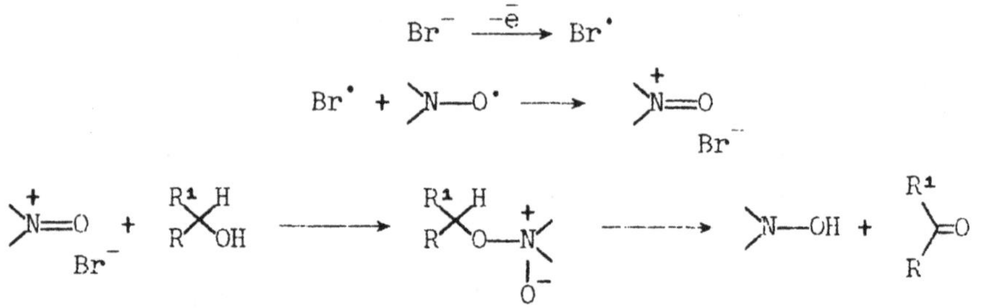

SCHEME 33.

C. DISPROPORTIONATION REACTIONS

The nitroxyl group shows weak basic properties (pK $\approx$ -5.5), due to which it may be protonated by strong acids: sulfuric acid, trifluoroacetic acid in the presence of sulfuric acid,[168] and AlCl$_3$ in nondehydrated methylene chloride.[169] When radical **54** reacts with hydrogen chloride in dry CCl$_4$, it is reduced to hydrochloride of the corresponding hydroxylamine.[113] This reaction may be due to the initial protonation of the nitroxyl group and further interaction of the radical-cation formed with one more nitroxide molecule to give the oxammonium salt and hydroxylamine. The oxammonium salt is reduced to hydroxylamine by water traces.

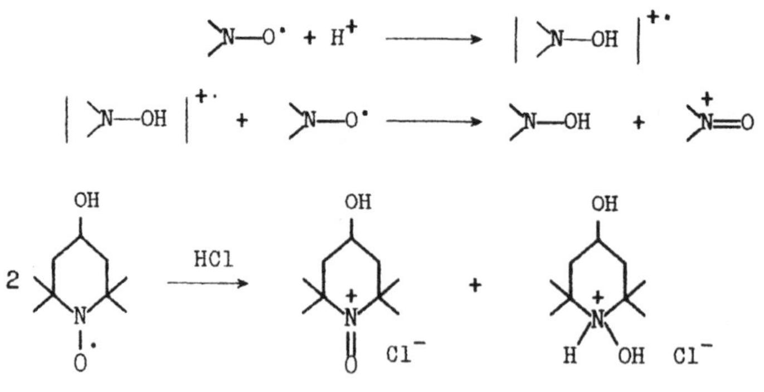

SCHEME 34.

Other piperidine nitroxides are disproportionated in a similar way by $HClO_4$, H_2SO_4, and HBF_4.[170]

In diluted solutions, disproportionation of piperidine nitroxides is reversible, but at pH $\leq$ 2 the equilibrium is almost entirely shifted toward the oxammonium salt and the protonated hydroxylamine. The introduction of electron-accepting substituents lowering the basicity both of the nitroxyl group and the hydroxylamine formed hinders the disproportionation reaction.[171]

Another factor hindering disproportionation reaction of nitroxides in acid media is the presence in a molecule of more basic groups than the nitroxyl group. Such nitroxides include the derivatives of 2- and 3-imidazoline, 3-imidazoline-3-oxide, and imidazolidine. In this case, protonation leads to formation of stable cation-radicals where the unpaired electron and the positive charge are located on different groups of atoms. Thus, 3-imidazoline derivatives are stable in the 0.1 N H_2SO_4 solution for more than 220 h.[172] The product composition of mixtures formed in protonation of 3-imidazoline and 3-imidazoline-3-oxide nitroxides in strong and superstrong acids have been studied by ^{1}H- and ^{13}C-NMR.[173]

Diarylnitroxides disproportionate in acid media in a slightly different way. No diarylhydroxylamines have been isolated, obviously, because of their instability; the reaction products were an oxammonium salt and diarylamine in the ratio 3:1.[97] Conversely, only a di-*tert*-butylhydroxylamine salt has been isolated from the products of acid disproportionation of di-*tert*-butylnitroxide.[174]

The reduction of the nitroxyl group by iodides in acid media is also supposed to proceed via the disproportionation stage. The oxammonium salt is reduced here by iodide-anion to the starting nitroxide which is again involved in the reaction.[175]

The nitroxides undergo disproportionation by allyl and benzyl bromides (the so-called homosolvolysis reaction) to form the bromides of oxammonium salts and the ethers of hydroxylamino derivatives.[176] A similar reaction takes place between di-*tert*-butylnitroxide and benzoyl chlorides, giving the *O*-benzoyl derivatives of hydroxylamine and oxammonium salts.[177]

An example of disproportionation of another type is formation of hydroxylamine **55** and aldehyde **56** by decomposition of nitroxide **57**[57] (Scheme 35).

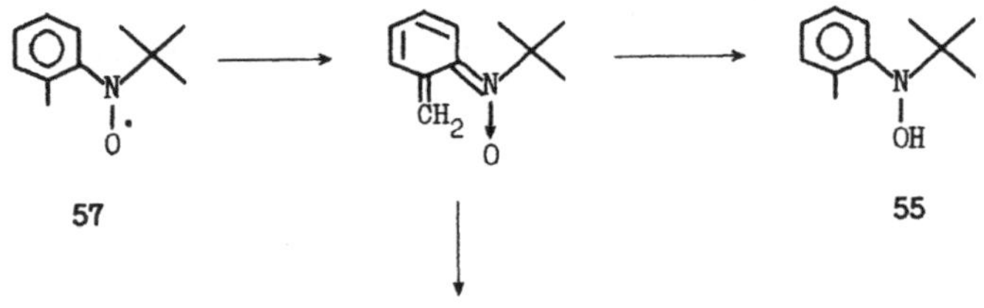

SCHEME 35.

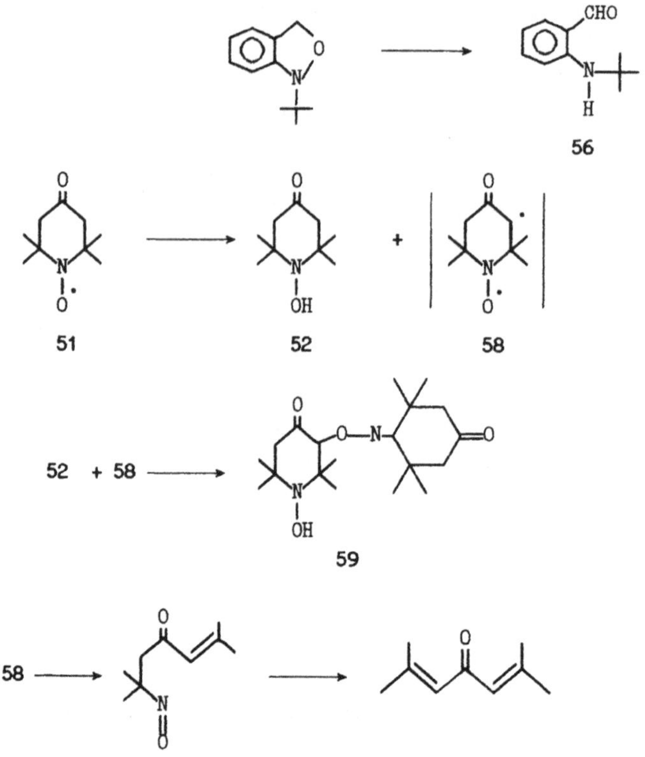

SCHEME 35 (continued).

Long storage of nitroxide **51** or its heating at 100°C or boiling it in benzene solution leads to its disproportionation as a result of hydrogen abstraction from another nitroxide molecule. The products of this reaction are the diamagnetic dimer **59** and a small amount of phorone.[178]

D. REACTIONS WITH RADICALS

The ability of nitroxides to recombine with short-lived radicals with formation of hydroxylamine ethers underlies their application as inhibitors of free radical processes.[7,97,179,180] Nitroxides readily react with free radicals in the oxidation of organometal compounds and substituted hydrazines.[118,122,123,129,130,181]

The interaction of imidazolidine nitroxides **60** or the piperidine nitroxide **51** with the 1-methyl-1-cyanoethyl radical formed in the thermolysis of azo-bisiso-butyronitrile yields stable hydroxylamine ethers.[182,183]

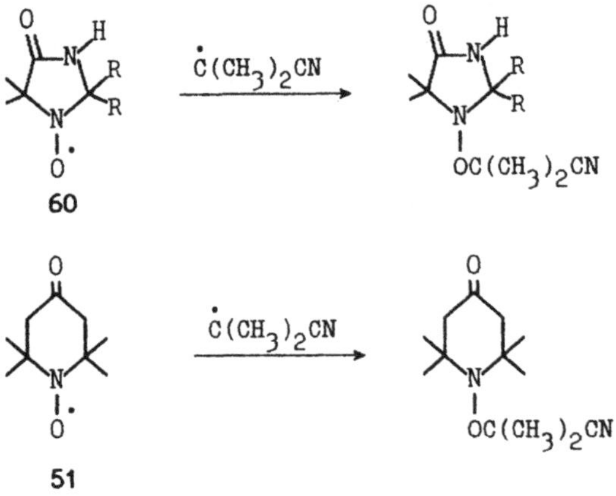

SCHEME 36.

It should be noted that nitroxide **51** is capable of adding the diphenylethyl radical to form an adduct of similar structure, which exists in solution as an equilibrium mixture of starting radicals.[184]

The main product in the copyrolysis of nitroxide **47** and acetylbenzoyl peroxide is the corresponding methoxy derivative.[185] The radical **47** easily reacts with inorganic radical-anions: $(CNS)_2^{\cdot}$, $Cl_2^{\cdot}$, $Br_2^{\cdot}$, and $CO_3^{\cdot}$ to produce diamagnetic adducts.[186] Bis(trifluoromethyl) nitroxide **9** is capable of forming stable adducts with nitrogen oxide and dioxide.[97]

In contrast to active, e.g., alkyl radicals, nitroxides generally do not enter recombinations with each other. This is due to the fact that the loss in the energy of delocalization of the three-electron, two-center bond is not compensated by the energy of formation of a weak O—O bond (compare to Chapter 1). Nevertheless, the sterically nonhindered nitroxide such as Fremy's salt **1**, bis(trifluoromethyl) nitroxide **9**, tetraphenylpyrrole-1-oxyl, 9-azabicyclo[3.2.1]nonane-3-1-9-oxyl and 8-azabicyclo[2.2.1]-octane-8-oxyl exist in the crystalline state as dimers.[57,187] However, in the 9-azabicyclo[3.2.1]nonane-3-1-9-oxyl dimer, the distance between nitroxyl groups has been estimated by X-ray structure analysis to be 2.278 Å. This rules out covalent bonding between these groups, i.e., the recombination does not occur.[188] Formation of sterically hindered nitroxide dimers has not been found at all.

Contrary to this, the arylalkyl- and diarylnitroxides can follow another pathway of recombination due to efficient delocalization of spin density in the aromatic moiety. This produces at a first stage the unstable dimers of type **61**, which decompose to form diamagnetic products.[97]

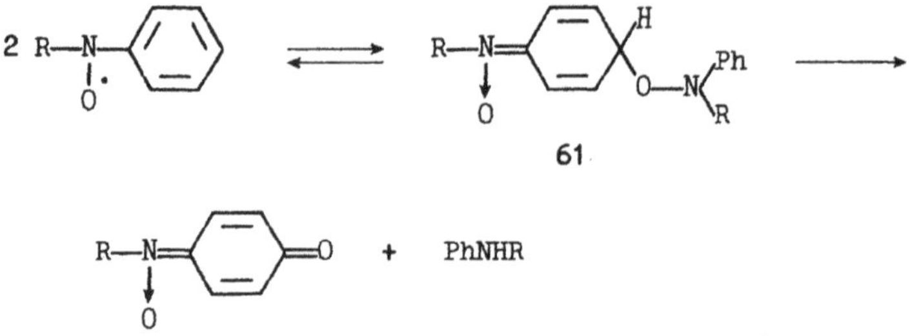

SCHEME 37.

This process is the reason for limited stability of nitroxides where the nitroxyl group is directly conjugated with the phenyl ring. Bulky substituents in the para- and, partly, meta-position can suppress this transformation.

E. PHOTOCHEMICAL DESTRUCTION OF THE NITROXYL GROUP

The additional reaction of free radicals at the nitroso group (Section I.D) may be reversible, especially in the case of the formation of stabilized C-radicals.

$$R-N=O + \dot{R}^1 \rightleftarrows R-N-R^1$$
$$|$$
$$O\cdot$$

$R^1 = Ph_3C\cdot,\ CH_2=CH-\dot{C}\ (CH_3)_2,\ CH\equiv C-\dot{C}(CH_3)_2$

The reaction proceeds especially readily in the case of bulky R substituents. Thus, nitroxide **62** is surprisingly unstable and decomposes even at 25°C, generating an allyl radical which recombines with the starting nitroxide to form a stable adduct **63**.[189] However, a nitroxide of related structure **64** proved to be quite stable and was recrystallized from boiling hexane without any pronounced decomposition.[190]

Above 125°C or when exposed to UV irradiation, di-*tert*-butylnitroxide under- goes similar cleavage of the C–N bond.[191] A recent example of such a reaction is the transformation of the extremely sterically hindered nitroxide **65** to azadiene proceeding in chromatographing on silica gel.[192]

Dialkylnitroxides in the ground state react very slowly with hydrocarbons, but under UV irradiation the reaction occurs easily. Thus, when the toluene solution of nitroxide **54** was irradiated, hydroxylamine and the *O*-benzyl derivative were formed in the ratio 1:1.[193] The hydrogen atom is abstracted from the alkane molecule by the photoexcited nitroxyl group in the n → π* excited state. The intramolecular abstraction of hydrogen atom by the photoexcited nitroxyl group is exemplified by the transformation of nitronyl nitroxide **66** to nitroxides **67** and **68**.[194]

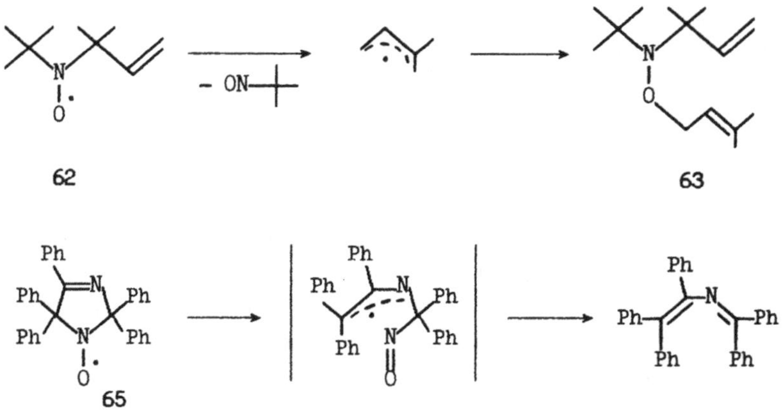

SCHEME 38.

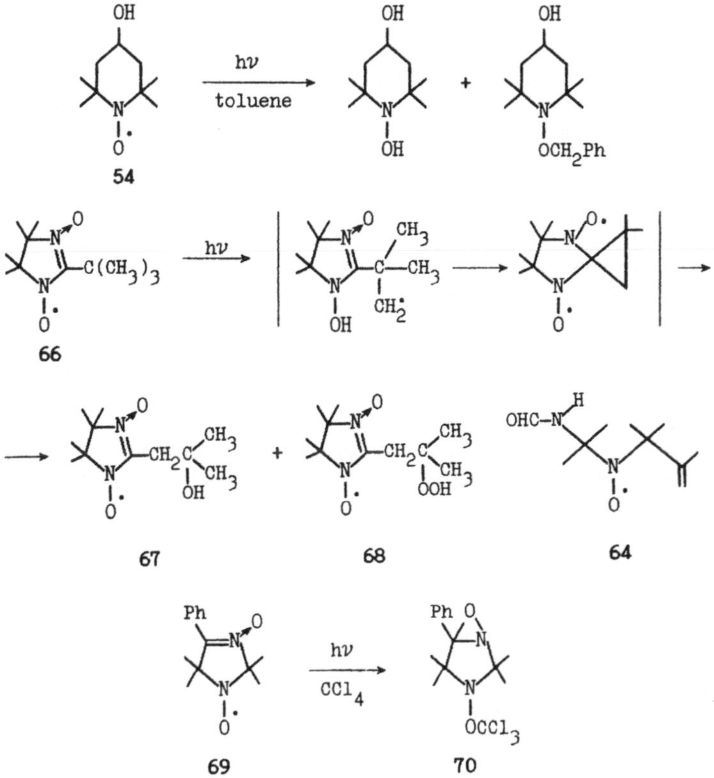

SCHEME 39.

In the photolysis of nitroxide **69** in CCl_4 solution, nitrone group isomerization not only occurs but also transformation of the nitroxyl group to the *O*-trichloromethylhydroxylamino group.[195]

REFERENCES

1. **Jansen, E. G. and Haire, D. L.**, Two decades of spin trapping, in *Advances in Free Radical Chemistry*, Vol. 1, JAI Press, CT, 1990.
2. **Piloty, O. and Graf Schwerin, B.**, Ueber die Existenz von Derivaten des vierwerthigen Stickstoffs, *Ber. Dtsch. Chem. Ges.*, 34, 2354, 1901.
3. **Sen', V. D., Golubev, V. A., and Rosenburg, A. N.**, Oxidation of amines and hydroxylamines to nitroxide radicals by H_2O_2. Kinetics and mechanism of reaction catalyzed by WO_4^{2-}, in *Symp. Stable Nitroxide Free Radicals: Synthesis and Application (Abstr.)*, Pècs, Hungary, 1979, 77.
4. **Keana, J. F. W., Norton, R. S., Morello, M., Van Engen, D., and Clardy, J.**, Mononitroxides and proximate dinitroxides derived by oxidation of 2,2,4,4,5,5-hexasubstituted imidazolidines. A new series of nitroxide and proximate dinitroxide spin labels, *J. Am. Chem. Soc.*, 100, 934, 1978.
5. **Volodarsky, L. B., Ed.**, *Imidazoline Nitroxides. Synthesis and Properties*, Vol. 1, CRC Press, Boca Raton, FL, 1988.
6. **Amitina, S. A. and Volodarskii, L. B.**, Synthesis of sterically hindered 1-hydroxy-2-acetyl-3-imidazoline-3-oxide and stable nitroxyl radicals based on them, *Izv. Akad. Nauk S.S.S.R., Ser. Khim.*, p. 2135, 1976; *Chem. Abstr.*, 86, 106467y.
7. **Rozantsev, E. G.**, *Free Nitroxyl Radicals*, Plenum Press, New York, 1970.
8. **Krinitskaya, L. A. and Volodarskii, L. B.**, Oxidation of di-*tert*-alkyl hydroxylamines to nitroxyl radicals using nitrous acid, *Izv. Akad. Nauk S.S.S.R., Ser. Khim.*, 391, 1983; *Chem. Abstr.*, 197967y.
9. **Schlude, H.**, Oxidation of hydroxylamines to nitroxyl radicals with Fremy's salt, *Tetrahedron*, 39, 4007, 1973.
10. **Dikanov, S. A., Grigor'ev, I. A., Volodarskii, L. B., and Tsvetkov, Yu. D.**, Oxidative properties of nitroxyl radicals in reactions between them and sterically hindered hydroxylamines, *Zh. Fis. Khim.*, 56, 2762, 1982; *Chem. Abstr.*, 98, 125151e.
11. **Zhukova, I. Yu., Kagan, E. Sh., and Smirnova, V. A.**, Synthesis and properties of 3,5-dibromo-4-oxo-2,2,6,6-tetramethylpiperidine-1-oxyl, *Khim. Geterotsikl. Soedin.*, 73, 1992.
12. **Booth, B. L. and Kosinski, E. D.**, Nitroxide chemistry. XXII. Oxidation of *N,N*-bis(trifluoromethyl)hydroxylamine, *J. Fluor. Chem.*, 37, 419, 1987.
13. **Krinitskaya, L. A. and Volodarskii, L. B.**, 3-Monohalo derivatives of triacetonamine, 1-hydroxy-2,2,6,6-tetramethyl-4-oxopiperidine, and 2,2,6,6-tetramethyl-4-oxopiperidine-1-oxyl, *Izv. Akad. Nauk S.S.S.R., Ser. Khim.*, 443, 1982; *Chem. Abstr.*, 96, 181113q.
14. **Lebedev, O. L. and Kazarnovskii, S. N.**, Catalytic oxidation of aliphatic amines with hydrogen peroxide, in *Tr. Khim. Tekhnol.*, Gorky, 649, 1959.
15. **Sholle, V. D., Krinitskaya, L. A., and Rosantsev, E. G.**, Unusual products from oxidation of tertiary amine, *Izv. Akad. Nauk S.S.S.R., Ser. Khim.*, 149, 1969; *Chem. Abstr.*, 70, 106295k.
16. **Abdallah, M. A., Andre, J.-J., and Biellman, J.-F.**, A new spin labelled analogue of nicotinamide dinucleotide, *Bioorg. Chem.*, 6, 157, 1977.
17. **Volodarskii, L. B., Martin, V. V., and Kobrin, V. S.**, U.S.S.R. Patent, 707914, 1980.
18. **Martin, V. V. and Volodarskii, L. B.**, Synthesis and some reactions of sterically-hindered 3-imidazoline 3-oxides, *Khim. Geterotsikl. Soedin.*, 103, 1979; *Chem. Abstr.*, 90, 186860s.

19. **Grigor'ev, I. A., Shchukin, G. I., and Volodarskii, L. B.**, Effect of a radical center on oxidative properties of the nitrone group in the reaction of nitroxyl radicals of 3-imidazoline-3-oxide with hydrazine, *Izv. Akad. Nauk S.S.S.R., Ser. Khim.*, 1140, 1983; *Chem. Abstr.*, 100, 6386k.

20. **Rozantsev, E. G. and Krinitskaya, L. A.**, U.S.S.R. Patent, 391137, 1973.

21. **Rozantsev, E. G. and Sholle, V. D.**, *Organic Chemistry of Free Radicals*, Khimia, Moscow, 1979; *Chem. Abstr.*, 93, 70865e.

22. **Sen', V. D., Golubev, V. A., and Efremova, N. N.**, Kinetics and mechanism of tungstate (2-)-catalyzed oxidation of di-*tert*-alkylamines and di-*tert*-alkylhydroxylamines by hydrogen peroxide to nitroxyl radicals, *Izv. Akad. Nauk S.S.S.R., Ser. Khim.*, 61, 1982; *Chem. Abstr.*, 96, 142057e.

23. **Briére, R., Lemaire, H., and Rassat, A.**, Nitroxydes. XV. Syntheseè et ètude de radicaux libres stables piperidiniques et pyrrolidiniques, *Bull. Soc. Chim. Fr.*, 3273, 1965.

24. **Chapelet-Letourneux, G., Lemaire, H., and Rassat, A.**, Nitroxydes. XVI. Mise en evidence de radicaux libres dans les oxydations par les peracides, *Bull. Soc. Chim. Fr.*, 3283, 1965.

25. **Sosnovsky, G. and Konieczny, M.**, Preparation of 4-hydroxy-2,2,6,6-tetramethylpiperidine-1-oxyl. A reinvestigation of methods using hydrogen peroxide in the presence of catalysts, *Z. Naturforsch.*, 31B, 1376, 1976.

26. **Martin, V. V., Volodarskii, L. B., and Vishnivetskaya, L. A.**, Interaction of sterically hindered imidazoline oxides which are precursors of stable nitroxyl radicals with organometallic reagents, *Izv. Sib. Otd. Akad. Nauk S.S.S.R., Ser. Khim. Nauk*, 4, 94, 1981; *Chem. Abstr.*, 96, 6648.

27. **Martin, V. V. and Volodarskii, L. B.**, Study of tautomerism and the chemical properties of β-oxonitrones, *Izv. Akad. Nauk S.S.S.R., Ser. Khim.*, 1336, 1980; *Chem. Abstr.*, 93, 204544k.

28. **Reznikov, V. A., Martin, V. V., and Volodarskii, L. B.**, Oxidative dimerization of heterocyclic nitrones derived from pyrroline and imidazoline, *Izv. Akad. Nauk S.S.S.R., Ser. Khim.*, 1398, 1990; *Chem. Abstr.*, 113, 190599z.

29. **Voinov, M. A., Martin, V. V., and Volodarskii, L. B.**, The reactions of aldonitrones of 3-imidazoline-3-oxide with isothiocyanates, *Izv. Akad. Nauk Russ.*, 2642, 1992.

30. **Espie, J.-C. and Rassat, A.**, Nitroxides. XLV. Radicaux libres azetidiniques (note preliminaire), *Bull. Soc. Chim. Fr.*, 4385, 1971.

31. **Rozantsev, E. G. and Kokhanov, Yu. V.**, A new individual free radical with primary amino group, *Izv. Akad. Nauk Russ., Ser. Khim.*, 1477, 1966; *Chem. Abstr.*, 66, 54758j.

32. **Reznikov, V. A., Vishnivetskaya, L. A., and Volodarskii, L. B.**, Synthesis and properties of heterocyclic β-nitronitrone-4-nitromethyl-1,2,2,5,5-pentamethyl-3-imidazoline-3-oxide, *Izv. Akad. Nauk Russ., Ser. Khim.*, in press.

33. **Kaliskà, V., Toma, Š., and Tkàč, A.**, Utilization of ultrasound for the synthesis of 4-substituted 2,2,6,6-tetramethylpiperidinyloxyls, *Chem. Pap. (ČSSR)*, 42, 243, 1988.

34. **Kaliskà, V., Toma, Š., and Lesho, J.**, Synthesis and mass spectra of piperidine and pyrazine *N*-oxyl radicals, *J. Coll. Czech. Commun.*, 52, 2266, 1987.

35. **Keana, J. F. W.**, Synthesis and chemistry of nitroxide spin labels, in *Spin Labeling in Pharmacology*, Holtzman, J. L., Ed., Academic Press, Orlando, FL, 1984, chap. 1.

36. **Keana, J. F. W., Keana, S. B., and Beetham, D.**, A new versatile ketone spin label, *J. Am. Chem. Soc.*, 89, 3055, 1967.

37. **Toda, T., Mori, E., and Murayama, K.**, Studies on stable free radicals. IX. Peroxy acid oxidation of hindered secondary amines to nitroxide radicals, *Bull. Chem. Soc. Jpn.*, 45, 1904, 1972.

38. **Cella, J. A., Kelley, J. A., and Kenehan, E. F.**, Oxidation of nitroxides by *m*-chloroperbenzoic acid, *Tetrahedron Lett.*, 2869, 1975.

39. **Golding, B. T., Ioannou, P. V., and O'Brien, M. M.**, A short-cut to spin-labelled amides, *Synthesis*, 462, 1975.

40. **Căproiu, M. T., Negoită, N., and Balaban, A. T.**, Aryl-nitrosoaminyloxide, aryl-tosyl-aminyloxide, and diaryl-aminyloxide, where the aryl is 3,5-di-*t*-butylphenyl, *Tetrahedron Lett.*, 1825, 1977.

41. **Pannell, J.,** Electron spin resonance spectra of radicals derived from secondary aromatic amines (I) diphenyl NO and the diphenylamino radical, *Mol. Phys.,* 5, 291, 1962.
42. **Coppinger, G. M. and Swalen, J. D.,** Electron paramagnetic resonance studies of unstable free radicals in the reaction of *t*-butyl hydroperoxides and alkylamines, *J. Am. Chem. Soc.,* 83, 4900, 1962.
43. **Buchachenko, A. L.,** Electron paramagnetic resonance (E.P.R.) spectra of some new radicals, *Opt. Spektrosk.,* 13, 795, 1962.
44. **Florin, R. E.,** Comment on "paramagnetic resonance of alkyl nitroxides", *J. Chem. Phys.,* 47, 345, 1967.
45. **Thomas, J. R.,** The identification of radical products from the oxidation of diphenylamine, *J. Am. Chem. Soc.,* 82, 5955, 1960.
46. **Adams, J. Q., Nickis, S. W., and Thomas, J. R.,** Paramagnetic resonance of alkyl nitroxides, *J. Chem. Phys.,* 45, 654, 1966.
47. **Frangopol, P. T., Frangopol, M., Negoita, N., and Balaban, A. T.,** Stability and equilibria of free radicals. IV. *Tert*-Butylcyano nitroxide, *Rev. Roum. Chim.,* 14, 385, 1969.
48. **Hudson, A. and Hussain, H. A.,** Electron spin resonance studies of aliphatic nitroxides. Part IV. The observation on conformational effects in medium sized rings, *J. Chem. Soc. (B),* 1346, 1968.
49. **Bruni, P. and Greci, L.,** Indoline nitroxides, *J. Heterocycl. Chem.,* 9, 1455, 1972.
50. **Neugebauer, F. A. and Fischer, P. H. M.,** Substituent effects in the electron spin resonance of 4,4'-disubstituted diaryl nitric oxides, *Z. Naturforsch.,* 21, 1036, 1966.
51. **Razumovskii, S. D., Buchachenko, A. L., Shapiro, A. B., Rozantsev, E. G., and Zaikov, G. E.,** Formation of nitroxyl radicals in the reaction between ozone and amines, *Dokl. Akad. Nauk S.S.S.R.* 183, 1106, 1968; *Chem. Abstr.,* 70, 95987j.
52. **Popova, E. G., Paramonova, V. I., Popov, L. K., Pankova, T. A., and Alekhina, N. B.,** U.S.S.R. Patent, 3884473, 1990.
53. **Murray, R. W. and Singh, M.,** A high yield one step synthesis of hydroxylamines, *Synth. Commun.,* 19, 3509, 1989.
54. **Murray, R. W. and Singh, M.,** U.S. Patent, 5001233, 1991.
55. **Bonvalet, C., Bourelle, F., Scholler, D., and Feigenbaum, A.,** An efficient preparation of crowded doxyl fatty esters using dimethyldioxyrane as an oxidizing agent, *J. Chem. Res. Synop.,* 348, 1991.
56. **Murray, R. W. and Singh, M.,** A convenient high yield synthesis of nitroxides, *Tetrahedron Lett.,* 29, 4677, 1988.
57. **Aurich, H. G. and Weiss, W.,** Formation and reactions of aminyloxides, *Top. Curr. Chem.,* 59, 65, 1975.
58. **Dikalov, S. I., Kirilyuk, I. A., Grigor'ev, I. A., and Volodarskii, L. B.,** The 2*H*-imidazole-*N*-oxides as spin traps, *Izv. Akad. Nauk. Russ., Ser. Khim.,* 1064, 1992.
59. **Iwamura, M. and Inamoto, N.,** Novel formation of nitroxide radicals by radical addition to nitrones, *Bull. Chem. Soc. Jpn.,* 40, 703, 1967.
60. **Kursakina, I. G., Starichenko, V. F., Kirilyuk, I. A., Grigor'ev, I. A., and Volodarskii, L. B.,** Electrochemical oxidation of 4*H*-imidazole-*N*-oxides, *Izv. Akad. Nauk Russ., Ser. Khim.,* in press.
61. **Kursakina, I. G., Starichenko, V. F., and Grigor'ev, I. A.,** Electrochemical oxidation of imidazoline *N*-oxides, *Izv. Akad. Nauk Russ., Ser. Khim.,* 2545, 1992.
62. **Grigor'ev, I. A., Volodarskii, L. B., Starichenko, V. F., and Kirilyuk, I. A.,** Synthesis of stable nitroxides with amino groups and fluorine atoms at α-carbon of the radical center, *Tetrahedron Lett.,* 30, 751, 1989.
63. **Grigor'ev, I. A., Volodarskii, L. B., Starichenko, V. F., Shchukin, G. I., and Kirilyuk, I. A.,** Rote to stable nitroxides with alkoxy groups at α-carbon — the derivatives of 2- and 3-imidazolines, *Tetrahedron Lett.,* 26, 5085, 1985.
64. **Grigor'ev, I. A., Kirilyuk, I. A., Starichenko, V. F., and Volodarskii, L. B.,** Oxidative alkoxylation of 4*H*-imidazole *N*-oxides as a new method of synthesis of stable 2- and 3-imidazoline nitroxyl radicals with alkoxy groups at the α-atom of the radical center, *Izv. Akad. Nauk S.S.S.R., Ser. Khim.,* p. 1624, 1989. *Chem. Abstr.,* 112, 55708y.

65. **Grigor'ev, I. A., Starichenko, V. F., Kirilyuk, I. A., and Volodarskii, L. B.**, Synthesis of stable nitroxyl radicals with amino group at the carbon α-atom to the radical center by oxidative amination of 4*H*-imidazole *N*-oxides, *Izv. Akad. Nauk S.S.S.R., Ser. Khim.*, 661, 1989; *Chem. Abstr.*, 111, 232661e, 1989.

66. **Shchukin, G. I., Starichenko, V. F., Grigor'ev, Dikanov, S. A., Gulin, V. I., and Volodarskii, L. B.**, Formation of methoxynitrones and stable nitroxyl radicals with gem-dimethoxy groups at α-carbon in oxidation of aldonitrones in methanol, *Izv. Akad. Nauk S.S.S.R., Ser. Khim.*, 125, 1987; *Chem. Abstr.*, 107, 236596c.

67. **Grigor'ev, I. A., Starichenko, V. F., Kirilyuk, I. A., and Volodarskii, L. B.**, Formation of nitroxides with α-fluorine atoms in the reaction of nitrones with xenone difluoride, *Izv. Akad. Nauk S.S.S.R., Ser. Khim.*, 933, 1989; *Chem. Abstr.*, 111, 214051b, 1989.

68. **Bakunova, S. M., Grigor'ev, I. A., Kirilyuk, I. A., Gatilov, Yu. V., Bagryanskaya, I. Yu., and Volodarskii, L. B.**, Synthesis of stable nitroxide radicals of the oxazolidine series with methoxy groups at α-carbons of radical centre, *Izv. Akad. Nauk. Russ. Ser. Khim.*, 966, 1992.

69. **Grigor'ev, I. A.**, private communication.

70. **Sutcliffe, H. and Wardale, H. W.**, Bis(trichloromethyl) nitroxide. A novel electron spin resonance spectrum, *J. Am. Chem. Soc.*, 89, 5487, 1967.

71. **Hartgerink, J. W., Engberts, J. B. F. N., Wajer, T. A. J. W., and De Boer, T. J.**, C-Nitroso compounds. IX. 1-Nitrosoadamantane. Trapping of trichloromethyl and trifluoromethyl radicals by monomeric C-nitrosoalkanes, *Rec. Trav. Chim.*, 88, 481, 1969.

72. **Xi-Kui J.**, Polyfluorinated nitroxides, *J. Pure Appl. Chem.*, 62, 189, 1990.

73. **Rassat, A. and Rey, P.**, Nitroxides: photochemical synthesis of trimethylisoquinuclidine-*N*-oxyl, *J. Chem. Soc. Chem. Commun.*, 1161, 1971.

74. **Lane, J. and Tauber, B.**, An electron spin resonance study of the reaction between 2-methyl-2-nitrosopropane and methyl-substituted vinyl monomers, *J. Chem. Soc. Perkin Trans. 2*, 1665, 1985.

75. **Motherwell, W. B. and Roberts, J. S.**, Inter- and intramolecular ''ene'' reactions of aliphatic nitroso compounds, *J. Chem. Soc. Chem. Commun.*, 329, 1972.

76. **Knight, G. T. and Loadman, M. J. R.**, The reaction of *p*-substituted nitrosobenzenes with *N*-alkyl-*N*-arylhydroxylamines, *J. Chem. Soc. (B)*, 2107, 1971.

77. **Mothewell, W. B. and Roberts, J. S.**, Nitroxide radical synthesis, *J. Chem. Soc. Chem. Commun.*, 328, 1972.

78. **Aurich, H. G.**, Nitroxides, in *Nitrones, Nitronates and Nitroxides*, Patai, S. and Rappoport, Z., Eds., John Wiley & Sons, New York, 1989, chap. 4.

79. **Hoffman, A. K., Feldman, A. M., Gelblum, E., and Hodgson, W. G.**, Mechanism of the formation of di-*t*-butylnitroxide from *t*-nitrobutane and sodium metal, *J. Am. Chem. Soc.*, 86, 639, 1964.

80. **Hoffman, A. K., Hodson, W. G., Maricle, D. L., and Jura, W. H.**, Cleavage reactions of tertiary nitro anion radicals, *J. Am. Chem. Soc.*, 86, 631, 1964.

81. **Brière, R. and Rassat, A.**, Nitroxydes. X., Synthèse magnèsienne du di-*t*-butylnitroxide, *Bull. Soc. Chim. Fr.*, 378, 1965.

82. **Lemaire, H., Rassat, A., and Ravet, J.-P.**, Nitroxydes. XII. Obtention du monophenyl nitroxide et du diphenyl nitroxyde par reductions a l'aluminohydrure de lithium, *Tetrahedron Lett.*, 3507, 1964.

83. **Lemaire, H., Marechal, Y., Ramasseul, R., and Rassat, A.**, Nitroxydes. IX. Rèsonance paramagnètique èlectronique de dèrivès du *t*-butylphènylnitroxyde substituès en para, *Bull. Soc. Chim. Fr.*, 372, 1965.

84. **Barbarella, G. and Rassat, A.**, Nitroxydes. XXXIV. Rèsonance paramagnètique èlectronique de dèrivès du *t*-butyl phènyl nitroxyde substituè, ècarts hyperfins dus à l'azote, *Bull. Soc. Chim. Fr.*, 2378, 1969.

85. **Brunel, Y., Lemaire, H., and Rassat, A.**, Nitroxydes. V. Le camphenyl-1 *t*-butyl nitroxyde, radical libre stable optiquement actif: un nouvean chromophore dichroique, *Bull. Soc. Chim. Fr.*, 1895, 1964.

86. **Ahrens, W. and Berndt, A.**, Ein einfacher weg zu Spinaddukten aus Nitrosoverbindungen und *t*-butylradikalen; ungewoehnlich stabile *N*-alkoxy-anilino-radikale, *Tetrahedron Lett.*, 4281, 1973.

87. **Roberts, J. R. and Ingold, K. U.**, Kinetic applications of electron paramagnetic resonance spectroscopy. X. Reactions of some alkylamino radicals in solution, *J. Am. Chem. Soc.*, 95, 3228, 1973.

88. **Meyer, K. H. and Reppe, W.**, Über die Reduktionsstafen von Arylderivaten der Salpetersäure, *Chem. Ber.*, 54, 327, 1921.

89. **Aurich, H. G., Hahn, K., Stork, K., and Weiss, W.**, Aminyloxide (nitroxide). XXIV. Empirische ermittlung der Spindichteverteilung in Aminyloxiden, *Tetrahedron*, 33, 969, 1977.

90. **Martin, V. V., Vishnivetskaya, L. A., Grigor'ev, I. A., Dikanov, S. A., and Volodarskii, L. B.**, Reactions of hydroxamic acid chloride derivatives of 3-imidazoline-3-oxide with nitrogen-containing nucleophilic agents and preparation of stable amidoxime-*N*-oxyl radicals, *Izv. Akad. Nauk. S.S.S.R., Ser. Khim.*, 1616, 1985; *Chem. Abstr.*, 104, 207200b.

91. **West, R. and Boudjouk, P.**, Bis(organosilyl) nitroxides, *J. Am. Chem. Soc.*, 93, 5901, 1971.

92. **Hankovszky, H. O., Hidek, K., and Lex, L.**, Nitroxyls. VII. Synthesis and reactions of highly reactive 1-oxyl-2,2,5,5-tetramethyl-2,5-dihydro-3-ylmethyl sulfonates, *Synthesis*, 914, 1980.

93. **Reznikov, V. A.**, unpublished results.

94. **Volodarskii, L. B., Grigor'ev, I. A., Kirilyuk, I. A., and Amitina, S. A.**, Nitration of 4-aryl- and 4-hetaryl-2,2,5,5-tetramethyl-3-imidazoline-3-oxides and corresponding nitroxides, *Zh. Org. Khim.*, 21, 443, 1985; *Chem. Abstr.*, 102, 166657u.

95. **Volodarsky, L. B., Grigor'ev, I. A., Grigor'eva L. N., and Kirilyuk, I. A.**, Nitration of imidazoline-*N*-oxide nitroxides containing the aryl nitrone group, *Tetrahedron Lett.*, 25, 5809, 1984.

96. **Hankovszky, H. O., Hidek, K., Lovas, M. J., and Jerkovich, G.**, Synthesis and reaction of *ortho*-fluoronitroaryl nitroxides, novel versatile synthons and reagents for spin-labeling studies, *Can. J. Chem.*, 67, 1392, 1989.

97. **Forrester, A. R., Hay, J. M., and Thomson, R. V.**, *Organic Chemistry of Stable Free Radicals*, Academic Press, London, 1968.

98. **Pokhodenko, A. R., Beloded, A. A., and Koshechko, V. G.**, *Oxidation-Reduction Reactions of Free Radicals*, Naukova Dumka, Kiev, 1977; *Chem. Abstr.*, 89, 188042e, 1978.

99. **Swartz, H. M.**, The use of nitroxides in viable biological systems: an opportunity and challenge for chemists and biochemists, *J. Pure Appl. Chem.*, 62, 235, 1990.

100. **Wieland, H. and Kögl, F.**, Über organische Radical mit vierwertigen Stikstoff (III), *Chem. Ber.*, 55, 1798, 1922.

101. **Volodarskii, L. B., Reznikov, V. A., and Kobrin, V. S.**, Preparation and properties of imidazolinium salts containing a nitroxyl radical center, *Zh. Org. Khim.*, 15, 415, 1979; *Chem. Abstr.*, 91, 5158w.

102. **Grigor'ev, I. A. and Volodarskii, L. B.**, Reaction of 4-bromomethyl derivatives of 2,2,5,5-tetramethyl-Δ^3-imidazoline 3-oxides with hydrazine and primary amines, *Zh. Org. Khim.*, 10, 118, 1974; *Chem. Abstr.*, 80, 108447s.

103. **Ramasseul, R., Rassat, A., and Rey, P.**, A useful protecting group in the preparation of amino nitroxides, *J. Chem. Soc. Chem. Commun.*, 83, 1976.

104. **Banks, R. E., Haszeldine, R. N., and Stevenson, M. J.**, Perfluoroalkyl derivatives of nitrogen. XXI. Some reactions of bistrifluoromethylnitroxide, *J. Chem. Soc. (C)*, 901, 1966.

105. **Makarov, S. P., Englin, M. A., Videiko, A. F., Tobolin, V. A., and Dubov, S. S.**, Reactions of hexafluoromethylnitrogen oxide, *Dokl. Akad. Nauk. S.S.S.R.*, 168, 344, 1966; *Chem. Abstr.*, 65, 8742a.

106. **Davydov, R. M.**, Interaction of iron (II) ions with an iminoxyl radicals, *Zh. Fis. Khim.*, 42, 2639, 1968; *Chem. Abstr.*, 70, 71434a.

107. **Schwartz, M. A., Parce, J. W., and McConnell, H. M.**, Hydrogen atom exchange between nitroxides and hydroxylamines, *J. Am. Chem. Soc.*, 101, 3592, 1979.

108. **Brackman, W. and Gaazbeek, C. J.**, Homogeneous catalysis. Radicals of the R$_2$NO type as catalysts for the oxidation of methanol by cupric complexes and as promoters for various oxidations catalyzed by copper, *Rec. Trav. Chim.*, 85, 221, 1966.

109. **Kornblum, N. and Pinnick, H. W.**, Reduction of nitroxides to amines by sodium sulfide, *J. Org. Chem.*, 37, 2050, 1972.
110. **Alper, H.**, Reaction of nitroxyl radicals with metal carbonyls, *J. Org. Chem.*, 38, 1417, 1973.
111. **Kozlova, L. M. and Litvin, E. F.**, Hydrogenation of stable nitroxides, in *Catalysis and Catalytic Processes in Chemistry Industry, Papers of Second Conference*, 1, 79, 1989.
112. **Paleos, C. M., Karyaunis, N. M., and Labes, M. N.**, Reduction of 2,2,6,6-tetramethylpiperidine nitroxide radical via complex formation with copper (II) perchlorate, *J. Chem. Soc. Chem. Commun.*, 195, 1970.
113. **Golubev, V. A., Rozantsev, E. G., and Neiman, M. B.**, Some reactions of free iminoxyl radicals with unpaired electron participation, *Izv. Akad. Nauk S.S.S.R., Ser. Khim.*, 1927, 1965; *Chem. Abstr.*, 64, 11164e.
114. **Levenson, J. L. and Kaplan, L.**, Deoxygenative reduction of nitroxyl and carbonyl groups, *J. Chem. Soc. Chem. Commun.*, 23, 1974.
115. **Lee, T. D., Birrel, G. B., and Keana, J. F. W.**, A new series of minimum steric perturbation nitroxide spin labels, *J. Am. Chem. Soc.*, 100, 1618, 1978.
116. **Grigor'ev, I. A., Martin, V. V., Shchukin, G. I., and Volodarskii, L. B.**, Reactions of α-haloalkylnitrones — derivatives of nitroxyl radicals of 3-imidazoline 3-oxide with nucleophilic agents, *Izv. Akad. Nauk S.S.S.R., Ser. Khim.*, 2711, 1979; *Chem. Abstr.*, 92, 198318a.
117. **Grigor'ev, I. A., Shchukin, G. I., and Volodarskii, L. B.**, 4-β-Dicarbonyl derivatives of stable nitroxyl radicals of 3-imidazoline, *Izv. Sib. Otd. Akad. Nauk S.S.S.R., Ser. Khim. Nauk.*, 4, 135, 1977; *Chem. Abstr.*, 88, 6796e.
118. **Rozantsev, E. G. and Golubev, V. A.**, Synthesis of heterocyclic analogs of hydroxylamine, *Izv. Akad. Nauk S.S.S.R., Ser. Khim.*, 891, 1966; *Chem. Abstr.*, 65, 10559e.
119. **Rozantsev, E. G. and Shapiro, A. B.**, A new stable free radical of the indole class — 2,2,4,4-tetramethyl-1,2,3,4-tetrahydro-γ-carboline-3-oxyl, *Izv. Akad. Nauk S.S.S.R., Ser. Khim.*, 1123, 1964; *Chem. Abstr.*, 61, 7000e.
120. **Rozantsev, E. G. and Burmistrova, R. S.**, Preparative synthesis of symmetrical di-*tert*-butylhydroxylamine and di-*tert*-butylamine, *Izv. Akad. Nauk S.S.S.R., Ser. Khim.*, 2364, 1968; *Chem. Abstr.*, 70, 28324c.
121. **Lee, T. D. and Keana, J. F. W.**, *In situ* reduction of nitroxide spin labels with phenylhydrazine in deuterochloroform solution. Convenient method for obtaining structural information on nitroxides using nuclear magnetic resonance spectroscopy, *J. Org. Chem.*, 40, 3145, 1975.
122. **Golubev, V. A. and Kobylyanskii, E. B.**, Reaction of oxo-piperidinium salts with organomagnesium compounds. Synthesis of *N*-alkoxy- and *N*-phenoxypiperidines, *Zh. Org. Khim.*, 8, 2607, 1972; *Chem. Abstr.*, 78, 84202p.
123. **Shchukin, G. I., Grigor'ev, I. A., and Volodarskii, L. B.**, Convenient method of converting nitroxyl radicals into *O*-methylhydroxylamines, *Izv. Akad. Nauk S.S.S.R., Ser. Khim.*, 2357, 1983; *Chem. Abstr.*, 100, 51514m.
124. **Mahoney, L. R., Mendenhall, G. D., and Ingold, K. U.**, Calorimetric and equilibrium studies on some stable nitroxides and iminoxy radicals. Approximate O–H bond dissociation energies in hydroxylamines and oximes, *J. Am. Chem. Soc.*, 95, 8610, 1973.
125. **Martin, V. V., Kobrin, V. S., and Volodarskii, L. B.**, Synthesis and properties of *N*-alkylhydroxylamino oximes. Stable nitroxyl radicals with α-ketoxime group, *Izv. Sib. Otd. Akad. Nauk S.S.S.R., Ser. Khim. Nauk.*, 2, 153, 1977; *Chem. Abstr.*, 87, 117517m.
126. **Reznikov, V. A. and Volodarskii, L. B.**, Recyclization of enaminoketones of imidazolidine to 1-pyrroline-4-on-oxides, *Khim. Geterotsikl. Soedin.*, 921, 1990.
127. **Reznikov, V. A., Vishnivetskaya, L. A., and Volodarskii, L. B.**, Synthesis and recyclization reactions of 4-*ortho*-phenyl substituted 3-imidazolines, *Izv. Akad. Nauk Russ.*, in press.
128. **Reznikov, V. A., Reznikova, T. I., and Volodarskii, L. B.**, Reaction of 1-hydroxy-2,2,4,5,5-pentamethyl-3-imidazoline and the appropriate nitroxyl radical with aldehydes, ketones and esters, *Zh. Org. Khim.*, 18, 2135, 1982; *Chem. Abstr.*, 98, 53775m.
129. **Whitesides, G. M. and Newirth, T. L.**, Reaction of *n*-butyllithium and 2,2,6,6-tetramethylpiperidine nitroxyl, *J. Org. Chem.*, 40, 3448, 1975.

130. **Sholle, V. D., Golubev, V. A., and Rozantsev, E. G.,** Interaction of nitroxyl radical with Grignard reagents, *Dokl. Akad. Nauk S.S.S.R.,* 200, 137, 1971; *Chem. Abstr.,* 76, 25049e.

131. **Kosman, D. J. and Piette, L. H.,** Nitroxide spin labels: new synthetic sequences, *J. Chem. Soc. Chem. Commun.,* 926, 1969.

132. **Yoshioka, T., Mori, E., and Murayama, K.,** Studies on stable free radicals. XIII. Synthesis and ESR spectral properties of hindered piperazine *N*-oxyls, *Bull. Soc. Chim. Jpn.,* 45, 1855, 1972.

133. **Rauckman, E. J. and Rosen, G. M.,** Synthesis of spin labeled probes; esterification and reduction, *Synth. Commun.,* 6, 325, 1976.

134. **Grigor'ev, I. A. and Volodarskii, L. B.,** Participation of nitroxyl radical in the oxidation of aldehyde and alcohol groups in 3-imidazoline-1-oxyls, *Izv. Akad. Nauk S.S.S.R., Ser. Khim.,* p. 208, 1978; *Chem. Abstr.,* 88, 170033x.

135. **Rozantsev, E. G.,** Selective reduction of carbonyl group in a free radical without participation of the unpaired electron, *Izv. Akad. Nauk S.S.S.R., Ser. Khim.,* 770, 1966; *Chem. Abstr.,* 65, 8712e.

136. **Berti, C., Greci, L., and Poloni, M.,** Stereoselective reduction of indoline nitroxide radicals, *J. Chem. Soc. Perkin Trans. 2,* 710, 1980.

137. **Malatesta, V. and Ingold, K. U.,** Kinetic applications of electron paramagnetic resonance spectroscopy. 37. The reaction of bis(trifluoromethyl) nitroxide with toluene. Evidence for quantum mechanical tunneling in an intramolecular hydrogen atom abstraction, *J. Am. Chem. Soc.,* 103, 3094, 1981.

138. **Banks, R. E., Haszeldine, R. N., and Myerscough, T.,** Nitroxide chemistry. I. Reactions of bistrifluoromethyl nitroxide with acetylene, 3,3,3-trifluoropropyne, perfluoropropyne, perfluorobut-2-yn, perfluorodiphenyl acetylene, and glyoxal, *J. Chem. Soc. (C),* 1957, 1971.

139. **Banks, R. E., Haszeldine, R. N., and Myerscough, T.,** Polyhalogeneoallenes. IX. Reaction of tetrafluoroallene with perfluoro-(2,4-dimethyl-3-oxa-2,4-diazapentane), *J. Chem. Soc. Perkin Trans.,* 1, 2336, 1972.

140. **Banks, R. E., Haszeldine, R. N., and Stephens, C. W.,** Free-radical attack on isocyanides: amino-oxyl and *t*-butyl or trifluoromethyl isocyanide, *Tetrahedron Lett.,* 3699, 1972.

141. **Banks, R. E., Haszeldine, R. N., and Murray, M. B.,** Nitroxide chemistry. XIX. Reaction of bistrifluoromethyl nitroxide with diphenylketene and related compounds (Ph$_2$CHOX, X = OH, Cl, NH$_2$), *J. Fluor. Chem.,* 17, 561, 1981.

142. **Makarov, S. P., Videiko, A. F., Tobolin, V. A., and Engline, M. A.,** Reactions of bis(trifluoromethyl) nitrogen oxide. I. Photolysis, pyrolysis, and some other reactions, *Zh. Obshch. Khim.,* 37, 1528, 1967; *Chem. Abstr.,* 68, 12361.

143. **Banks, R. E., Choudhury, D. R., and Haszeldine, R. N.,** Nitroxide chemistry. IV. Reactions of bistrifluoromethyl nitroxide with aldehydes, *J. Chem. Soc. Perkin Trans. 1,* 80, 1973.

144. **Banks, R. E., Choudhury, D. R., and Haszeldine, R. N.,** Nitroxide chemistry. V. Reactions between bistrifluoromethyl nitroxide and alkylbenzenes; mechanism of formation of carbonyl ⋅compounds, *J. Chem. Soc. Perkin Trans. 1,* 1092, 1973.

145. **Banks, R. E., Haszeldine, R. N., and Justin, B.,** Nitroxide chemistry. II. Reaction of bistrifluoromethyl nitroxide with some alkanes and alkenes; free radical dehydrogenation of alkanes to alkenes, *J. Chem. Soc. (C),* 2777, 1971.

146. **Banks, R. E., Birchall, J. M., Brown, A. K., Haszeldine, R. N., and Moss, F.,** Nitroxide chemistry. VIII. Abstraction of allylic hydrogen from isobutene by bistrifluoromethyl nitroxide, *J. Chem. Soc. Perkin Trans. 1,* 2033, 1975.

147. **Buchahenko, A. L. and Vasserman, A. M.,** *Stable Radicals. Electronic Structure Reactivity and Use,* Khimia, Moscow, 1973; *Chem. Abstr.,* 80, 113018g.

148. **Shchukin, G. I., Ryabinin, V. A., Grigor'ev, I. A., and Volodarskii, L. B.,** Electrochemical oxidation of nitroxides, *Zh. Obshch. Khim.,* 56, 855, 1986; *Chem. Abstr.,* 190330q.

149. **Golubev, V. A., Zhdanov, R. I., and Rozantsev, E. G.,** Interaction of iminoxyl radicals with chlorine, *Izv. Akad. Nauk S.S.S.R., Ser. Khim.,* 184, 1970; *Chem. Abstr.,* 72, 111229n.

150. **Zhdanov, R. I., Golubev, V. A., and Rozantsev, E. G.,** Synthesis and structure of 1-oxopiperidinium tribromides, *Izv. Akad. Nauk S.S.S.R., Ser. Khim.,* 186, 1970; *Chem. Abstr.,* 72, 111227k.

151. **Golubev, V. A., Voronina, G. N., and Rozantsev, E. G.,** *N*-Oxopyrrolidinium salts, *Izv. Akad. Nauk S.S.S.R., Ser. Khim.,* 161, 1972; *Chem. Abstr.,* 77, 34231f.

152. **Golubev, V. A., Voronina, G. N., and Rozantsev, E. G.,** Oxammonium salts of the pyrrolidine series, *Izv. Akad. Nauk S.S.S.R., Ser. Khim.,* 2605, 1970; *Chem. Abstr.,* 74, 111841x.

153. **Golubev, V. A., Solodova, V. V., Aleinikov, N. N., Korsunskii, B. L., Rozantsev, E. G., and Dubovitskii, F. I.,** Oxidation of nitroxyl radicals by xenone difluoride, *Izv. Akad. Nauk S.S.S.R., Ser. Khim.,* 572, 1978; *Chem. Abstr.,* 89, 43237u.

154. **Zhdanov, R. I., Golubev, V. A., Gida, V. M., and Rozantsev, E. G.,** Interaction of iminoxyl radicals with antimony pentachloride, *Dokl. Akad. Nauk S.S.S.R.,* 196, 856, 1971; *Chem. Abstr.,* 74, 141470t.

155. **Golubev, V. A. and Voronina, G. N.,** Synthesis and structure of *N*-oxopyrrolinium hexachloroantimonates, *Izv. Akad. Nauk. S.S.S.R., Ser. Khim.,* 164, 1972; *Chem. Abstr.,* 77, 34232g.

156. **Golubev, V. A. and Voronina, G. N.,** *N,N*-Dialkyl-*N*-oxo-ammonium salts, *Izv. Akad. Nauk. S.S.S.R., Ser. Khim.,* 2089, 1972; *Chem. Abstr.,* 78, 3608a.

157. **Mar'in V. A., Kornienko, G. K., Matkovskii, P. E., Shebaldova, A. D., and Khidekel, M. L.,** Products of the reaction of nitroxyl radicals with tungsten hexachloride and their catalytic properties in the disproportionation of 2-pentene, *Izv. Akad. Nauk. S.S.S.R., Ser. Khim.,* 2269, 1980; *Chem. Abstr.,* 94, 64812q.

158. **Shchukin, G. I., Grigor'ev, I. A., and Volodarskii, L. B.,** Recyclization of the imidazoline ring in nitroxyl radicals of 3-imidazoline derivatives in oxidation by halogens and nitrous acid, *Izv. Akad. Nauk. S.S.S.R., Ser. Khim.,* 1581, 1981; *Chem. Abstr.,* 95, 187150a.

159. **Nelson, J. A., Chou, S., and Spencer, T. A.,** Reaction of 4,4-dimethyloxazolidine-*N*-oxyl (doxyl) derivatives with nitrogen dioxide: a novel and efficient reconversion into parent ketones, *J. Chem. Soc. Chem. Commun.,* 1580, 1971.

160. **Chou, S., Nelson, J. A., and Spencer, T. A.,** Oxidation and mass spectra of 4,4-dimethyloxazolidine-*N*-oxyl (doxyl) derivatives of ketones, *J. Org. Chem.,* 39, 2356, 1974.

161. **Bailey, P. S. and Heller, J. E.,** Ozonation of amines. V. Di-*t*-butyl nitroxide, *J. Org. Chem.,* 35, 2782, 1970.

162. **Golubev, V. A., Zhdanov, R. I., Gida, V. M., and Rozantsev, E. G.,** Interaction of iminoxy radicals with triphenylcarbonyum salts, *Izv. Akad. Nauk. S.S.S.R., Ser. Khim.,* 2815, 1970; *Chem. Abstr.,* 74, 111324f.

163. **Bobbitt, J. H. and Flores, M. C.,** Organic nitrozonium salts as oxidants in organic chemistry, *Heterocycles,* 27, 509, 1988.

164. **Yamaguchi, M., Miyazawa, T., Takata, T., and Endo, T.,** Application of redox system based on nitroxides to organic chemistry, *J. Pure Appl. Chem.,* 62, 217, 1990.

165. **Anelli, P. Y., Banfi, S., Montanari, F., and Quici, S.,** Oxidation of diols with alkali hypochlorites catalyzed by oxammonium salts under two phase conditions, *J. Org. Chem.,* 54, 2970, 1989.

166. **Ganem, B.,** Biological spin labels as organic reagents. Oxidation of alcohols to carbonyl compounds using nitroxyls, *J. Org. Chem.,* 40, 1998, 1975.

167. **Inokuchi, T., Matsumoto, S., and Torii, S.,** Indirect electooxidation of alcohols by a double mediatory system with two redox couples of $[R_2^+ N{=}0]/R_2NO$ and $[Br^{\cdot}$ or $Br^+]/Br^-$ in an organic-aqueous two-phase solution, *J. Org. Chem.,* 56, 2416, 1991.

168. **Malatesta, V. and Ingold, K. U.,** Protonated nitroxide radicals, *J. Am. Chem. Soc.,* 95, 6404, 1973.

169. **Hoffman, B. M. and Eames, T. B.,** Protonated nitroxide free radical, *J. Am. Chem. Soc.,* 91, 2169, 1969.

170. **Golubev, V. A., Zhdanov, R. I., Gida, V. M., and Rozantsev, E. G.,** Interaction of iminoxyl radicals with some inorganic acids, *Izv. Akad. Nauk. S.S.S.R., Ser. Khim.,* 853, 1971; *Chem. Abstr.,* 75, 63571f.

171. **Golubev, V. A., Sen', V. D., Kulyk, I. V., and Aleksandrov, A. L.,** Mechanism of the acid disproportionation of di-*tert*-alkyl-nitroxyl radicals, *Izv. Akad. Nauk. S.S.S.R., Ser. Khim.,* 2235, 1975; *Chem. Abstr.,* 84, 42801d.

172. **Usvyatsov, A. A., Medvedeva, I. M., and Volodarskii, L. B.,** Anodic polarography and amperometric titration of imidazoline nitroxides, *Izv. Sib. Otd. Akad. Nauk S.S.S.R., Ser. Khim. Nauk,* 4, 138, 1980; *Chem. Abstr.,* 93, 194379u.

173. **Grigor'ev, I. A., Shchukin, G. I., and Volodarskii, L. B.,** ESR, ^{1}H and ^{13}C NMR studies of nitroxides transformations in strong and super strong acids, *Izv. Akad. Nauk. S.S.S.R., Ser. Khim.,* 2277, 1986; *Chem. Abstr.,* 107, 197363c.

174. **Hoffman, A. K.,** A new stable free radical: di-*t*-butylnitroxide, *J. Am. Chem. Soc.,* 83, 4671, 1961.

175. **Sen', V. D., Golubev, V. A., and Kosheleva, T. M.,** Mechanism of redox reactions of oxo-piperidinium salts and piperidine oxyl radicals with iodides and iodine, *Izv. Akad. Nauk. S.S.S.R., Ser. Khim.,* 747, 1977; *Chem. Abstr.,* 87, 67600k.

176. **Low, H., Paterson, I., Tedder, J. M., and Walton, J.,** Homosolvolysis, *J. Chem. Soc. Chem. Commun.,* 171, 1977.

177. **Smith, J. and Tedder, J. M.,** Homolysis. Part 3. The reaction of aromatic acid chlorides with di-*t*-butylnitroxide, *J. Chem. Soc. Perkin Trans. 2,* 895, 1987.

178. **Yoshioka, T., Higashida, S., Morimura, S., and Murayama, K.,** Studies on stable free radicals. V. Reactivity of a stable free radical, 2,2,6,6-tetramethyl-4-oxopiperidine-1-oxyl, *Bull. Chem. Soc. Jpn.,* 44, 2207, 1971.

179. **Bolsman, T. A. B. M., Blok, A. P., and Frijns, J. H. G.,** Catalytic inhibition of hydrocarbon autoxidation by secondary amines and nitroxides, *Recl. Trav. Chim.,* 97, 310, 1978.

180. **Bolsman, T. A. B. M., Blok, A. P., and Frijns, J. H. G.,** Mechanism of catalytic inhibition of hydrocarbon auto-oxidation by secondary amines and nitroxides, *Recl. Trav. Chim.,* 97, 313, 1978.

181. **Khrushch, N. E., Chukanova, O. M., D'yachkovskii, F. S., and Golubev, V. A.,** Mechanism of reaction of methyltitanium chloride with 2,2,6,6-tetramethylpiperidine-1-oxyl, *Izv. Akad. Nauk. S.S.S.R., Ser. Khim.* 1239, 1981; *Chem. Abstr.,* 95, 131937v.

182. **Murayama, K., Morimura, S., and Yoshioka, T.,** Studies on stable free radicals. II. Reactivity of stable nitroxide radicals and NMR spectra of reactions products, *Bull. Chem. Soc. Jpn.,* 42, 1640, 1969.

183. **Rozantsev, E. G. and Sholle, V. D.,** Synthesis and reactions of stable nitroxyl radicals. II. Reactions, *Synthesis,* 401 1971.

184. **Howard, J. A. and Tait, J. C.,** 2,2,6,6-Tetramethyl-4-oxo-1-(1,1-diphenylethoxy)piperidine: synthesis and thermal stability, *J. Org. Chem.,* 43, 4279, 1978.

185. **Rykov, S. V. and Sholle, V. D.,** Chemical polarization of nuclei in the reaction of a stable nitroxyl radical with acetyl-benzoyl peroxide, *Izv. Akad. Nauk. S.S.S.R., Ser. Khim.,* 2351, 1971; *Chem. Abstr.,* 76, 45448n.

186. **Willson, R. L.,** Pulse radiolysis studies on reaction of triacetonamine-*N*-oxyl with radiation-induced free radicals, *Trans. Faraday Soc.,* 67, 3008, 1971.

187. **Mendenhall, G. D. and Ingold, K. U.,** Reversible dimerization and some solid-state properties of two bicyclic nitroxides, *J. Am. Chem. Soc.,* 95, 6390, 1973.

188. **Capiomont, A., Chion, B., and Lajzerowicz, J.,** Refinement of the structure of the dimerized nitroxide radical; 9-azabicyclo[3.2.1]nonan-3-on-9-oxy, *Acta Cryst., (B),* 27, 322, 1971.

189. **Craig, R. L. and Roberts, J. S.,** Decomposition of an allyl-substituted nitroxide radical, *J. Chem. Soc. Chem. Commun.,* 1142, 1972.

190. **Reznikov, V. A., Vishnivetskaya, L. A., and Volodarskii, L. B.,** Reactions of the heterocyclic enamides of imidazolidine nitroxyl radicals with electrophilic and nucleophilic reagents, *Khim. Geterotsikl. Soedin.,* 620, 1988; *Chem. Abstr.,* 110, 135142h.

191. **Anderson, D. R. and Koch, T. H.,** Upper excited state photochemistry of di-*tert*-butyl nitroxide, *Tetrahedron Lett.,* 3015, 1977.

192. **Reznikov, V. A. and Volodarskii, L. B.,** Reaction of 2*H* (4*H*)-imidazole oxides with lithium organic compounds — a new route to stable nitroxides of 2(3)-imidazoline series, *Izv. Akad. Nauk. Russ. Ser. Khim.,* in press.

193. **Keana, J. F. W., Dinerstein, R. J., and Baitis, F.,** Photolytic studies on 4-hydroxy-2,2,6,6-tetramethylpiperidine-1-oxyl, a stable nitroxide free radical, *J. Org. Chem.,* 36 209, 1971.
194. **Call, L. and Ullman, E.,** Stable free radicals. XI. Photochemistry of a nitronyl nitroxide, *Tetrahedron Lett.,* 961, 1973.
195. **Shchukin, G. I., Grigor'ev, I. A., and Volodarskii, L. B.,** Photochemical reactions of nitroxyl radicals derivatives of 3-imidazoline 3-oxide and 3-imidazoline, *Izv. Akad. Nauk. S.S.S.R., Ser. Khim.,* 1421, 1980; *Chem. Abstr.,* 93, 186241c.

Chapter 3

SYNTHETIC CHEMISTRY OF STABLE NITROXIDES

I. INTRODUCTION

The aim of synthetic chemistry of stable nitroxides is the selective design of various structures containing the nitroxyl group in a molecule. The presence of the nitroxyl group in a molecule is thus the only feature defining the object of this field and distinguishing this field from general synthetic chemistry regarding its aims and tasks. The definition of the object is therefore indiscriminative, because it covers the most diverse organic compounds, both cyclic and acyclic in which structure may in principle include any groups. For nitroxides, however, such a definition of an object is justified by the fact that investigation of the chemistry of nitroxyl-containing molecules is not an end in itself though it is of interest. The necessity of particular functionalized stable nitroxides is determined first of all by their utility in solving various scientific and applied problems using EPR spectroscopy. Thus, the choice of nitroxide structure is largely determined by social (practical or scientific) needs. On the other hand, the development of synthetic chemistry of nitroxides leads to new applications of nitroxides or provides new information on traditional objects. In this connection, before discussing synthetic possibilities of nitroxides and synthetic strategy for various nitroxyl-containing structures, we must consider briefly the main applications of these compounds.

A. MOLECULAR BIOLOGY, BIOCHEMISTRY, AND BIOPHYSICS

Wide use of nitroxides to study various biological systems and processes in living organisms is associated with their unique property, paramagnetism.[1-3] The design of molecules with a nitroxyl group and another reactive group or groups that ensure covalent binding with a substrate makes it possible to obtain paramagnetic spin-labeled analogs of various biogenic molecules in which metabolism, migrations, distribution in systems, conformations, etc. may be investigated using EPR. These compounds were named as spin labels and their use, as the spin-label method. This approach allows, for example, investigation of drug metabolism in organisms. The behavior of a spin-labeled substrate in a surrounding medium gives us knowledge of its properties, such as viscosity, structure, diffusivity, electric potentials, etc.

The design of nitroxide molecules with given physicochemical properties and not covalently bonded with a substrate — spin probes — allows them to be used to study the structure and functions of active centers of enzymes, the structure and properties of biological membranes.[3-4] Using spin probes in the spin echo method, one can judge the structure of the nearest nitroxyl group environment.[5-7]

The ability of nitroxides to undergo reversible protonation or deprotonation at the groups in close proximity to the nitroxyl group leading to variation in the EPR spectrum underlies the method of medium pH determination in biological micro-objects: cells, micelles, and organelles.[8]

A comparatively new field of application of nitroxides is the determination of free oxygen concentration in various substrates in the range of $10^{-3} - 10^{-7} M$.[9,10] This method takes advantage of EPR line broadening as a result of interaction of paramagnetic oxygen molecules of the nitroxide molecule with the magnitude that is directly related to oxygen concentration. This method is highly sensitive, exact, and simple to use. In particular, it was used to study cell perspiration, oxygen transport through lipid bilayer, mechanism of photosynthesis, etc.[10]

Owing to paramagnetism, nitroxide molecules can affect relaxation times T_1 and T_2 of the nuclei nearest to the nitroxyl group. Due to this, stable nitroxides may be used as contrasting agents in the NMR tomography.[11] They are preferable to the high-spin metal ions commonly used for that purpose, because they may impart the properties that ensure localization in a given volume and in a given organ. In this respect, the EPR tomography method which allows determination of nitroxide distribution in a sample volume should be mentioned. This method is currently at a primary stage of development,[12] but one can suggest its use in the future to study spin probe distribution in living organisms.[13] One of the questions arising in this connection is: what is the maximal depth in an object under study at which the paramagnetic probe may be recorded and its distribution investigated? Another potential problem is the ability of the nitroxyl group to be reduced in these conditions to the diamagnetic hydroxylamino group. It is interesting that the reduction ability of various types of nitroxides practically does not correlate with electrochemical reduction potential, which may be due to the enzymatic character of the reduction process.[13] On the other hand, reduction of the nitroxyl group to the diamagnetic hydroxylamino group may be used for investigating electron transfer processes in the oxidative enzyme chains,[14] for example, to study enzymatic hydroxylation of various substrates.[15]

It should be noted that to raise the sensitivity of the spin probe to the environment, molecules with a minimal number of hydrogens must be used because splitting on them leads to broadening of the EPR spectrum lines. One of the possible approaches to the design of high-sensitive spin labels and probes is substitution of hydrogen in a molecule by deuterium, and the nitroxyl ^{14}N atom by ^{15}N isotope.[16] This approach was used to create a high-sensitive spin probe for the EPR oximetry with one hydrogen atom in a molecule whose splitting is rather sensitive to oxygen concentration.[17]

A number of special spectral methods have been developed using isotope-substituted nitroxides to give information on the structure, geometry, and mobility of biogenic structures, membrane permeability, conformational changes of protein molecules, and other fast processes.[3]

B. COORDINATION AND ANALYTICAL CHEMISTRY

The development of the chemistry of coordination compounds with paramagnetic ligands is stimulated by the fact that the nitroxyl group can impart peculiar chemical, physical, and spectral properties to complex compounds. Due to the presence of the nitroxyl group in a ligand, the EPR method may be used to obtain information on the structure of complex compounds, their geometry, and electronic configuration.[18] Paramagnetism of complexes with nitroxyl-containing ligands underlies the radiospectroscopy method of quantitative determination both of paramagnetic and diamagnetic metal ions.[19] Other applications are the study of sorption, extraction, and flotation processes,[19] as well as transmembrane transport of metal ions.[20]

Complex compounds of high-spin metal ions with paramagnetic ligands may possibly be used as contrasting agents in the NMR tomography (compare to Reference 11).

Because chelate complexes with paramagnetic ligands are the polyspin molecules, they present a convenient model to study spin exchange between the same (nitroxyl-nitroxyl) and different (metal-nitroxyl) fragments.[21-22] In the series of these compounds one can expect materials with unusual physical properties, in particular, organic ferromagnets and antiferromagnets (Chapter 4).[23-24] The highly efficient spin exchange and easily reversible reduction of the nitroxyl group to the hydroxylamino group are the promising qualities for the design of catalytic systems for one-electron redox processes widely occurring in nature, which may be used, for example, for atmospheric nitrogen fixation.

C. OXIDATION CATALYSTS

The application of nitroxides as catalysts of redox processes is possible due to reversibility of oxidation of the nitroxyl group to the oxammonium group and reduction to the hydroxylamino group. Oxammonium salts are strong oxidants which can, for example, selectively convert alcohols to carbonyl compounds. The hydroxylamino group is oxidized in alkaline media by atmospheric oxygen to superoxide ion capable of catalyzing the hydrolysis of nitriles to amides.[25]

It should be noted that reversibility of conversion of the hydroxylamino group anion to the oxammonium group serves as a basis for creation of an accumulator. The electromotive force of the cell: $(-)$ $Pt/>N-O^-/>N-O^{\cdot}$, CH_3CN, $LiBr//LiBr$, CH_3CN, $>N=O/>N-O^{\cdot}$, $Pt(+)$ is 0.85 V.[26]

D. STABILIZERS OF POLYMER AND MONOMER MATERIALS, ANTIOXIDANTS OF BIOCHEMICAL PREPARATES

Application of stable nitroxides as stabilizers and antioxidants is based on recombination of short-lived radicals involving the nitroxyl group and leading to chain termination in free-radical processes.[27-29] The presence of functional groups in a nitroxide molecule is not as important in this case. Essential factors are nitroxyl group stability and physical properties of a molecule: hydrophilicity, geometry, and dimension.

E. OIL FIELD GEOLOGY

A comparatively new field of application of nitroxides is their use as tracers for underground oil and water fluids.[30] This method is based on paramagnetism, which is a unique property for organic compounds providing easy detection of a tracer with the help of EPR spectroscopy. A condition for using nitroxides as tracers is their stability within a wide pH and temperature range and against various chemical agents.

More efficient tracers are the diamagnetic precursors of nitroxides, which are often more stable and oxidized to nitroxides in the analysis of the samples.[30] The limitation of this method is the similarity of EPR spectra of nitroxides of different classes and the impossibility of distinguishing between them using a routine EPR spectrometer. This has posed a task to synthesize nitroxides (or the respective diamagnetic precursors) with essentially different spectral properties. This may be done by varying substituents in the vicinity of the nitroxyl group.

In discussing potential applications of nitroxides, we consider here only those fields of scientific knowledge where nitroxides are already used or may find use in the near future. It is likely that creation of polyspin systems with spatially ordered nitroxyl groups with or without metal ion coordination may give rise to materials with unusual physical properties: ferromagnets or high-temperature organic super-conductors. The highly efficient spin exchange in these systems covering large distances may lead to the application of these compounds as "molecular wire" and to the design of catalytic redox systems functioning similarly to natural enzyme systems. The paramagnetism of nitroxides and ease of their detection by high-sensitive radiospectroscopic methods which make it possible to obtain information about the systems under study will also serve to solve other problems in different fields of scientific knowledge.

II. SYNTHETIC STRATEGY FOR FUNCTIONALIZED NITROXIDES

Potential applications of stable nitroxides are determined, on the one hand, by the presence of a nitroxyl group in a molecule responsible for its paramagnetism, and on the other hand, by structural peculiarities of the whole molecule. To choose a nitroxide structure for any specific application, it is generally considered which reactive group or groups are desirable in a molecule, as well as molecular geometry, lipophilicity or hydrophilicity, mobility of the nitroxyl-containing fragment relative to the macromolecule involved, the distance between the nitroxyl fragment and the substrate studied and the degree of its bonding with it, and, finally, nitroxyl group stability within any specific pH range against oxidants or reducing agents and to UV irradiation. Generally, it is desirable to have as simple an EPR spectrum as possible. It may or may not depend on medium pH, the ionic strength of the solvent, and the presence of any specific ions in a medium. The statement of variable and desirable characteristics of a candidate for a spin-label, probe, or chelating spin-labeled reagent demonstrates the necessity to create stable nitroxides with different chemical and physical properties to solve different scientific problems. Moreover, with development of the chemistry of stable nitroxides, new fields of application may emerge.

There are three main approaches to the design of a molecule of stable nitroxide:

1. The nitroxyl fragment or its precursor containing a spatially hindered secondary or tertiary amino, hydroxylamino, or nitroxyl group is constructed using acyclic bifunctional diamagnetic compounds and a molecule with properties required to solve the given problem. This method is most often used to obtain spin probes and labels from simple bifunctional compounds and molecules of biogenic origin.
2. The nitroxyl group or, more frequently, its nearest precursor, the hydroxylamino group, is incorporated into a molecule with necessary properties simultaneously with frame construction.
3. A comparatively small molecule containing the nitroxyl group or its precursor is modified in such a way that it could be covalently bound with the macromolecule or contain functional or other groups imparting the necessary qualities for use as a spin label, probe, or chelating reagent. These modifications are generally confined to conversion of one or two reactive groups by usual organic synthetic procedures. The nitroxyl group is regarded in this case as one more functional group not involved in the transformations. In the synthesis of functionally substituted nitroxides, it is common practice to provide such conditions of transformations in which the nitroxyl group would remain intact or could be easily regenerated in further treatment.

Such a classification of methods to construct a nitroxide molecule is rather conventional and we have adopted it to define the conception of the chemistry of stable nitroxides.

A. DIRECT INTRODUCTION OF THE NITROXYL-CONTAINING FRAGMENT INTO A BIOGENIC MOLECULE

In this chapter, methods of nitroxide design that construct the heterocyclic nitroxyl-containing fragment using simple acyclic bifunctional compounds and a functional group of a comparatively more complex molecule is discussed. This approach requires substrate stability at the stage of building a heterocycle containing the nitroxyl group precursor and further oxidation. The most frequently used substrates are the molecules containing ketone or aldehyde groups. The use of the keto group to introduce the nitroxyl-containing fragment is exemplified by the synthesis of spin-labeled sterines **1-9** by interaction of subsequent ketones **10** with α-aminocarbinol **11**, 2,3-diamino-2,3-dimethylbutane **12**, and hydroxylamino ketone **13** in the presence of ammonium acetate (Scheme 1).

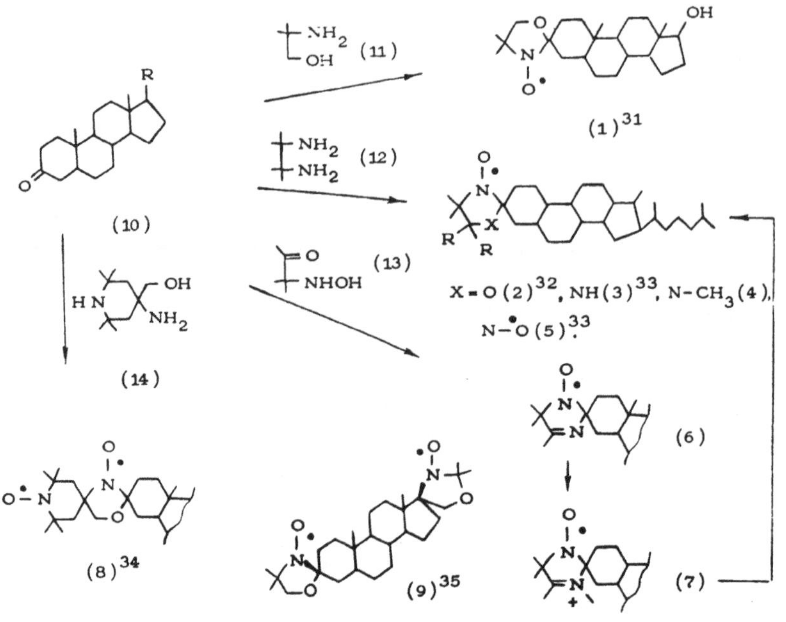

SCHEME 1.

Diamine **12** may be used for the synthesis of biradical spin probe **5**.[33] In the reaction of aminocarbinol **14** with ketone **10** and subsequent oxidation, a highly anisotropic spin probe **8** is formed.[34] Alkylation of imidazoline-containing spin probe **6** with dimethyl sulfate leads to salt **7** where reduction with sodium borohydride gives compound **4** with an imidazolidine nitroxyl-containing fragment in the molecule.

A similar approach is used in the synthesis of spin-labeled aliphatic carboxylic acids and other structures containing a long aliphatic chain with or without other functional groups. Thus, condensation of aminocarbinol **11** with ketostearic acids with subsequent oxidation affords spin-labeled stearic acids with a nitroxyl-containing fragment in different positions of the aliphatic chain (Scheme 2).[36-37]

The carboxy group in a molecule of compound **15** is reduced to the alcoholic group by the reaction with borane-dimethyl sulfide complex.[38] The nitroxyl group remains intact here.

Condensation of 1,2-hydroxylaminoketone **13** with esters of ketostearic acids and subsequent oxidation in mild conditions leads to spin-labeled ethers **18** which are readily hydrolyzed in alkaline media to form acids **20**.[39,40] Compounds **18** react with dimethyl sulfate to form imidazolinium salts **19**. It strikes one as unusual that in the reaction of acids **20** with an equimolar amount of lithium aluminium hydride, the carboxylic group is reduced to the alcoholic one with preservation of the nitroxyl group.[40] The resulting alcohol **21** was further converted to other functional derivatives **22**.[40]

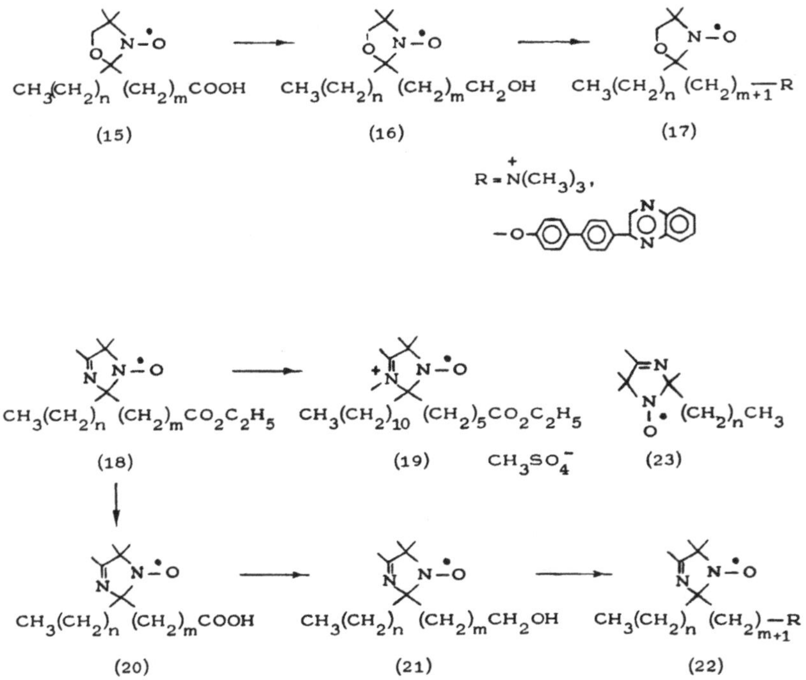

SCHEME 2.

Hydroxylaminoketone **13** seems to be the most convenient synthon of the nitroxyl-containing heterocyclic fragment, since the condensation reaction proceeds in mild conditions (room temperature or boiling for 3 to 5 h in methanol solution). The diamagnetic hydroxylamino derivatives formed are oxidized to the respective nitroxides under very mild conditions: by treatment with manganese or lead dioxides in organic solvents or simply with atmospheric oxygen without a catalyst.[41] In particular, compound **13** was used in the synthesis of nitroxides **23** containing alkyl substituents with different lengths of aliphatic chain in the 2-position of the heterocycle.[42]

Spin-labeled crown-ether **25** was obtained by condensation of ketone **24** with diamine **12** and subsequent oxidation.[43]

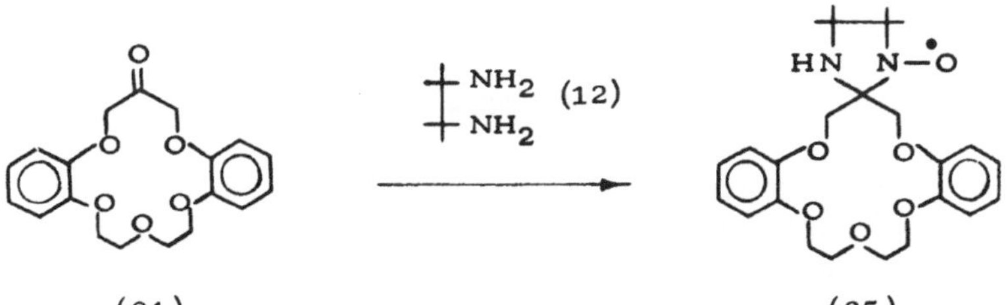

To introduce a spin label at the aldehyde group, condensation with 2,3-dimethyl-2,3-bishydroxyaminobutane **26** is used. This approach was used to synthesize compounds **27–29**. The derivatives of dihydroxyimidazolidine **30** formed at the first stage are oxidized by lead dioxide in mild conditions to nitroxides (Scheme 3).[44]

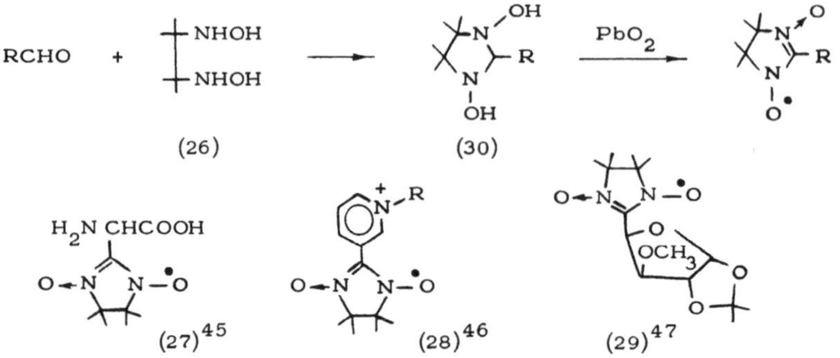

R = adenosine diphosphoribosyl

SCHEME 3.

The carboxy group can also be used for the direct introduction of the nitroxyl-containing fragment. Thus, the reaction of carboxylic acid **31** with aminocarbinol **11** produces oxazoline derivative **32** in which oxidation with a peracid and rearrangement of the resulting oxaziridine **33** leads to nitrone **34**. The reaction with organomagnesium reagents and subsequent oxidation leads to nitroxide **35** (Scheme 4).[48,49]

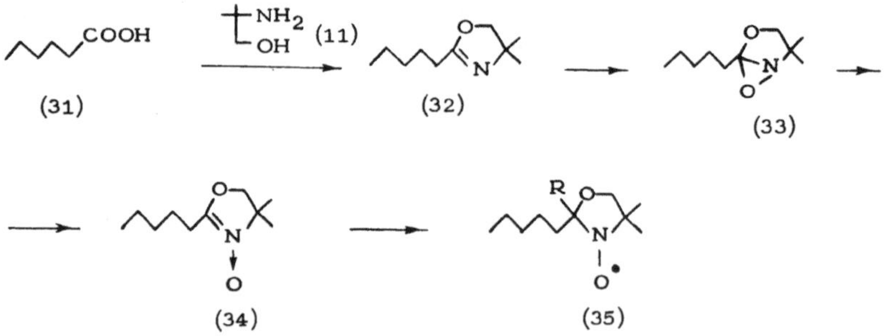

SCHEME 4.

A variation of this approach is an introduction into a substrate of one or, more frequently, two functional groups which are further used to build a nitroxyl-containing fragment. In particular, this allows spin-label introduction at the C=C bond. At a primary stage, iodoisocyanates **36** are obtained which are hydrolyzed to aziridines **37**. Subsequent interaction with acetone and oxidation leads to nitroxides **38**. Condensation with acetone proceeds nonregioselectively to form an isomeric mixture. Spin probes **38** contain the hydrogen atom at the nitroxyl α-carbon atom, due to which they are unstable and are not isolated in a free state (Scheme 5).[50]

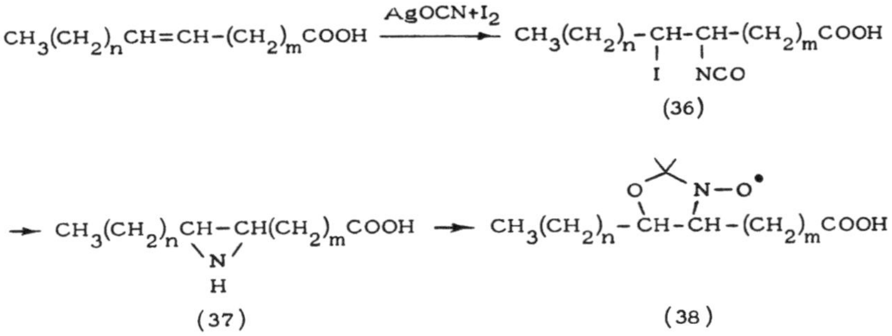

SCHEME 5.

As the nitro and nitroso groups may be transformed to the nitroxyl group (Chapter 2), they can also be regarded as synthons of the nitroxyl-containing fragment.

Due to ease of the transformation of the vicinal aminocarbinol group to the nitroxyl-containing fragment, the molecules with this group may be transformed to spin probes. This is exemplified by the synthesis of the structural spin-labeled analog of progesterone **39** (Scheme 6).[51]

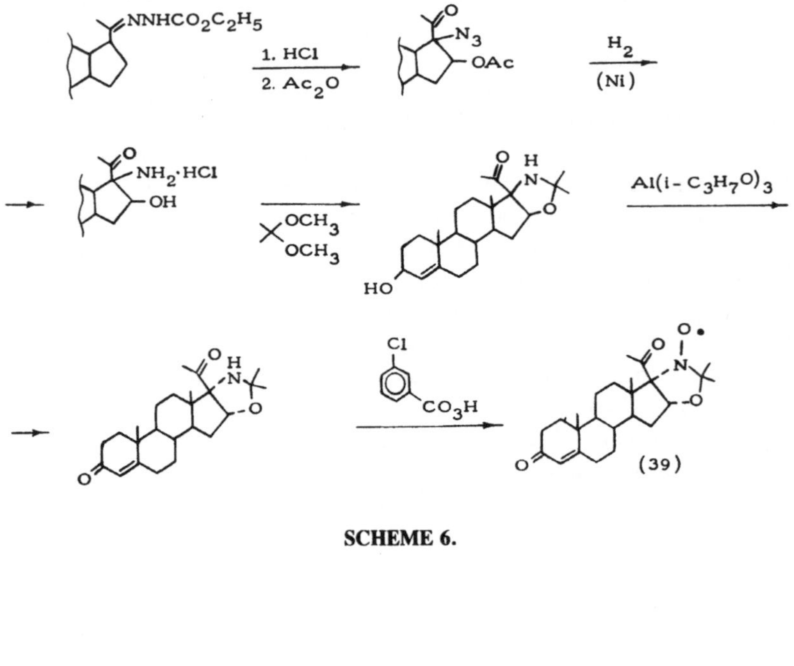

SCHEME 6.

Heterocyclic nitrones and methoxyimines can be easily transformed to nitroxides (compare to Chapter 2). This approach was used to synthesize the paramagnetic structural analog of androstane **40** (Scheme 7).[52]

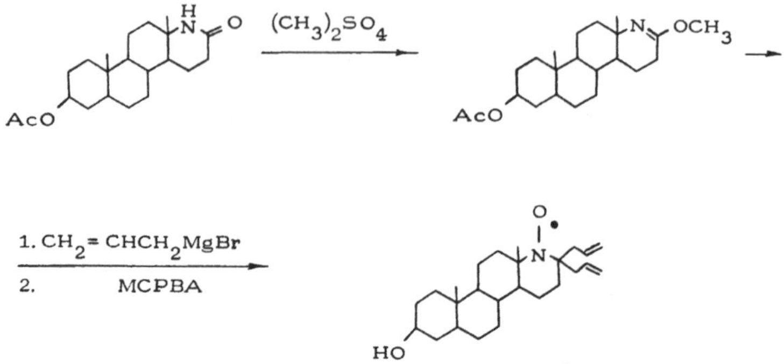

SCHEME 7.

B. INTRODUCTION OF THE NITROXYL GROUP OR ITS PRECURSOR WITH SIMULTANEOUS FRAME CONSTRUCTION

The following approach to the synthesis of spin labels, probes, and spin-labeled chelating reagents involves initial frame construction for the molecule with given properties and containing a group that is a precursor of the nitroxyl group. The nitroxyl group is generated at the last stage of synthesis or directly during frame construction. Examples of such an approach are syntheses based on the use of pyrroline-*N*-oxide derivatives **41**. Compounds **41** are obtained from available γ-nitroketones **42**.[53,54] The possibility of varying substituents R¹–R³ over a wide range allows synthesis of pyrrolines **41** with various substituents. Subsequent interaction with organometal compounds and oxidation with atmospheric oxygen in the presence of Cu^{2+} salts leads to nitroxides **43** (Scheme 8).[55,57]

Possible complication for this transformation is self-condensation of alkyl-nitrones occurring as a result of metallation of α-position of the alkylnitrone group.[58,59] This reaction is observed more often for lithium derivatives, although it also occurred in the presence of methylmagnesium iodide.[60]

In the case of addition of an organometallic reagent to the nitrone group to form hydroxylamino derivatives with hydrogen atoms at the α-carbon atom, further oxidation under the same conditions leads to more substituted nitrones. Repeating this sequence of transformations, one can introduce up to four various substituents into the nearest environment of the nitroxyl group. Substituents introduced into the heterocycle at the stage of reaction with the organometallic reagent can contain at the end of aliphatic chain the hydroxy group in the form of the tetrahydropyranyl derivative. Using a sequence of transformations: O–THP → OH → OMs → X → CN → COOH, one can obtain the carboxy group at the end of the aliphatic chain.

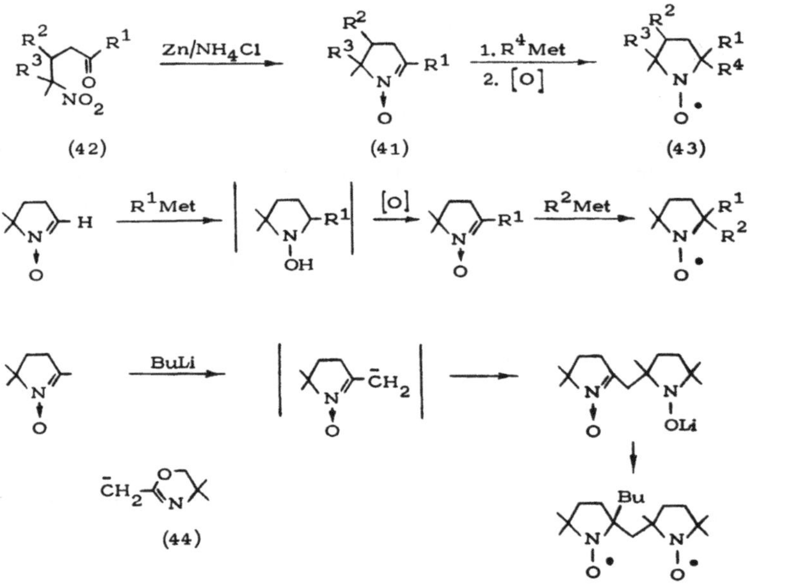

SCHEME 8.

Modification of the method is interaction of the respective halide with carbanion **44**. Subsequent alkylation and hydrolysis also lead to carboxylic acids.[61] Another method of introducing the carboxy group at the end of the aliphatic chain of the substituent is based on the fact that the substituent introduced by the reaction with an organometallic compound contains oxidation of the terminal ethylene group in the system $RuO_4/NaJO_4/CCl_4/CH_3CN$ that leads to carboxylic acids of the pyrrolidine-1-oxyl **45-47**.[57,62,63]

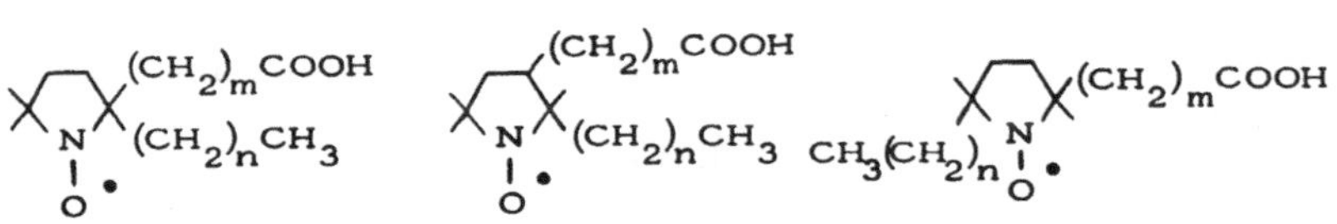

Such an approach to the synthesis of nitroxides of the pyrrolidine series is rather fruitful for the preparation of spin labels and probes. It allows construction of a molecule with a given steric location of substituents. In particular, this allows synthesis of *cis*-2,5-disubstituted derivatives, which reduces the perturbations caused by the spin probe in the native system under study.[55] Stereoselectivity of addition of the organometallic reagent to the trisubstituted nitrone, the pyrroline derivative, is determined by the fact that the reagent attack takes place from the least sterically hindered side. By varying the order of introduction of the most bulky substituent, one can obtain predominantly either *trans*- or *cis*-2,5-disubstituted derivatives.[55,59,64,65]

The stereoselectivity of the addition of organometallic reagents to the nitrone group was used in the synthesis of spin-labeled crown-ethers **48** with the nitroxyl group located in close proximity to the coordination sphere of the crown-ether (Scheme 9).[66] In this synthesis, the relative lability of nitroxyl with respect to acids was taken into account. A more obvious approach to synthesis of compounds **48** could involve the use of diphenol **49**, but the hydrolysis of methoxy groups in the presence of the nitroxyl group failed. Therefore, at first the macrocycle **50** was initially constructed which contains in the pyrroline heterocycle the nitrone group further converted to the nitroxyl group by reaction with methyllithium and subsequent oxidation.[66]

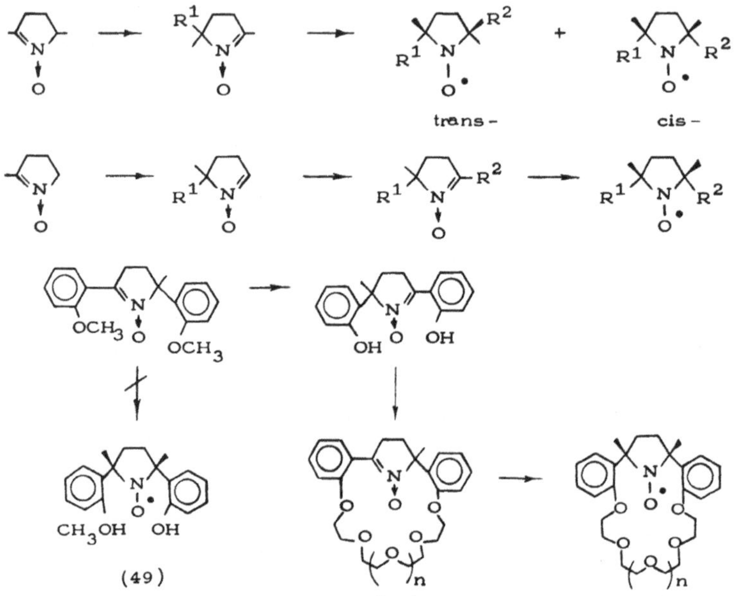

SCHEME 9.

To obtain the spin-labeled analog of cholesterol **51** a procedure based on the following sequence was used. The starting nitro derivative **52** was converted to pyrroline **53** and then to the nitroxyl radical **51** structurally related to cholesterol.[67]

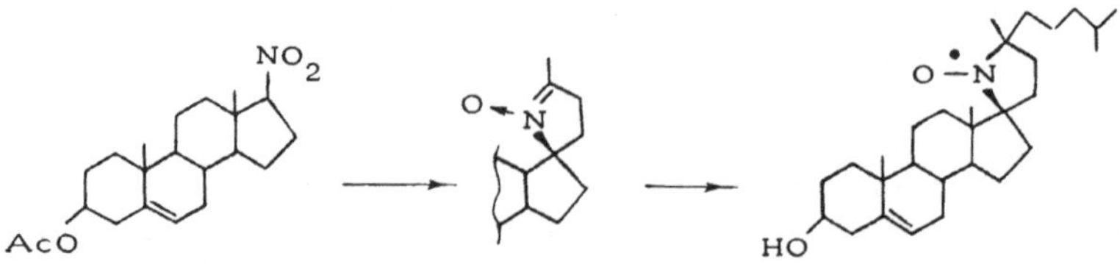

In general, the principle of structure similarity of the spin-labeled analog to its diamagnetic precursor is of great importance for the use of spin probes to study native systems. It is important that the probe should be geometrically similar to its diamagnetic precursor, and it is highly desirable that the nitroxyl group-containing fragment is not constructed with the use of functional groups in starting a diamagnetic precursor, as one cannot rule out the meaning of their specific interaction with the medium. Only meeting these two conditions can ensure that the information obtained using a spin probe will adequately reflect the real state of the system.[55]

An unusual approach to synthesis of bifunctional 2,5-disubstituted derivatives of stable nitroxides was realized using the bicyclic derivatives **54** (Scheme 10).[68]

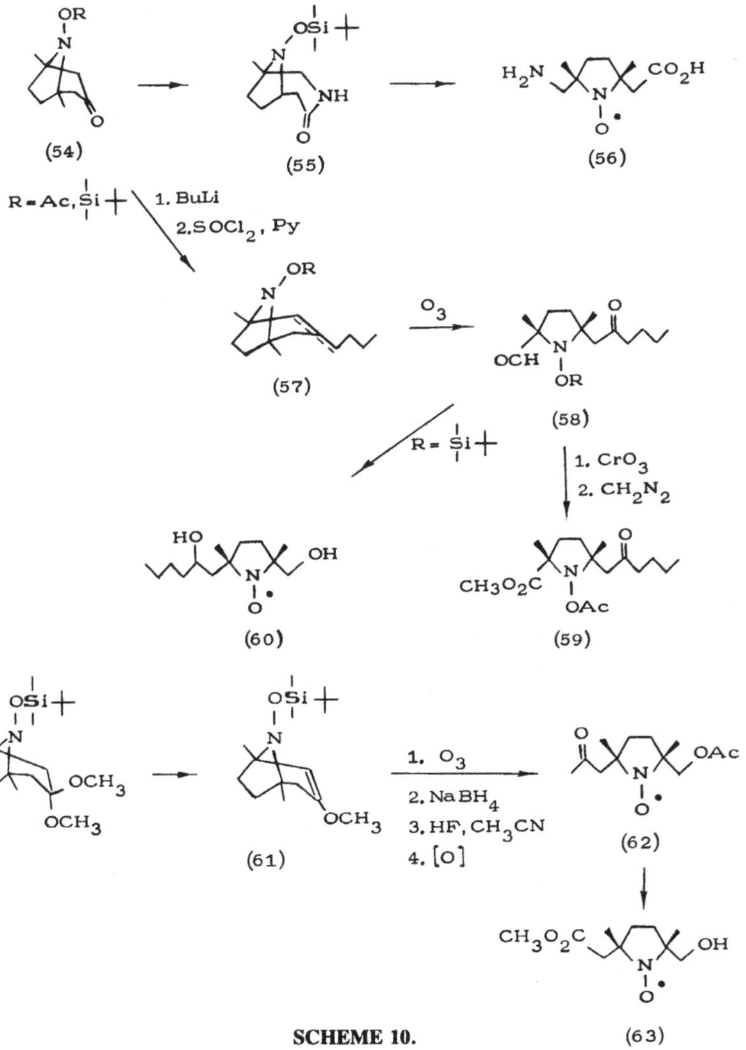

SCHEME 10.

One possible application of compound **54** is stereospecific cleavage of the lactame cycle of molecule **55** to form the 2,5-*cis*-disubstituted amino acid **56**. Another possibility of cleaving the bicyclic system is to obtain a mixture of olefins **57** which is subjected to ozonolysis. The reductive cleavage of the ozonide formed leads to aldehyde **58**. Oxidation of the aldehyde group led to an acid that was isolated as methyl ether **59**, and the reduction of compound **58** produced a mixture of diastereomeric alcohols that may be separated. Subsequent removal of the protecting acyl group and oxidation lead to radical **60**. Ozonolysis of enol ether **61** also leads to pyrrolidine derivatives which were used to prepare nitroxides **62-63**.[68]

The following structure presents synthesis of macrocyclic nitroxide **64** in which the nitroxyl group is directly incorporated into the coordination sphere of the crown-ether formed.[20]

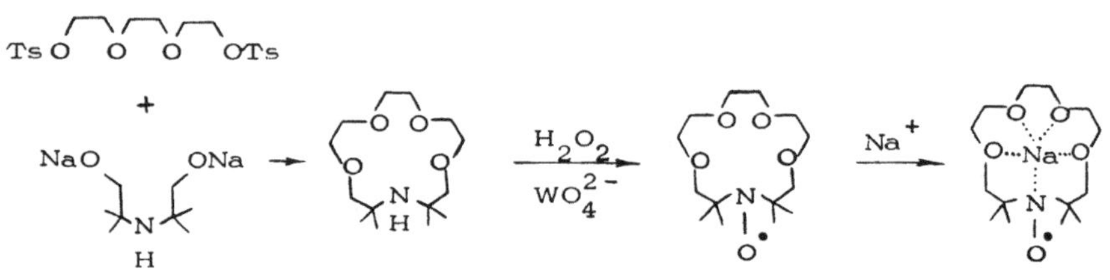

(64)   (65)

The complex of nitroxide **64** with Na[+] **65** has the spectral data which differ from those of the starting crown-ether. Thus, compound **64** is a single spin-labeled crown-ether where the EPR spectrum changes on complexation.[20]

III. SYNTHETIC CHEMISTRY OF STABLE NITROXIDES

The most widespread approach to the synthesis of functionally substituted nitroxides is as follows. A comparatively small molecule with the nitroxyl group or its precursor, the hydroxylamino group, the secondary or tertiary amino group, is modified by introducing the appropriate groups enabling it to further bind covalently or in any other way with the macromolecule studied, or in such a way that it contained functional or other groups imparting the necessary properties of a spin label, probe, or chelating agent. These modifications are generally confined to transformations of one or two reactive groups using conventional organic synthesis methods. The nitroxyl group is regarded as a more functional group that is not involved in the transformations. In synthesis of functionally substituted nitroxides, the transformations are performed under such conditions that the nitroxyl group remains intact or may be easily regenerated in the course of further treatment. Usually the nitroxyl group does not produce a specific effect ensuing from its paramagnetism on the reactions occurring at remote reaction centers. Exceptions are some reactions of 3-imidazoline-3-oxide nitroxides that will be discussed later. In view of these considerations, it seems paradoxical that at first sight the notion "synthetic chemistry of nitroxides" should be defined as a set of chemical transformations of molecules with the nitroxyl group which is not directly involved in these transformations. Nevertheless, such a definition seems to us well justified, as it clearly defines the range of chemical transformations considered. The main strategic task of the synthetic chemistry of nitroxides is creation of stable nitroxides for practical use. This aim is achieved by the search for and the use of such transformation and reaction conditions where the nitroxyl group is preserved in reaction products or may be easily regenerated.

A. PROTECTING GROUPS IN SYNTHESIS OF FUNCTIONAL DERIVATIVES OF NITROXIDES

In view of the fact that conditions of generation (or regeneration) of nitroxyl groups largely depend on the nature of the precursor employed, in choosing a synthetic scheme one must take into account the stability of the groups introduced and choose an appropriate sequence of introducing functional and nitroxyl groups. In some cases it is necessary to employ protecting groups for nitroxyl or its precursors. The simplest protecting group for the nitroxyl group with respect to acids is the hydroxylamino group. Under the reaction conditions, however, the hydroxylamino group may be transformed to a nitroxyl group (compare with Reference 69). Another circumstance to be taken into account is that the hydrolytic stability of the diamagnetic derivative may be essentially lower than that of the respective nitroxide, especially in the case of heterocyclic compounds with a 1,3-position of heteroatoms. The use of the hydroxylamino derivatives as a protecting group may be exemplified by the synthesis of aldehyde **66** from triacetonamine.

Condensation of triacetonamine with methoxymethylene triphenylphosphonium chloride in the presence of phenyllithium and subsequent heating in the methanol solution of HCl leads to acetal **67**. To obtain aldehyde **66** from **67** one may imagine three possible consecutive stages of hydrolysis-oxidation represented by the previous structure. Oxidation of the amino derivative **68** to aldehyde **66** may not be performed because of rigid conditions of amino group oxidation to the nitroxyl group, in which the aldehyde group is also oxidized. On the other hand, the nitroxyl group is unstable under conditions of acidic hydrolysis of the acetal group. Due to this, acetal **67** was initially oxidized to the radical **70**, and then reduced to the hydroxylamino derivative **71**. Subsequent hydrolysis and oxidation by Fremy's salt under mild conditions leads to aldehyde **66** in high yield.[70]

In such an approach to protection of the nitroxyl group by conversion to the hydroxylamino group, its acetyl or tetrahydropyranyl derivatives may be rather efficient for the hydrolytically stable compounds: nitroxides of the piperidine, pyrroline, and pyrrolidine series. It is of less importance for heterocyclic nitroxides with a 1,3-position of heteroatoms (oxazolidine and imidazolidine derivatives), because of the readily occurring acid-catalyzed hydrolytic cleavage of the heterocycle. In particular, imidazolidine derivatives are regarded as a protecting group for carbonyl and are easily removable in acid media. However, this approach may also be applied to imidazoline and imidazolidine derivatives. Thus, in the synthesis of isocyclic biradical **72**, acetyl protection for the hydroxylamino group is used.[33]

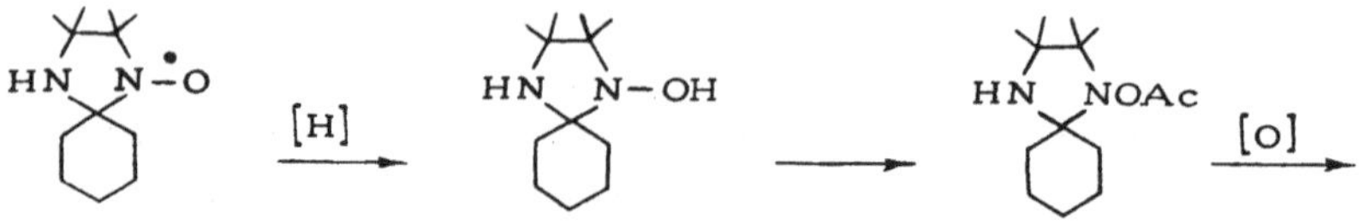

(73)

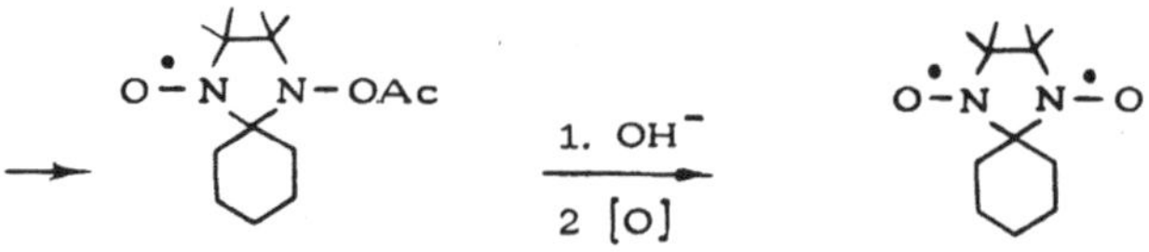

(72)

Acylation of the hydroxylamino derivative **73** was fulfilled without its isolation in view of its hydrolytic lability. The use of acetyl protection at the stage of oxidation of the sterically hindered secondary amino group in heterocycle is necessary because of the limited stability of the nitroxyl group toward the *m*-chloroperbenzoic acid oxidation conditions, though this oxidant is a rather widespread reagent for nitroxyl group generation.

Some attempts were made to obtain the isocyclic biradical **74** using 3-imidazoline-3-oxide derivatives **75-77**. The reaction of nitroxide **77** with excess methylmagnesium iodide was shown to occur primarily at the nitroxyl group to form a mixture of 1-hydroxy- **78** and 1-methoxy derivatives **79**,[70] where interaction with methylmagnesium iodide and subsequent oxidation give, respectively, the acyclic nitroxides **80** and methoxy derivatives **81** (Scheme 11).[72,73]

In the reaction of 1-hydroxy derivative **78** with organomagnesium compounds, an acyclic nitroxide **80** is formed after oxidation in a high yield. Attempted transformation of the *N*-methoxy group to the nitroxyl group leads to imidazolidine heterocycle cleavage.[74] At the same time, the use of tetrahydropyranyl or ethoxy-

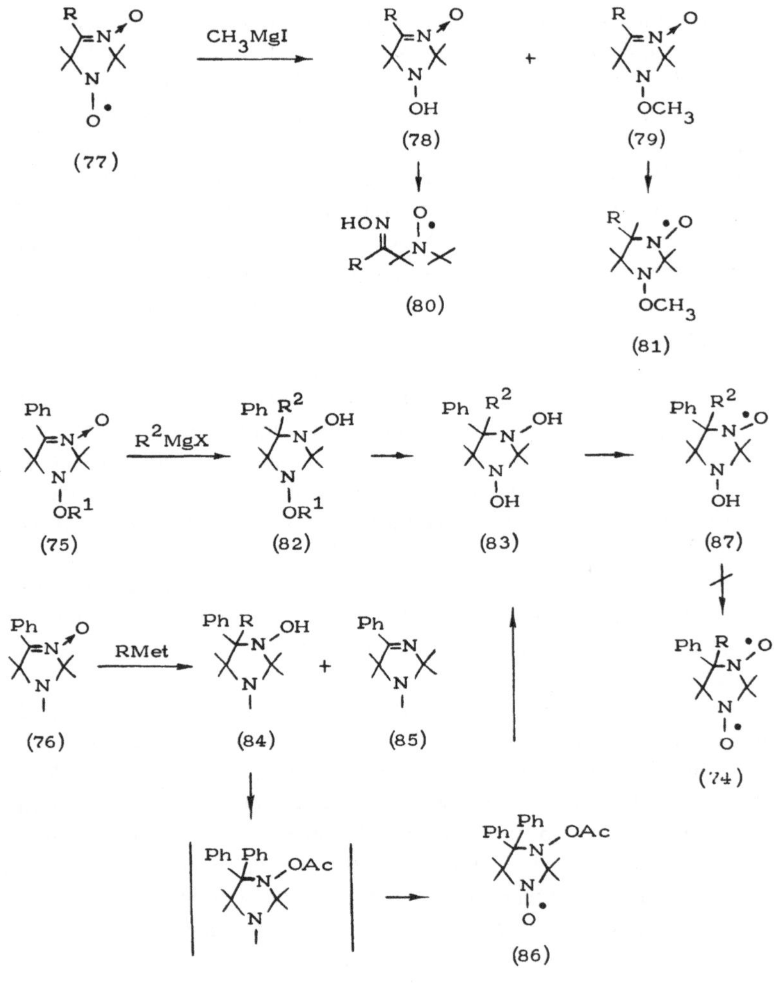

R = tetrahydropyranyl , $C_2H_5OCHCH_3$

SCHEME 11.

ethyl protection for the 1-hydroxy group in the reaction of 3-imidazoline-3-oxide derivatives **75** with organomagnesium and -lithium compounds leads to formation of imidazolidine derivatives **82** in high yields, where the protecting group is removed by treatment with dilute hydrochloric acid to form the 1,3-dihydroxy derivatives **83**.

The interaction of 1-methyl-substituted derivative **76** with organometallic reagents leads to a mixture of the products of addition at the nitrone group **84** and deoxygenation **85**.[73,76] The yield of deoxygenation product decreases from methylmagnesium iodide to phenylmagnesium bromide, the yield of the addition product increases accordingly. When phenyllithium is used, the yield of compound **84** approaches the quantitative one. Acylation of imidazolidine **84** and subsequent oxidation of the methylamino group to the nitroxyl group by hydrogen peroxide in the presence of tungstate leads to nitroxide **86** in which the acetyl group is hydrolyzed and further reduction brings about the dihydroxy derivatives **83**. Thus, the use of various protecting groups for the nitroxyl group in the reaction of 3-imidazoline-3-oxide derivatives with organometallic compounds affords the precursors of imidazolidine isocyclic biradicals **83**; however, attempts to oxidize these compounds to respective biradicals failed, the only paramagnetic products isolated were monoradicals **87**. It seems there must be a spiro group between two nitroxyl groups in one heterocycle to obtain a stable isocyclic biradical.[74]

Though it is possible to use the tetrahydropyranyl and ethoxyethyl protecting groups, conditions of their removal are much more rigid than in the case of protected alcohols. Together with hydrolytical lability of the imidazolidine heterocycle, this restricts applicability of these protecting groups for this class of nitroxides. Possibly, these groups will be better applied to other classes of nitroxides. Of the related protecting groups, we can mention successful use of silyl derivatives as protecting groups (Scheme 10).

The tertiary amino group has a limited applicability as a protecting group or the nitroxyl group precursor due to rather rigid conditions of its oxidation to the nitroxyl group. A previously mentioned example seems to be the only example where such an approach is justified.

In view of the aims and tasks of the synthetic chemistry of nitroxides, the possibilities of modification and functionalization of the nitroxide molecule are determined by the presence of an additional functional group in the molecule. Thus, synthetic possibilities of any class of nitroxides are generally already determined at the stage of construction of the molecular frame with the nitroxyl group or its precursor. In this sense the chemical properties of the given class of nitroxides coincide to a certain extent with those of the functional group of the molecule. One must take into account three circumstances due to the presence of the nitroxyl group in a molecule:

1. The limited stability of the nitroxyl group, which determines the conditions of transformations or necessity of the use of protecting groups
2. The electron-accepting effect of the nitroxyl group, which takes place when the nitroxyl group and reaction center are close to each other
3. The possibility of a specific effect of the nitroxyl group on the remote reaction center due to paramagnetism of the nitroxyl group

These circumstances determine the strategy of synthesis of functionally substituted nitroxide derivatives. In the following section specific classes of nitroxides, methods of their construction, and synthetic possibilities will be discussed. Emphasis will be made on stable nitroxides and their transformations leading in turn to new stable nitroxides.

B. ACYCLIC NITROXIDES

Under acyclic nitroxides we will understand the compounds with nitroxyl groups outside the heterocycle. This class of compounds includes analogs of di-*tert*-butylnitroxide (**A**), aryl-(**B**), and diarylnitroxides (**C**), acylnitroxides and their derivatives (**D**), and vinylnitroxides (**E**).

The general method for the synthesis of compounds **A-C** involves the reaction of *tert*-alkyl- or aryl nitroso compounds with *tert*-alkyl or arylmagnesium- or -lithium compounds (compare to Reference 77). This method was used to prepare many acyclic nitroxides **A-C**, both with and without hydrogen atoms at the α-carbon of the nitroxyl group, many of which were isolated in individual form. Aryl-*tert*-alkylnitroxides may also be obtained by oxidation of the respective secondary aryl-*tert*-alkylamines.[78] It should be noted that the method was designed by Berti for the preparation of symmetric diarylnitroxides, by the reaction of arylmagnesium bromides with amyl nitrite.[79]

As a result of their specific synthesis, compounds **A-C** generally do not contain additional functional groups. Due to this, practically all their reactions proceed with participation of a nitroxyl group. The reactions involving nitroxyl groups are most typical for aryl- and diarylnitroxides (**B,C**) owing to efficient spin density delocalization over the aromatic system. This is also the reason for the limited stability of such compounds (compare to Chapter 1). Thus, aryl-*tert*-butylnitroxides dimerize as a result of nitroxyl oxygen attack on the *para*-position of another molecule if this position is not blocked by a substituent.[80,81] Dimerization is hindered by bulky substituents in the *ortho*-position which disturb co-planarity of nitroxyl groups with the aromatic ring, thus decreasing the efficiency of spin density delocalization.[82]

One of the few examples of the reactions with aryl-*tert*-butylnitroxides proceeding with preservation of the nitroxyl group is the reaction of compound **88** with singlet oxygen leading to endo-epoxide **89**.[83]

As this transformation leads to marked variation in the EPR spectrum due to the changed spin density distribution in the aromatic system, it was immediately used as a basis for the method of singlet oxygen determination by EPR.[83]

Acyclic nitroxides containing additional functional groups in a molecule **80, 92** may be obtained by the interaction of 3-imidazoline-3-oxide **78** and imidazolidine **91** derivatives with organometal reagents (Scheme 12).

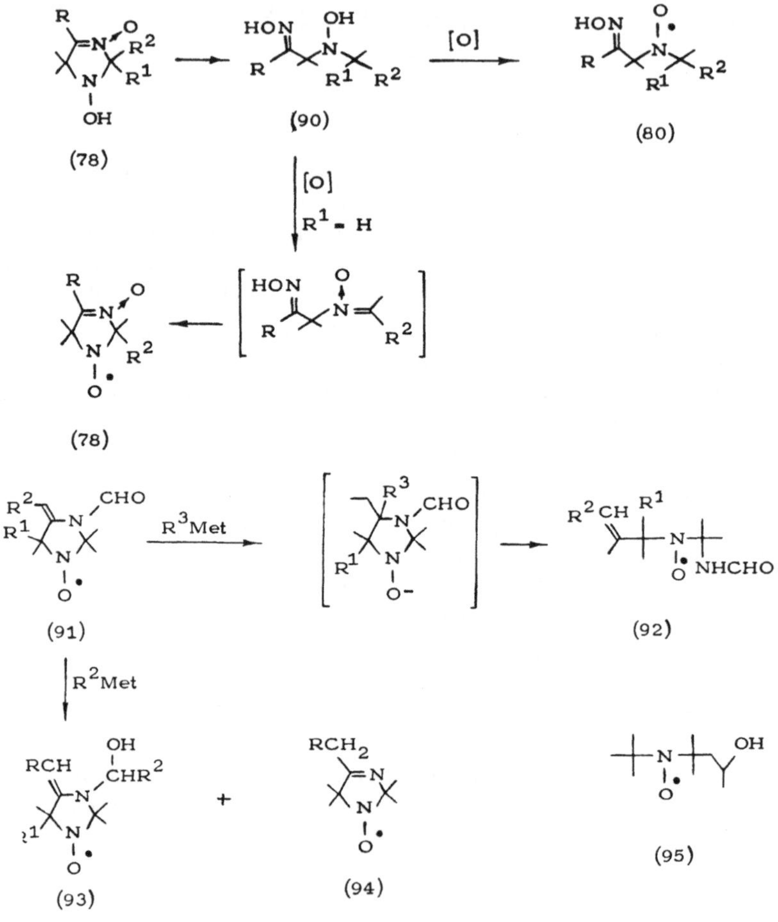

SCHEME 12.

In the reaction of 3-imidazoline-3-oxide derivatives with methylmagnesium iodide, the hydroxylaminooxime derivatives **90** are initially formed. With R^1 and R^2 being other than hydrogen they are oxidized by air oxygen in the presence of copper salts to nitroxides **80**. If one of the substituents is hydrogen, the oxidation leads to nitroxide **78**.[72]

The character of reaction products of imidazolidine derivatives **91** with organ-ometal reagents is determined by the presence of two reaction centers in a molecule: the C^4 atom in heterocycle and the carbonyl group. In the former case, acyclic nitroxides **92** are formed, while an attack at the carbonyl is the reason for the formation of compounds **93** and **94**.[84]

In the reaction of the tetrahydrooxazine-*N*-oxide derivative with methyllithium, an acyclic nitroxide **95** is formed as one of the products.[49]

Oxidation of triacetonamine with hydrogen peroxide in alkaline media or with potassium permanganate leads to the hydroxyamino acid **96**.[85,86] Its oxidation with H_2O_2 in the presence of sodium tungstate gives a stable acyclic nitroxide **97** con-taining two functional groups in a molecule.[86]

A series of acyclic diglycosyl nitroxides was synthesized. Synthesis of one of them **(98)** is represented here.[47]

A method for the synthesis of acyclic nitroxide **99** with carboxyl at the nitroxyl α-carbon atom has been developed.[87] It has been shown that free acids **100** are unstable and may be isolated only in the form of salts **99**.

Acyl nitroxides (**D**, X = O) are formed as a result of one-electron oxidation of *N*-substituted hydroxamic acids. In the case of *N-tert*-alkyl-substituted deriva-tives, the acyl nitroxides **D** may be isolated. Thus, the X-ray structure analysis of compound **101** has been carried out and the nitroxyl fragment has been shown to be practically planar with anti-position of oxygen atoms.[88]

Despite the fact that many acyl nitroxides may be isolated individually, they are much more reactive than di-*tert*-alkyl nitroxides. This is due to the fact that the lone electron to a larger extent is localized on nitroxyl oxygen than in the case of dialkyl nitroxides.[89]

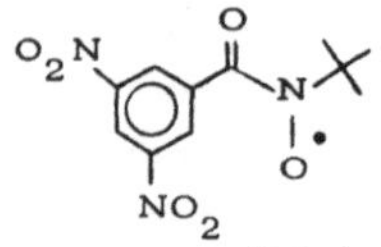

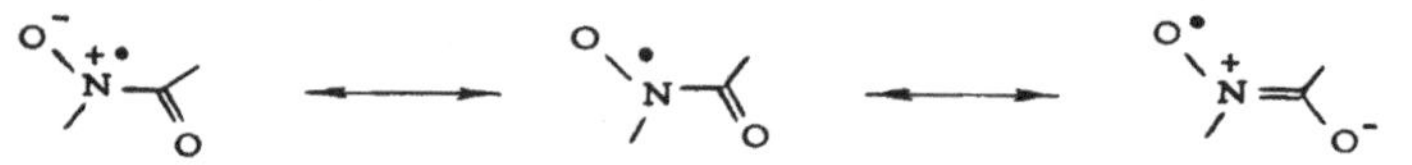

These data are in agreement with the fact that the a_N constant in acyl nitroxides is about twice as low as in the case of dialkyl nitroxides. Besides, the presence of a strong intramolecular hydrogen bond stabilizes the corresponding hydroxy form — hydroxamic acid. Due to this, the most typical reactions of acyl nitroxides are the reactions of hydrogen abstraction from various substrates by the nitroxyl fragment. In detail, the reactivity of these compounds is discussed in the review.[89] No data are available on the reactions of acyl nitroxides where the nitroxyl group is not involved in the reactions or is at least preserved in the products. Consequently, functionalization of these compounds or their modification is unlikely, and the structure of acyl nitroxides is unambiguously determined by the structure of the starting hydroxamic acids.

A number of stable acyl nitroxides containing an additional function in a molecule was recently obtained from 1,2-hydroxylaminoketones **102** (Scheme 13).[90]

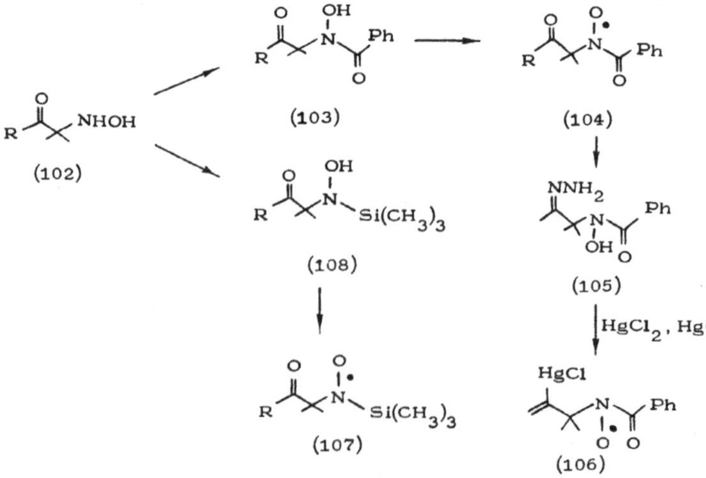

SCHEME 13.

Oxidation of keto hydroxamic acids **103** with lead dioxide gives stable acyl nitroxides **104**. The reaction of hydrazone **105** with HgCl$_2$ and HgO in ether afforded the stable organomercurous radical **106**. The organosilicon analogs of acylhydroxamic acids **107** are formed in a small yield in the oxidation of compounds **108** by lead dioxide.[90]

Vinylnitroxides **D** are generated by oxidation of enolizable nitrones or the corresponding enhydroxylamines. Ease of nitrone oxidation increases if there are substituents with the −M effect at the nitrone α-carbon, which is associated with greater enolization of such compounds.[91]

Spin density delocalization over the conjugated π-system leads to a high reactivity of these compounds. A usual reaction route for these compounds is recombination to form the C–C or C–O dimers. Due to this, vinylnitroxides are rather unstable compounds, which may not be isolated individually and in some cases may not even be recorded by EPR. Formation of these compounds and their structures are determined by the structure of the dimers formed or by the EPR spectra of spin adducts formed in the oxidation of the respective nitrones in the presence of spin traps.[91] Due to this, chemical transformations involving vinylnitroxides with preservation of the nitroxyl group seem unlikely.

Acyclic stable nitroxides are practically not used in the synthetic chemistry of stable nitroxides and their structure is unambiguously determined by the structure of their precursors. Modification of the molecule seems to be difficult or altogether impossible.

C. CYCLIC NITROXIDES WITHOUT AN ADDITIONAL HETEROATOM IN HETEROCYCLE

1. Three- and Four-Membered Nitroxides

While investigating the properties of bishydroxylamine **26**, it has been found that keeping this compound in air at room temperature results in an oily mixture of products with the EPR spectrum typical for nitroxides. A colorless substance was isolated from this mixture having the EPR spectrum and the band M$^+$ 114 in the mass spectrum, due to which it was assigned the structure of aziridine derivative nitroxide **109**. The reduction of this compound with lithium aluminium hydride leads to 2-amino-2-methylbutane, which also confirms the structure of compound **109**.[92]

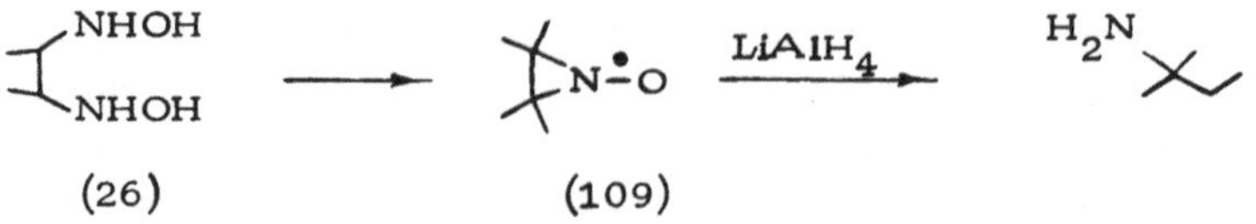

Another research group tested the results of this work in identical conditions. They isolated a colorless compound with the same melting point, giving a signal in the EPR spectrum. However, its subsequent purification led to the disappearance of paramagnetism and the compound was identified as acetone oxime.[93] Thus, the existence of aziridine nitroxides has not been proved.

The first four-membered heterocyclic azetidine nitroxides **110, 111** were obtained according to Scheme 14.[94]

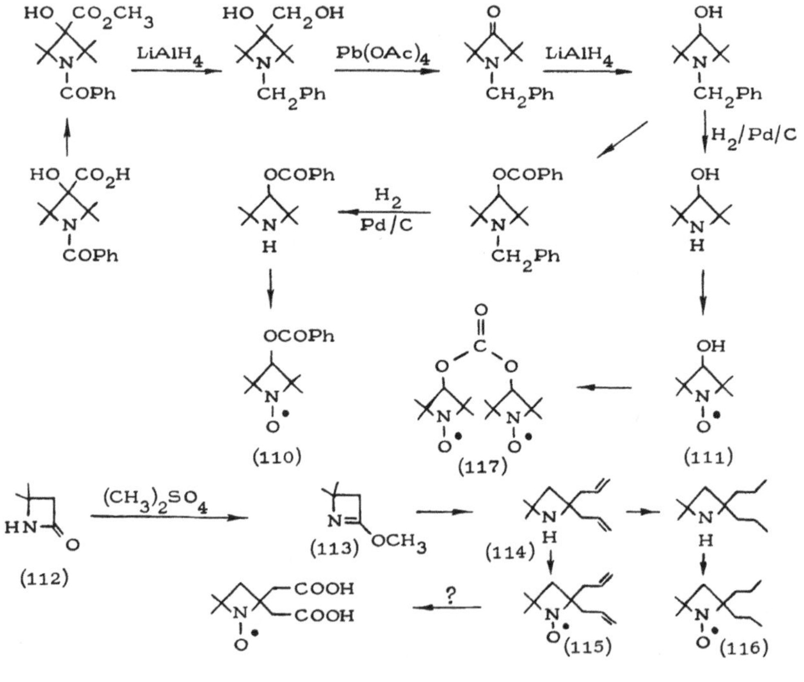

SCHEME 14.

Later on, the same research group developed another method for the preparation of azetidine nitroxides which involved alkylation of the four-membered lactame **112** leading to iminoether **113**. Interaction of the latter with two equivalents of allylmagnesium bromide and subsequent oxidation of sterically hindered secondary amine **114** leads to nitroxide **115**. The preliminary reduction of the allyl C=C bonds and subsequent oxidation leads to radical **116** (Scheme 14).[95] This method of heterocyclic amide group transformation to the nitroxyl group seems to be of general character, as demonstrated for the six-membered nitroxide-sterine analog (compare to Scheme 7).

A distinguishing feature of the four-membered nitroxide azetidine derivatives is their small size, which must minimize the perturbations in the medium under study when these compounds are used as spin probes.

The chemical modifications of compound **111** are possible owing to the presence of the hydroxy group in a molecule, due to which this compound may be a synthon for a number of spin labels and probes similar to 1-hydroxy-2,2,6,6-tetramethyl-piperidine-1-oxyl (compare to References 27 and 96). The only transformation in which compound **111** was involved was synthesis of the biradical **117**.[91] The possibilities for functionalization of azetidine (**115**) are determined by the presence of terminal C=C bonds in substituents in the 2-position of the heterocycle. For the pyrroline derivatives with such substituents, the possibility of converting the terminal ethylene bonds to carboxyl groups has been shown (compare to Section II.B). One can suggest that such transformations can also be performed for the azetidine derivative **115**. The dicarboxylic acids obtained in this way could in turn be used as starting compounds in the synthesis of various functional and bifunctional derivatives of nitroxides, including the ''cross-linking'' spin labels. The azetidine nitroxides **111** and **115**, evidently, have not been used thus far in the synthesis of functionally substituted nitroxide derivatives, and the chemical properties of these compounds have remained uninvestigated.

2. Pyrrolines and Pyrrolidines

Two approaches to the synthesis of stable pyrroline and pyrrolidine nitroxides are known. The approach that was the first to appear involves a rearrangement of triacetoneamine dibromo derivative **118** in the presence of aqueous ammonia.[27] Later it was shown that such transformation is also possible for nitroxide **119**, and instead of ammonia other nitrogen-containing nucleophiles may be used.[97] Amide **121**, R = H, may also be obtained directly from radical **122** by treatment with iodine in an alkaline medium.[98]

Hydrogenation of amide **120** leads to the pyrrolidine derivative **123** in which oxidation with hydrogen peroxide in the presence of sodium tungstate gives the nitroxide **124**. Under conditions of the Hofmann reaction, ketone **126** was obtained from pyrroline **125**, and amine **127** from pyrrolidine **124**.[27] Amides **124**, **125** are the starting compounds in the synthesis of esters **128**, **129**.[27] Thus, an available starting compound, triacetoneamine, was used to obtain a number of nitroxide derivatives containing the amide, ester, keto, or amino groups in addition to the nitroxyl group, which reactivity served as a basis for the development of synthetic chemistry of this class of nitroxides. These compounds were later used for the synthesis of other mono- and bifunctional derivatives of stable nitroxides by ordinary synthetic procedures.

Scheme 16 shows synthesis of some synthetically useful nitroxide derivatives. These compounds may be used as synthons to prepare spin labels or are spin labels themselves. These transformations proceed with participation of only one functional group of nitroxides.

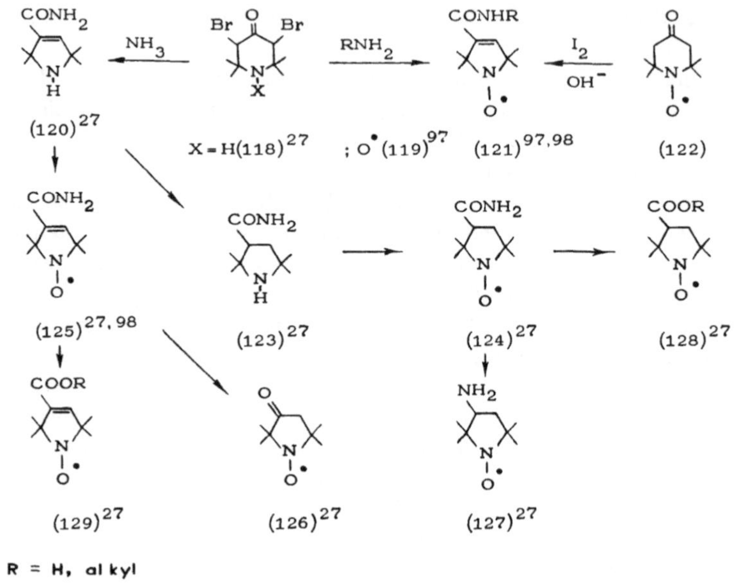

SCHEME 15.

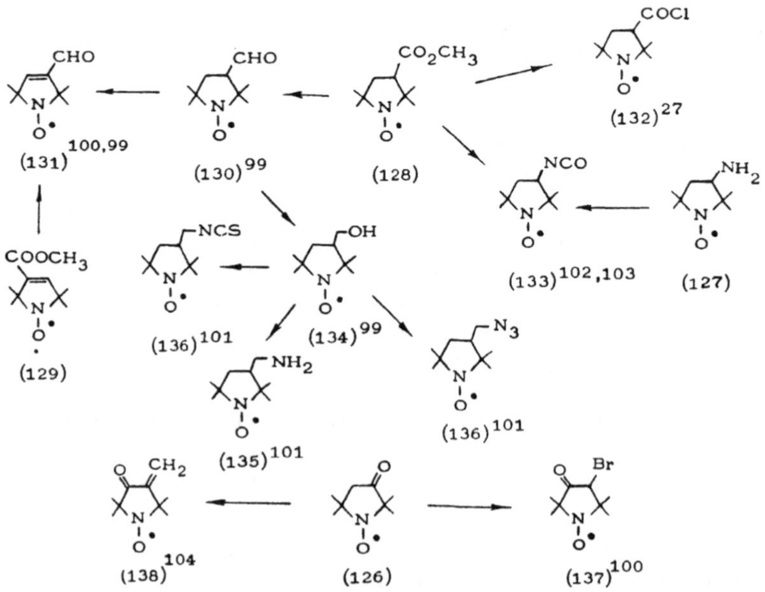

SCHEME 16.

Ketone **126** is a starting compound in the synthesis of bromoketone **137** and α-methyleneketone **138**, which can also be regarded as a spin label because of easy formation of 1,4-addition products in the reactions with nucleophilic reagents.[105]

Scheme 17 shows some nitroxide derivatives that may be obtained from nitrile **139** formed in the dehydration of amide **125**.[106] This scheme illustrates the methodological approach developed by Hideg's group for the preparation of bifunctional nitroxide derivatives, which is based on the conjugated 1,4-addition of nucleophiles to activated olefins (the Michael addition).[56,57]

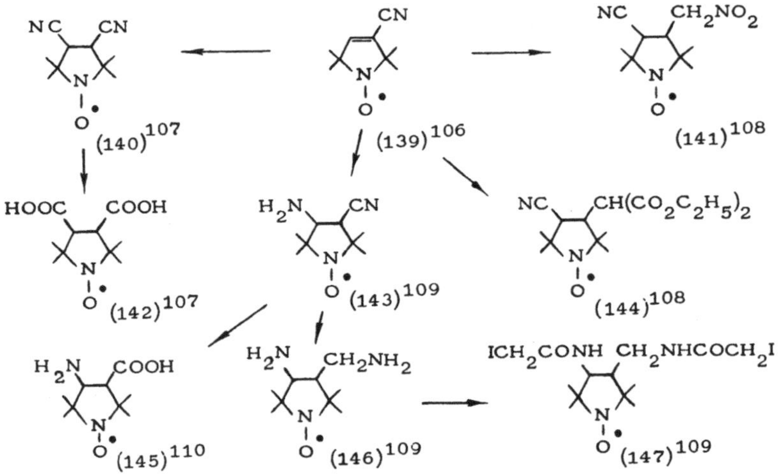

SCHEME 17.

The nucleophilic addition to the activated C=C bond is a rather promising synthetic technique to prepare the bi- and polyfunctionally substituted nitroxides that are used as the cross-linking spin labels. The addition proceeds nonstereoselectively, leading to the formation of a mixture of *cis-* and *trans*-3,4-disubstituted pyrrolidine derivatives, which generally may be easily separated. It seems most advantageous to use the α,β-unsaturated ester **129** as a starting compound in the synthesis of various bifunctionally substituted nitroxides. Schemes 18 and 19 show some derivatives obtained using this compound.

The nucleophilic addition here is also nonstereospecific. Thus, addition of phenylmagnesium bromide to ester **129** in the presence of copper salts leads regiospecifically to a mixture of two diastereomers **151** which is easily separated. The phenyl-substituted derivatives **151** were transformed to azides **155** via the stage of the respective nitro derivatives, their subsequent reduction to amines, diazotization, and interaction with sodium azide. The authors of Reference 111 maintain that in the nitration reaction under normal conditions (HNO$_3$ + H$_2$SO$_4$ concentrated), the radical center remains intact. It seems more likely that it is oxidized to the oxammonium group or undergoes disproportionation; the nitroxyl group is

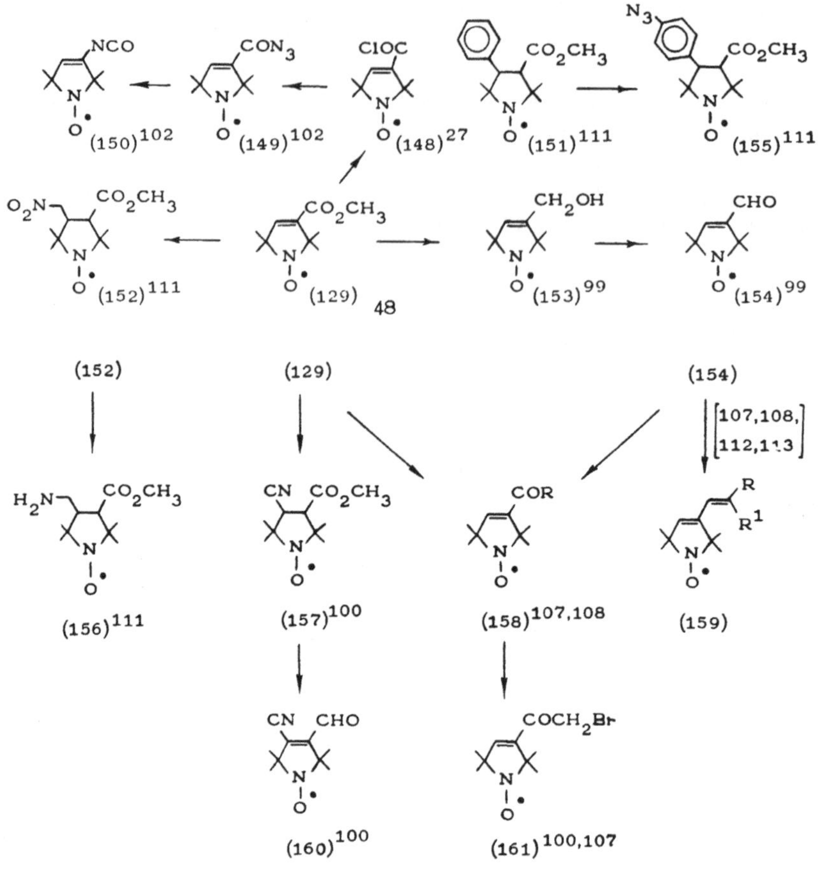

SCHEME 18.

regenerated in the process of further treatment (pouring the reaction mixture into water) (compare to Reference 114). The coexistence of the thiol and nitroxyl groups in one molecule of the mercapto derivative **162** (Scheme 19) is explained in our view by the low oxidative potential of the nitroxyl group in the pyrroline heterocycle. The data on the oxidative ability of the nitroxyl group in various heterocyclic systems given in References 115 and 116 suggested the possibility of predicting the stability of molecules containing a nitroxyl group and a group capable of oxidation with respect to intramolecular redox processes.

The synthetically useful compounds also include unsaturated alcohol **153** and halides **163, 164** which may also be used as spin labels. Thus, the high nucleophilic mobility of the halogen atom in the nitroxide molecule **163** due to its allyl position allows synthesis of various functional derivatives (Scheme 19).

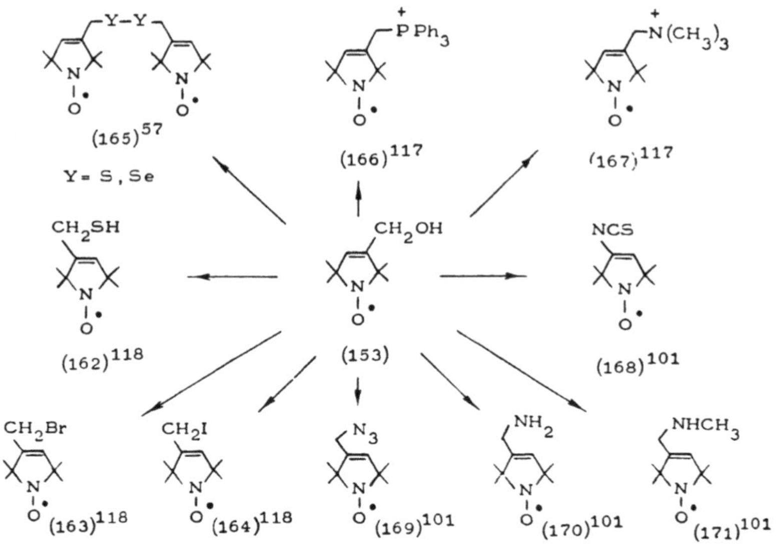

SCHEME 19.

As seen from the previous examples, pyrroline and pyrrolidine derivatives possess a wide spectrum of chemical properties that allow functionalization of these compounds. A number of the derivatives of this class of nitroxides synthesized according to this procedure are reviewed in References 55 to 57. A certain limitation of the synthetic potential of these compounds is practical unfeasibility or, at least, difficulty of varying substituents in the 2,5-positions of the heterocycle, that is, in the nearest vicinity of nitroxyl, because of the specific character of a method of generating precursors of this class of nitroxides. This demerit is absent in the method developed by Keana's group[55,58] and recently widely used by Hideg's group,[56,57] which consists of the reaction of pyrroline-*N*-oxide derivatives with organometallic reagents (compare to above). Availability of the starting pyrrolines with various substituents in the 2,3- and 5-positions of the heterocycle, on the one hand, and the possibility of using various, including functionally substituted organometallic derivatives (with the terminal C=C bond or the hydroxy group in the form of tetrahydropyranyl derivative) opens up wide possibilities for the synthesis of 2,2,3,5,5-substituted derivatives of pyrrolidine nitroxides.[55-58] In this approach, the functionalization of pyrrolidine nitroxides is associated exclusively with the presence of functional groups in the side chain. Some of the possibilities were discussed in Section II.B. Scheme 20 shows examples of spin labels and probes synthesized in terms of this approach (compare to References 55 and 58).

A limitation of this approach is the impossibility of introducing functional groups directly into the pyrrolidine heterocycle.

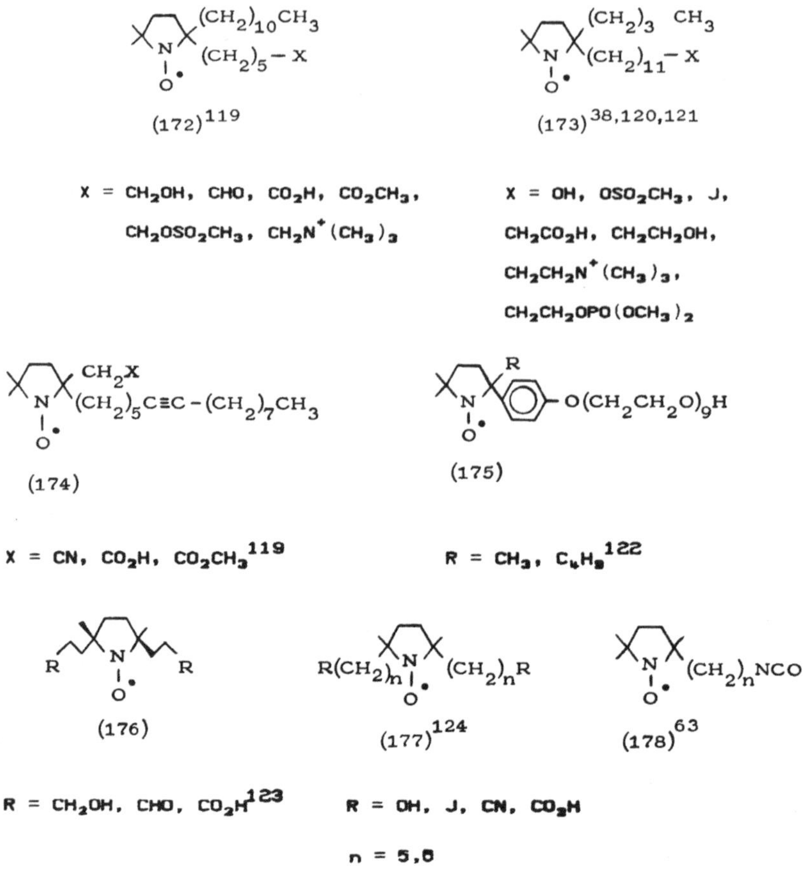

$$(172)^{119} \qquad\qquad (173)^{38,120,121}$$

X = CH_2OH, CHO, CO_2H, CO_2CH_3,

$CH_2OSO_2CH_3$, $CH_2N^+(CH_3)_3$

X = OH, OSO_2CH_3, J,

CH_2CO_2H, CH_2CH_2OH,

$CH_2CH_2N^+(CH_3)_3$,

$CH_2CH_2OPO(OCH_3)_2$

(174)   (175)

X = CN, CO_2H, $CO_2CH_3^{119}$ R = CH_3, $C_4H_9^{122}$

(176) $(177)^{124}$ $(178)^{63}$

R = CH_2OH, CHO, CO_2H^{123} R = OH, J, CN, CO_2H

n = 5,6

SCHEME 20.

Recently, the authors of this book have worked out a method for generating pyrrolidine nitroxides having to some extent advantages of both traditional approaches and based on the reaction of organometallic reagents with pyrroline derivatives **179, 180** with oxo and nitrone groups in the heterocycle. Compounds **179, 180** are formed as a result of recyclization of imidazolidine enaminoketones **181** (Scheme 21).[125-128]

The reaction of pyrroline derivatives **179** with an excess of organometallic reagent was found to proceed exclusively as an addition at the nitrone group forming after subsequent oxidation nitroxides **182**. Thus, on the one hand, one can vary substituents in the 2-position of heterocycle, and on the other hand, carbonyl is preserved in the heterocycle, which can also be used for further functionalization. Nitroxyl-containing chelating reagents **183** and **184** were synthesized in this way. The reaction of the chloro derivative **185** with methylmagnesium iodide leads to α-chloroketone **186** in a good yield.[125]

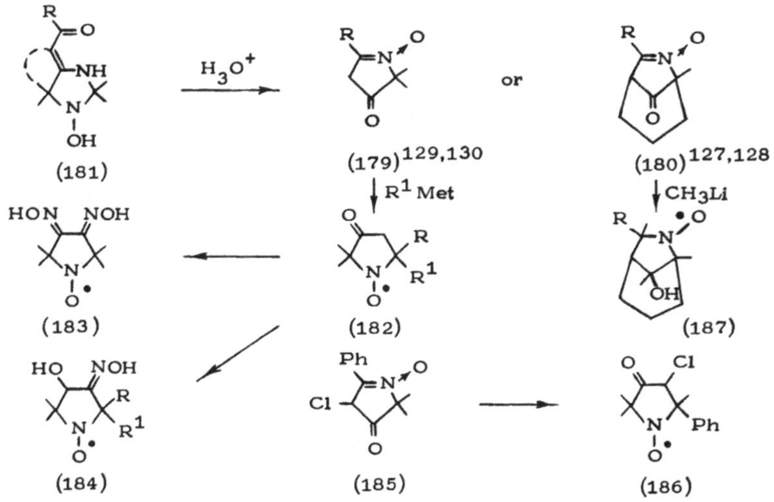

SCHEME 21.

The bicyclic derivatives **180** react with an excess of methylmagnesium iodide or methyllithium to form nitroxides **187** after oxidation, so that here addition occurs at both reactive groups, but primarily at the carbonyl group.[127-128] Scheme 22 shows some stable nitroxides containing the pyrroline heterocycle with a nitroxyl group.

Out of these compounds, synthetically useful compounds are those with carbonyl groups in a molecule. In particular, it has been shown that ketone **190** may

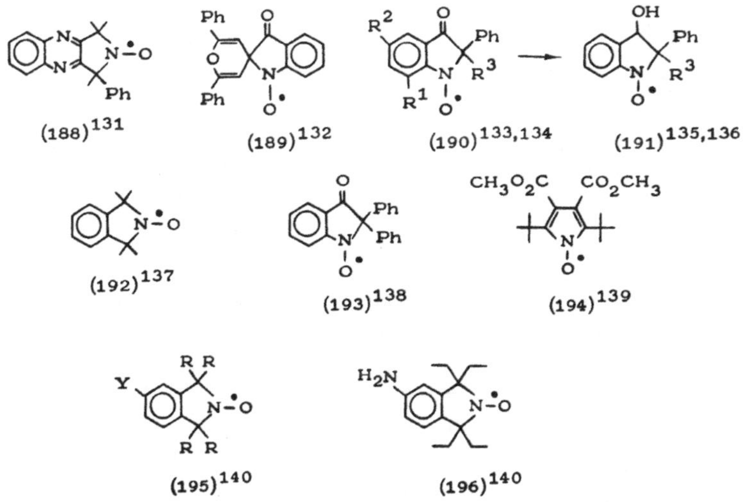

SCHEME 22.

be reduced to alcohol **191** by sodium borohydride.[135] Compound **196** is of interest in connection with the possibility of carrying out reactions at the amino group in the aromatic ring.[141]

Reference 142 has reported the possibility of direct acyloxylation and methoxylation of nitroxides (**195**, Y = H) in the aromatic ring. The authors postulated initial oxidation of the nitroxyl group to the oxammonium group and subsequent nucleophilic substitution in the aromatic ring. At the same time, in an earlier work, they suggested the homolytic mechanism of the reaction of compound **195** with aroyl peroxides.[143] In both cases, regioselectivity of the reactions is the same, which may indicate similar or even identical mechanisms of these reactions.

Due to great synthetic possibilities of pyrroline and pyrrolidine nitroxides, many of them are the commercially available spin labels and probes.

3. Piperidines

Piperidine nitroxides are one of the first representatives of stable nitroxides that were synthesized. Due to this, and the availability of the starting triacetoneamine, synthetic chemistry of these compounds was best developed. This is illustrated by the large number of commercially available spin labels and probes — piperidine, pyrroline, and pyrrolidine derivatives — that are also usually synthesized from triacetoneamine (compare to Section III.C.2.). In this section, reactions of piperidine derivatives and nitroxides on their basis, proceeding with preservation of the heterocycle, are discussed.

The nitroxyl group in the piperidine heterocycle is usually generated by oxidation of the secondary amino group with hydrogen peroxide in the presence of sodium tungstate or without a catalyst. Oxidation of sterically hindered amino derivatives **197** to nitroxides **198** is effectively proceeded in the system H_2O_2/WO_4^{2-}, only by exposure to ultrasound.[144]

(197)   (198)

Recently, it was suggested that sterically hindered secondary amines be oxidized to hydroxylamine derivatives by the reaction with dimethyldioxyrane.[145] In turn, the hydroxylamino group in the piperidine heterocycle may be oxidized to the nitroxyl group under mild conditions by various oxidants.

In the synthesis of functionally substituted piperidine nitroxide derivatives, 2,2,6,6-tetramethyl-substituted derivatives are used almost exclusively. Variation of substituents in the nearest vicinity of the nitroxyl group is difficult due to the specific character of the heterocycle construction method. A number of stable nitroxides have been synthesized containing the condensed piperidine ring with nitroxyl, as well as frame compounds that may be regarded as piperidine derivatives. Examples of such compounds are given later. Here we will only note that these compounds are not practically used in the synthesis of functional-containing derivatives. They are used as spin probes in which the structure is determined by the method of heterocyclic system design.

It is also worthwhile to mention an approach to generating the piperidine fragment with the nitroxyl group in heterocycle, consisting in transforming the lactame group to the nitroxyl group in the sterine series,[52] as mentioned earlier (Scheme 7). Such an approach evidently may be used for the generation of piperidine nitroxides of simpler structure. In this case, various substituents may be introduced into the nearest environment of a nitroxyl group, including those that may further be used for functionalization.

(199)   (200)

Another possibility for generating piperidine nitroxides may be the use of heterocyclic nitrones **199** as the starting compounds, as shown for the pyrrolidine derivatives. However, this has not been tried yet, possibly because compounds **200** have no advantages over pyrrolidine derivatives and differ from them only by one "extra" methylene group. This leads to some band broadening in the EPR spectrum due to splitting on additional hydrogen atoms, and increases mobility of the nitroxyl-containing fragment, thus making interpretation of the immobilized EPR spectra more difficult. Finally, this leads to an increased size of the nitroxyl-containing fragment, which can enhance the perturbations caused by the spin probe in the system under study.

Thus, the main objects of chemical transformations of piperidine nitroxides are triacetoneamine and the respective nitroxide. The possibilities of functionalizing these compounds arise from the presence of a carbonyl group in heterocycle, while the order of introducing functional groups and oxidation of the amino group to the nitroxyl group is determined by the rigidity of conditions of generating the latter. The groups that are labile in these conditions are either introduced into the molecule already having nitroxyl or are protected before the oxidation stage, as in the case of the 4-amino-derivative **207**. The functionally substituted derivatives of this class of nitroxides are usually synthesized from nitroxyl-containing heterocyclic compounds by selecting the appropriate conditions of transformations. The hydroxylamino derivatives may also be used, in which oxidation to nitroxides proceeds in mild conditions with many other functional groups remaining intact. Scheme 23 shows some of the derivatives most useful for the synthesis of spin labels and probes, which may be obtained from triacetoneamine or the corresponding nitroxide. Scheme 23 represents the transformations proceeding, at least at the first stage, at the carbonyl group.

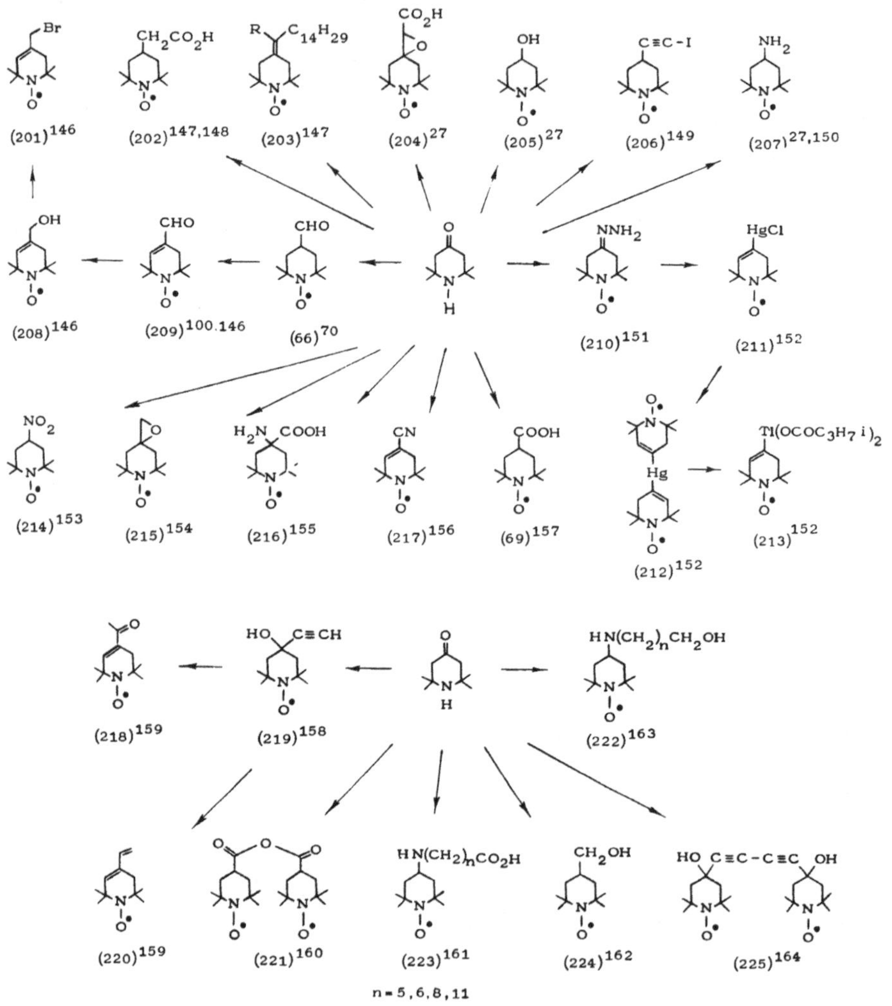

n = 5, 6, 8, 11

SCHEME 23.

The acid **202** may be obtained in two ways: by using the modified Wittig reagent,[147] or lithiated trimethylsilylacetate.[148]

Compounds that are rather prospective in the synthesis of bi- and polyfunctional derivatives of nitroxides are α,β-unsaturated aldehyde **209**, ketone **218**, and nitrile **217**, for which the reaction of conjugative addition must be feasible, effectively used by the Hideg's group in the series of pyrroline nitroxides (Schemes 17 and 18). Aldehyde **209** is a starting compound in the synthesis of such useful synthons as alcohol **208** and allyl bromide **201**. The alcohol **205** and the amine **207** are useful starting compounds in the synthesis of other functional derivatives (Schemes 25 and 26) as well as polyradicals and paramagnetic models of biologically active compounds.[27,96] A convenient method has been developed for the preparation of *N*-alkyl-substituted derivatives **224**, **225** based on the reductive amination reaction and consisting in the condensation of amine **207** with aldehydes or the ketone **122** with amines with simultaneous reduction with NaBH₃CN.[161,163,165-167] The nitroxyl group remains intact in these conditions. The product of acetylene addition at the carbonyl group **219** is a starting compound in the synthesis of α,β-unsaturated ketone **218**, as well as of diene **220**. Compound **220** may act as the diene component in the Diels-Alder reactions which afford polycyclic compounds with the nitroxyl-containing fragment in a molecule and polyfunctionally substituted nitroxides.[159]

It was reported that polymerization of diacetylene alcohol **225** at 80 to 100°C leads to a mixture of mono- and polycrystalline polymers which possess spontaneous magnetization.[168] The value of this magnetization is small, which was attributed to the fact that polymerization proceeds only to an insignificant extent on the monocrystalline matrix (topochemical polymerization). This leads to a small content (~0.1%) of ferromagnetic crystals of the polymer. In a similar work for diacetylene biradical **226**, a polymer with the content of ferromagnetic fraction ~0.7% was obtained.[169]

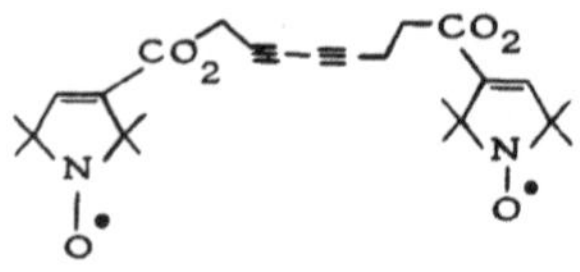

(226)

Verification has shown that no polymerization of compounds **225, 226** takes place under these or other conditions.[170,171] On heating, the starting biradicals decomposed producing compounds with fewer nitroxyl groups in a molecule, and the ferromagnetism was not reproduced.

The next scheme shows the reactions of triacetoneamine derivatives or the corresponding nitroxide **122** proceeding at the α-position to carbonyl group.

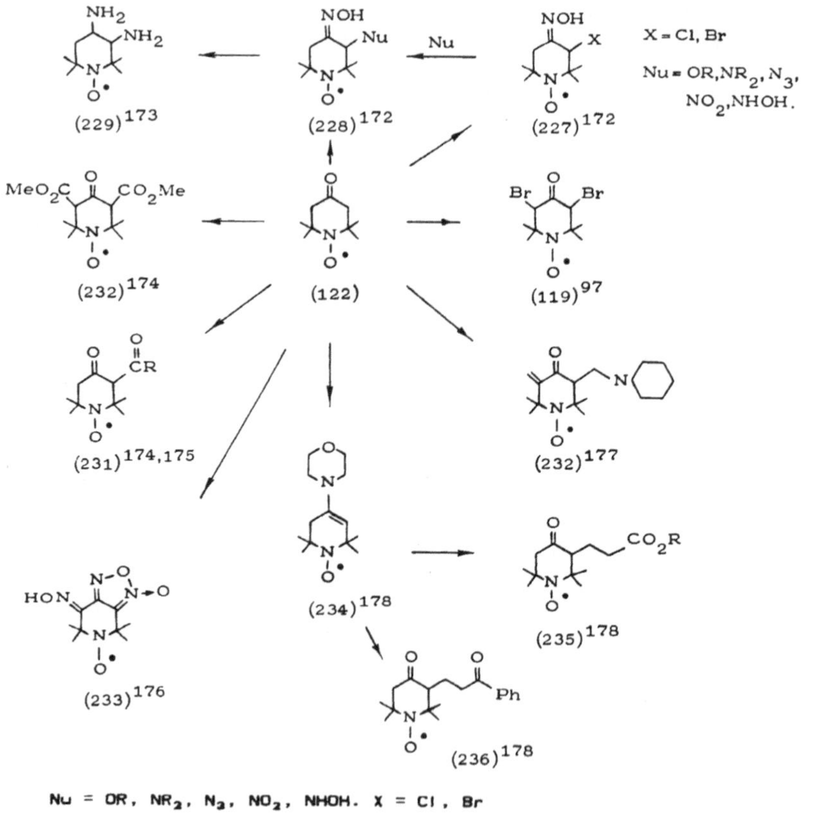

Nu = OR, NR$_2$, N$_3$, NO$_2$, NHOH. X = Cl, Br

SCHEME 24.

Of certain interest among the compounds presented in the scheme is bromo oxime **227**, in which the bromine atom has a high nucleophilic mobility allowing this compound to be used in the synthesis of various α-disubstituted bifunctional

derivatives **228**, in particular, α-hydroxylaminooxime (Nu = NHOH) and amino-
oxime (Nu = NH$_2$) which reduction leads to diamine **229**. Compounds **228, 229**
are capable of forming complexes with various metals, in particular, with Pt^{2+}.[173,179]
Hydroxylaminooxime (**228**, Nu = NHOH) may also be used for the synthesis of
condensed bicyclic nitroxides.[180] The β-dicarbonyl compounds **223** can form complexes with a number of metals, which allow them to be used as spin-labeled
chelating reagents. An interesting method for the preparation of esters (**230, 231,**
R = OCH$_3$) suggested in Reference 174 consists of carboxylation of nitroxide **122**
with CO$_2$ in the presence of potassium phenoxide. Depending on the reagent ratio,
one or two carboxyl groups may be introduced into the α-position to carbonyl. A
convenient method to alkylate triacetoneamine derivatives is to go over to the highly
reactive enamine **234** that reacts with electrophilic olefins with subsequent hydrolysis in mild conditions to form the alkylation products **235, 236**. The high
reactivity of enamines[181] suggests that compound **234** has many more synthetic
potentialities.

The following schemes (25 and 26) represent synthetically useful functional
derivatives and spin labels based on alcohol **205** and amine **206**.

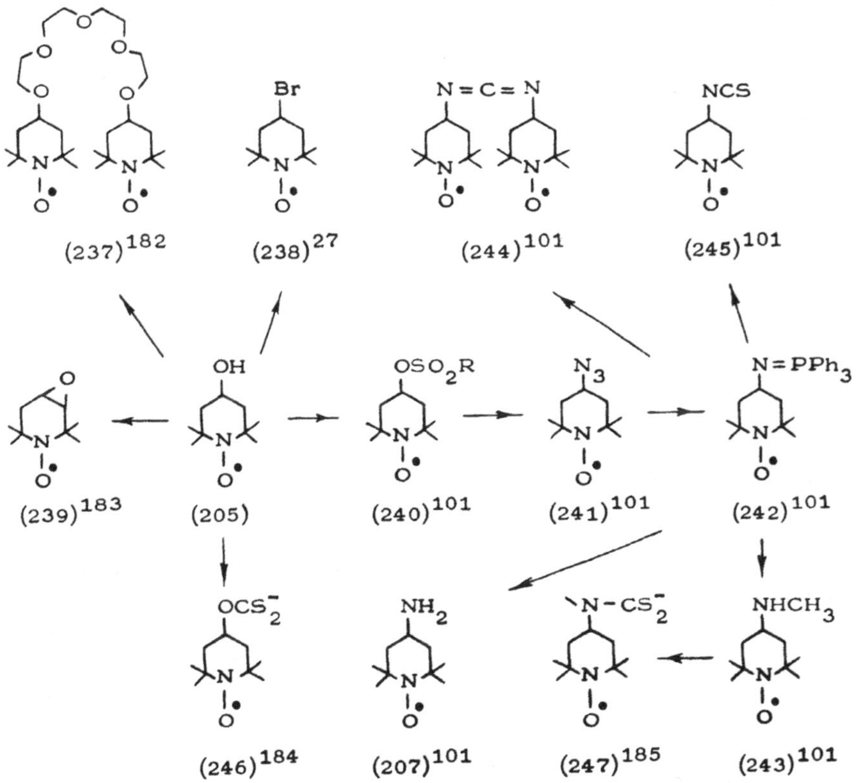

SCHEME 25.

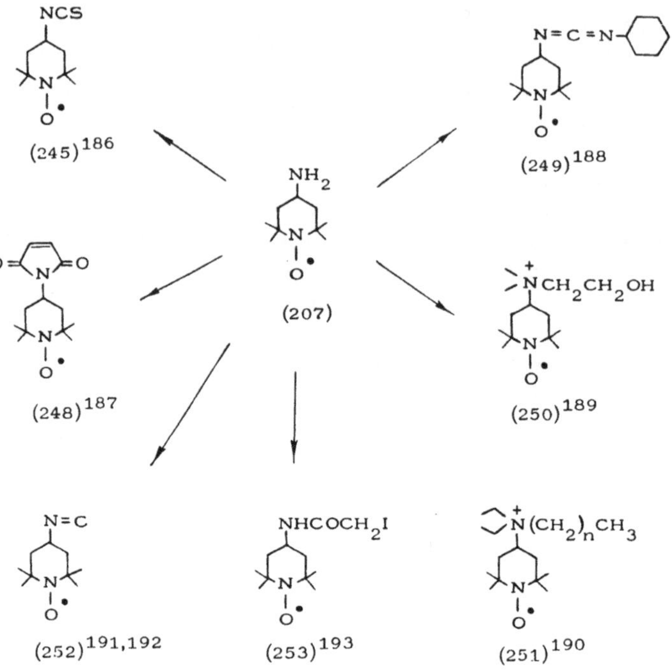

SCHEME 26.

Scheme 27 shows some stable nitroxides with molecules which include the piperidine nitroxyl-containing fragment. These compounds are mainly used as spin probes in spectral studies.

Owing to the presence of ester or alcohol groups, compounds **265, 266** may probably be converted to the highly anisotropic biradical spin labels with a rigid mutual position of the nitroxyl groups.

D. HETEROCYCLIC NITROXIDES WITH AN ADDITIONAL HETEROATOM IN HETEROCYCLE

1. Five-Membered Heterocyclic Nitroxides

A common feature of heterocyclic nitroxides, especially of those containing an additional nitrogen atom in heterocycle, is their stability in acids. This is due to the fact that an additional heteroatom is first protonated rather than the nitroxyl group. The possibility of reversible protonation of stable nitroxides at an additional nucleophilic center to form stable radical cations where the positive charge and lone electron are located on different groups of atoms underlies the radiospectroscopic method of pH determination in biological microobjects.[8] At the same time, the corresponding amino and hydroxylamino derivatives with 1,3-position of heteroatoms in heterocycle are much less stable to hydrolytic cleavage. The same refers to the oxidized diamagnetic form of nitroxides — the oxammonium salts — which

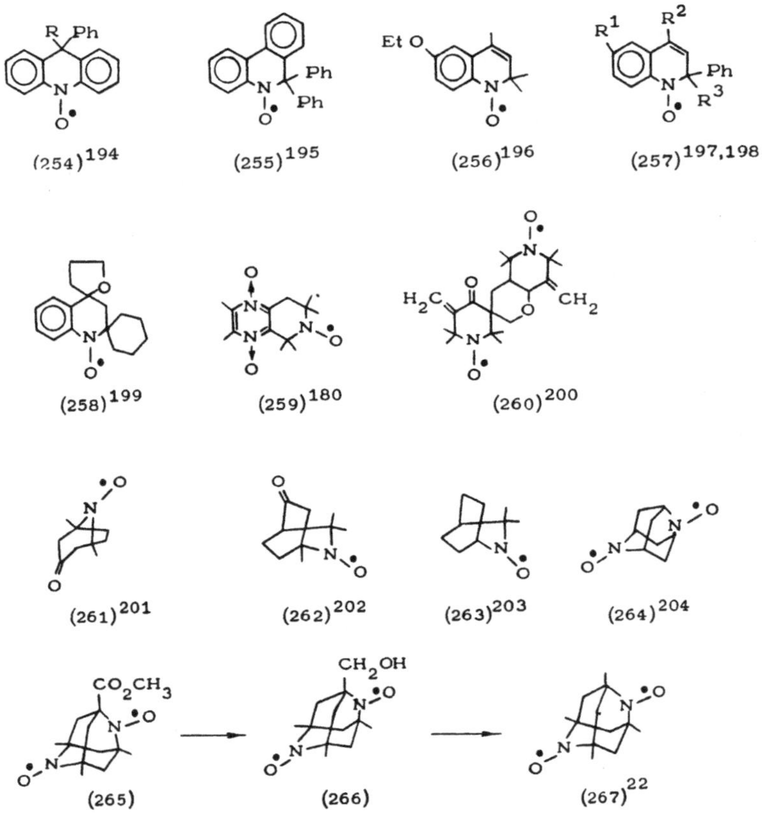

SCHEME 27.

in the case of 1,3-position of heteroatoms undergo the irreversible hydrolytic decomposition. An example of such a transformation is the oxidative destruction of the oxazolidine derivatives by *m*-chloroperbenzoic acid, bromine, or NO_2.[205-207]

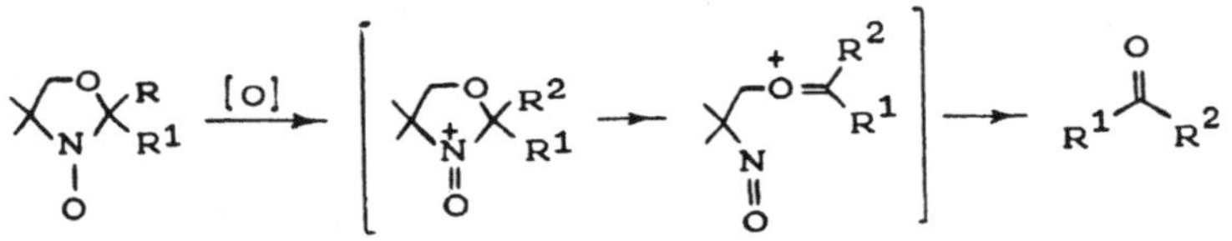

This seems to be the reason for the fact that in the synthesis of isocyclic biradical **72**, it was necessary to protect the nitroxyl group at the stage of oxidation of the amino group in imidazolidine heterocycle. (Section III.A). Nevertheless, for heterocyclic nitroxides with 1,3-position of heteroatoms — 3-imidazoline-3-oxide derivatives — the reversible formation of oxammonium salts is also possible (compare to Scheme 42).

a. Oxazolidines

Synthesis of oxazolidine nitroxides is generally used to introduce spin labels into different molecules with the carbonyl or carboxy groups (compare to section II.A). The method consists of boiling the benzene, toluene, or xylene solution of ketone with 2-amino-2-methyl-propanol-1 **11** in the presence of the catalytic amount of *p*-toluenesulfonic acid with constant removal of water as an azeotrope mixture or its binding with anhydrous potassium carbonate. Sometimes an excess of aminocarbinol **11** is used as a solvent.[32,55,208] Subsequent oxidation of sterically hindered oxazolidines **268** with *m*-chloroperbenzoic acid leads to the formation of nitroxides (doxyls) **269** (Scheme 28).[55]

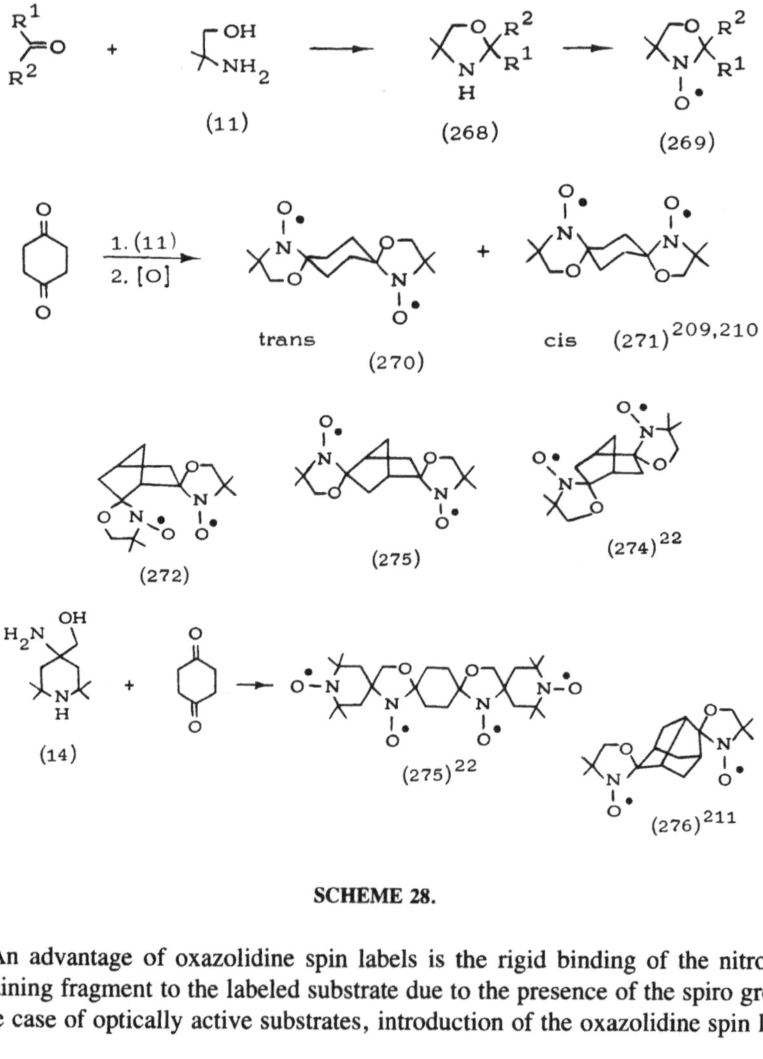

SCHEME 28.

An advantage of oxazolidine spin labels is the rigid binding of the nitroxyl-containing fragment to the labeled substrate due to the presence of the spiro group. In the case of optically active substrates, introduction of the oxazolidine spin label leads to racemates that may be separated by recrystallization already at the stage of diamagnetic derivatives.[55] Rigid fusion of the oxazolidine ring to the hydrocarbon frame of the starting ketone allows synthesis of bi- and polyradicals with a certain mutual orientation of nitroxyl groups. A series of such nitroxyl polyradicals **270–276** was synthesized by Rassat's group to investigate the magnetic properties of these compounds and the character of interaction of nitroxyl groups depending on their mutual orientation.[22]

Synthetic possibilities of oxazolidine nitroxides are due exclusively to the presence of functional groups in the molecule of the starting ketone. Therefore, as starting compounds for the synthesis of spin labels and probes, the ketocarboxylic acids or their esters are used. Further modification of the carboxylic group affords other functional derivatives (Scheme 2)

In Reference 212 a method has been suggested for the synthesis of oxazolidine nitroxyls with the ammonium group at the end of the aliphatic chain in the 2-position of the heterocycle from ketone **277** (Scheme 29).

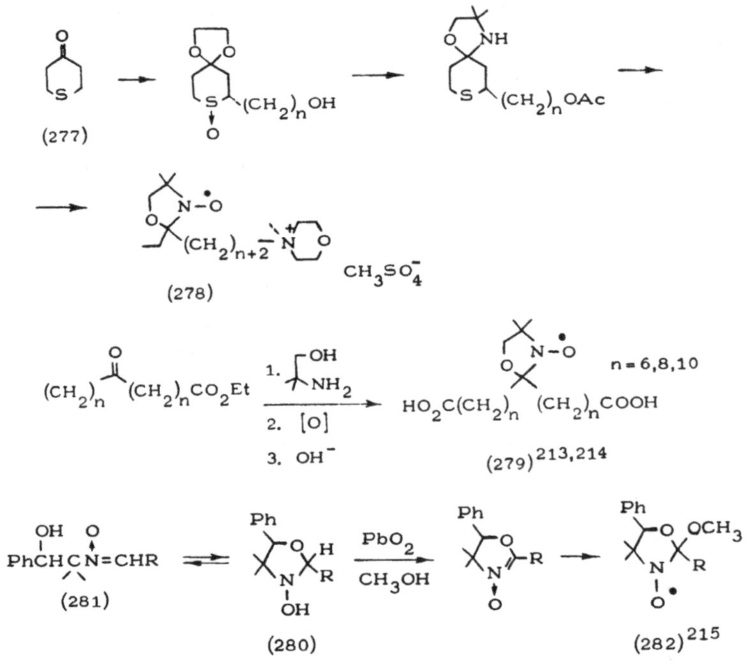

SCHEME 29.

The reaction of aminocarbinol **11** with esters of ketodicarboxylic acids is a key stage in the synthesis of bifunctional derivatives of isoxazolidine nitroxides **279**. The transformation of ester groups to succinimide derivatives leads to the cross-linking diacylating spin labels.[213,214]

An alternative method to prepare oxazolidine nitroxides consists of oxidation of oxazoline derivatives **32** to oxaziridines **33**, which are subsequently rearranged to nitrones **34** upon chromatographing on silica gel. Subsequent interaction with organometallic reagents and oxidation with atmospheric oxygen leads to radicals **35** (Scheme 4). This method has two advantages over the one based on condensation of aminocarbinol **11** with ketones. First, the starting carboxylic acids are much more available than the functionally substituted ketones, due to which the oxazoline derivatives **32** are readily available compounds.[61,216,217] Second, since the substituent R^1 is introduced at a stage following the oxidation with *m*-chloroperbenzoic acid, it may contain the groups that are unstable under the oxidation conditions, which then may be transformed to other functional groups.

Recently, a method has been worked out for the synthesis of oxazolidine nitroxides with the methoxy group at the nitroxyl α-carbon atom (Scheme 29) based on the reaction of oxidative methoxylation of oxazolidines **280**. It is interesting to note that oxazolidines **280** are in a tautomeric equilibrium with acyclic hydroxynitrones **281**. Still, irrespective of the state of tautomeric equilibrium, the oxidation leads to nitroxides **282**. The reaction is of a general character and since the oxidation occurs in mild conditions, the R substituent may contain functional groups that are labile to oxidation.[215]

b. Imidazolidines

A first representative of stable heterocyclic nitroxides in general and in particular imidazolidine nitroxides, is compound **283** synthesized back in 1901 (see Chapter 1). In 1963, formation of sterically hindered imidazolidine **284** was reported as a result of self-condensation of 1-amino-1-cyanocyclohexane **285** in the presence of sodium ethoxide.[218] The mechanism of this reaction involves partial hydrolysis of aminonitrile **285** to cyclohexanone and its condensation with aminonitrile.[219] A similar transformation is described in Reference 220 (Scheme 30).

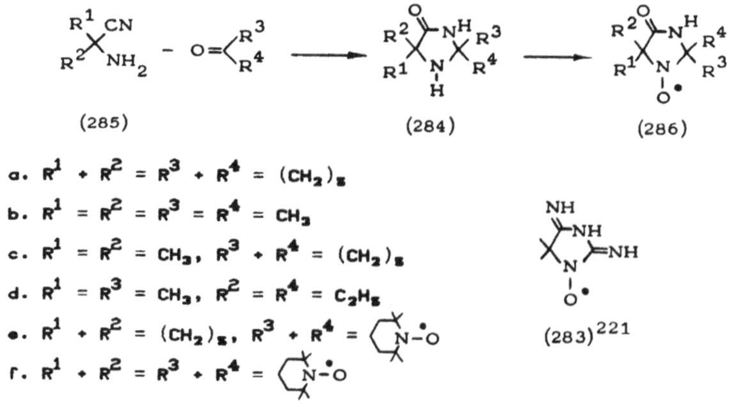

(285)   (284)   (286)

a. $R^1 + R^2 = R^3 + R^4 = (CH_2)_5$

b. $R^1 = R^2 = R^3 = R^4 = CH_3$

c. $R^1 = R^2 = CH_3$, $R^3 + R^4 = (CH_2)_5$

d. $R^1 = R^3 = CH_3$, $R^2 = R^4 = C_2H_5$

e. $R^1 + R^2 = (CH_2)_5$, $R^3 + R^4 = $

f. $R^1 + R^2 = R^3 + R^4 = $

(283)[221]

SCHEME 30.

The condensation of various ketones with aminonitriles affords imidazolidinone derivatives **284** containing the sterically hindered amino group in heterocycle in which oxidation with hydrogen peroxide in the presence of sodium tungstate leads to the respective nitroxides **286**.[222] The only information known about the chemical properties of these compounds is that they possess high stability and have been patented as thermal and light stabilizers of polymer materials.[223]

A general method for the synthesis of imidazolidine nitroxides **287** consists of the reaction of 2,3-dimethyl-2,3-diaminobutane **12** with ketones and subsequent oxidation of the resulting imidazolidines **286** with *m*-chloroperbenzoic acid.[33,55]

The reaction is of a sufficiently general character and allows spin label introduction at the carbonyl group of various compounds: steroids, crown ethers, etc. (Section II.A). The possibilities of chemical modification of these compounds arise from the presence of an additional nitrogen atom in heterocycle. These reactions include the alkylation, acylation, and oxidation of the endocyclic amino group. Evidently, the use of functionally substituted ketones, as was done for oxazolidine and 3-imidazoline derivatives, could expand synthetic possibilities of this class of nitroxides, but no data are currently available regarding this.

Recently, formation of stable imidazolidine nitroxides with two methoxy groups at the nitroxyl α-carbon has been reported.[224] This method is based on the oxidation of heterocyclic aldonitrones **289** with lead dioxide in methanol. An intermediate product in the reaction is the highly reactive methoxy nitrone **290** (Scheme 31).

On oxidation of nitrones **292, 289** by xenon difluoride, nitroxides with fluorine atoms at the nitroxyl α-carbon are generated. The nitroxide (**293**, R = C_6H_5) is stable and was isolated individually.[225, 226] The other fluorine-containing nitroxides **293** and α,α-difluoronitroxides **294** are unstable. Their formation was registered by the characteristic splitting of the lone electron on fluorine atoms. The splitting constant α_F values vary from 20 to 60 G.[225,226] It should be noted that the compound (**293**, R = C_6H_5) may be involved in the nucleophilic substitution reactions proceeding with preservation of the radical center. With ammonia and methylamine these reactions lead to amines **295**; with sodium methoxide, to the methoxy derivative **296**.[226]

Formation of stable bicyclic nitroxides — the derivatives of spiroimidazoisoxazole **297** — was reported in the oxidation of oxime **298** and semicarbazone **299** — the derivatives of 3-imidazoline-3-oxide β-oxonitrones — with manganese dioxide. Reduction of nitroxide **297a** with zinc leads to the starting 3-imidazoline-3-oxide derivative.[227]

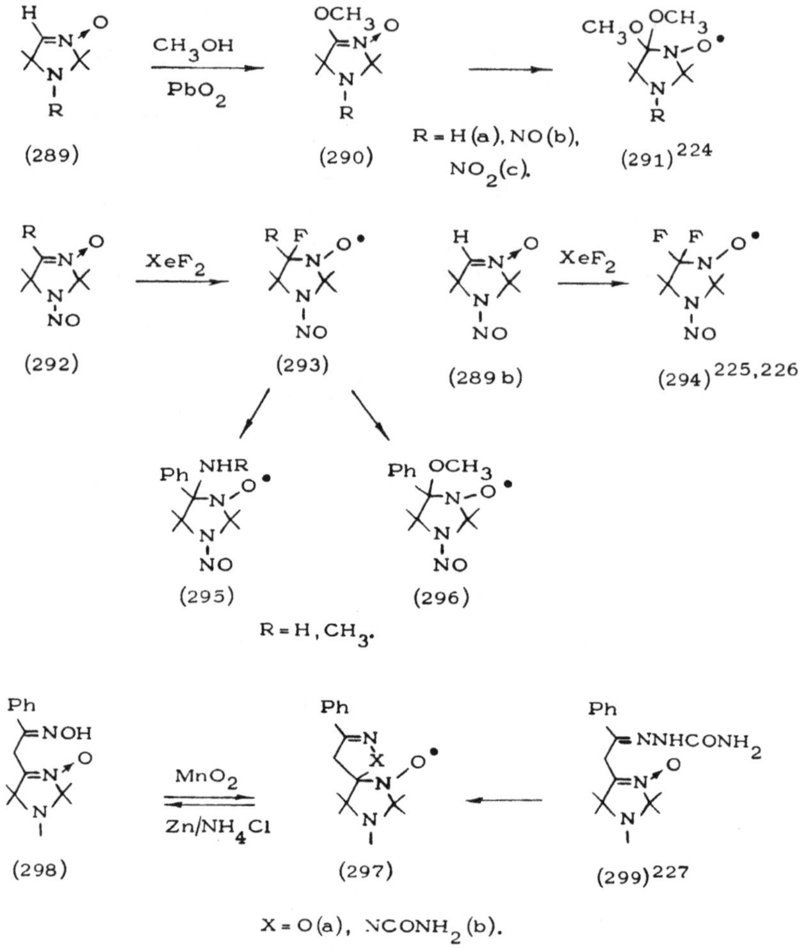

SCHEME 31.

It should be noted that a number of functionally substituted imidazolidine nitroxides were obtained by modification of 3-imidazoline derivatives (compare to Section III.D.1.d).

c. 2-Imidazolines and 2-Imidazoline-3-Oxides

Stable 2-imidazoline-3-oxide nitroxides — nitronyl nitroxides **302** — were first obtained by Ullman's group, by the reaction of aryl- or alkyl-substituted aldehydes with 2,3-bis-hydroxylamino-2,3-dimethylbutane **26** and subsequent oxidation of intermediate 1,3-dihydroxyimidazolidines **30** with lead dioxide.[228,229] Oxidation may be stoped at the stage of imidazoline **300**, which is sensitive to oxidants. The reaction with dialdehydes after oxidation leads to biradicals **302**.[44]

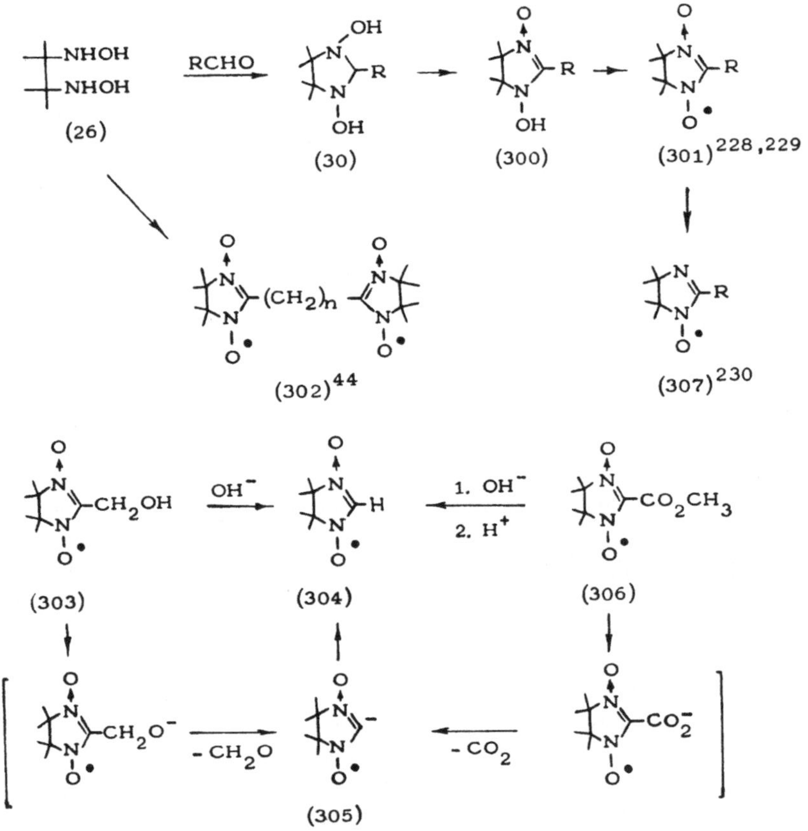

SCHEME 32.

Nitronyl nitroxides usually have sufficiently high stability, the radical **303** is less stable. Upon standing, it is transformed to radical **304**. The transformation is quicker in the presence of bases. This allowed the authors in Reference 229 to suggest that in alkaline media, the formaldehyde molecule is eliminated to form the resonance-stabilized radical-anion **305**. A similar transformation is observed on alkaline hydrolysis of ester **306** and subsequent neutralization that leads to compound **304**. The possibility of the formation of carbanion **305** and its stability are supported by deuterium exchange of hydrogen at C-2 proceeding in alkaline media.[229]

When nitronyl nitroxides **301** are treated with triphenylphosphine or nitric acid, imino nitroxides **307** are formed, which are the derivatives of 2-imidazoline.[230] The EPR spectrum of these compounds markedly differs from that of the starting nitronyl nitroxides **301**: the spectrum of the starting compounds has five lines, while that of the products has nine lines (splitting on two nonequivalent nitrogen atoms). This reaction of compound **301** (R = C_6H_5) underlies the method of nitrogen oxide determination in air.[231,232]

Synthetic possibilities of nitronyl and imino nitroxides are due to the presence of the functional group in substituent R or the possibility of electrophilic substitution in the 2-position of heterocycle when there is no substituent there (R = H). Thus, condensation of bishydroxylamine **26** with halo-substituted acetaldehydes gaves haloalkyl derivatives **308**, where the halogen atom possesses high nucleophilic mobility, the reactivity of halogen increasing in the series Cl < Br < J. The thermal stability of halo-derivatives changes in the reverse order, the radical **308** (R = J) may only be obtained in the reaction of chloro or bromo derivatives **308** with KJ.[44]

The halomethyl derivatives **308** may not be obtained by halogenation of the 2-methyl-substituted derivative. The reaction of compound **301** (R = CH$_3$) with halogens, hypohalogenites, or *N*-bromosuccinimide produces a complex mixture of products.[44]

The halomethyl derivatives **308** react with water, alcohols, and amines to form subsequent substitution products.[44] The amino derivatives **309** are stable only in the presence of strong electron-accepting substituents.

Ease of generation of anion **305** (compare to Reference 235) allows halogenation of compound **304** which proceeds by its reaction with hypohalogenites or bromo- or iodocyan in aqueous bicarbonate.[233] The halogen atom in the molecule **310** also has high nucleophilic mobility, as shown in the case of the reactions of halo derivatives **310** with NaCN and NaOH, leading to nucleophilic substitution products.[233] The anion **305** is readily oxidized by atmospheric oxygen or potassium ferrocyanide to form bis-nitronyl nitroxide **311**, which can also be obtained by treatment of the iodo derivative **310** (X=J) with copper in DMFA.[234]

The hydrogen atom in the 2-position of the imino nitroxide **312** heterocycle is also likely to experience deuterium exchange. However, in view of the fact that the reaction rate remains constant under medium pH variation, it was suggested that the anion is not formed, the reaction proceeding via the stage of zwitterion **313**.[230]

For the imino nitroxide **312**, the electrophilic substitution reactions at the C-2 atom are unknown, but the bromo derivative **314** may be obtained by deoxygenation of the subsequent nitronyl nitroxide **310**. It is interesting to note that keeping compound **312** in aqueous alkali leads to stable carbamoyl nitroxide **315** in a small yield. In a high yield it may be obtained in similar conditions from the bromo derivative **314**. In the reaction of compound **314** with amines, stable amidinonitroxides **316** are formed, which are moderately strong bases (pK 6.4 to 6.8) and are capable of reversible protonation leading to a marked change of the EPR spectrum.[230]

Evaluating the synthetic possibilities of nitronyl and imino nitroxides in general, it should be noted that they are limited, because of the presence of only one reaction center in the molecule of this class of nitroxides, which may be used for functionalization. Due to this, bifunctional derivatives may only be synthesized by modifications of a substituent in a side chain. The limited applicability of these compounds as spin labels and probes arises from the fact that these compounds have a rather complex EPR spectrum in which interpretation upon immobilization may be difficult.

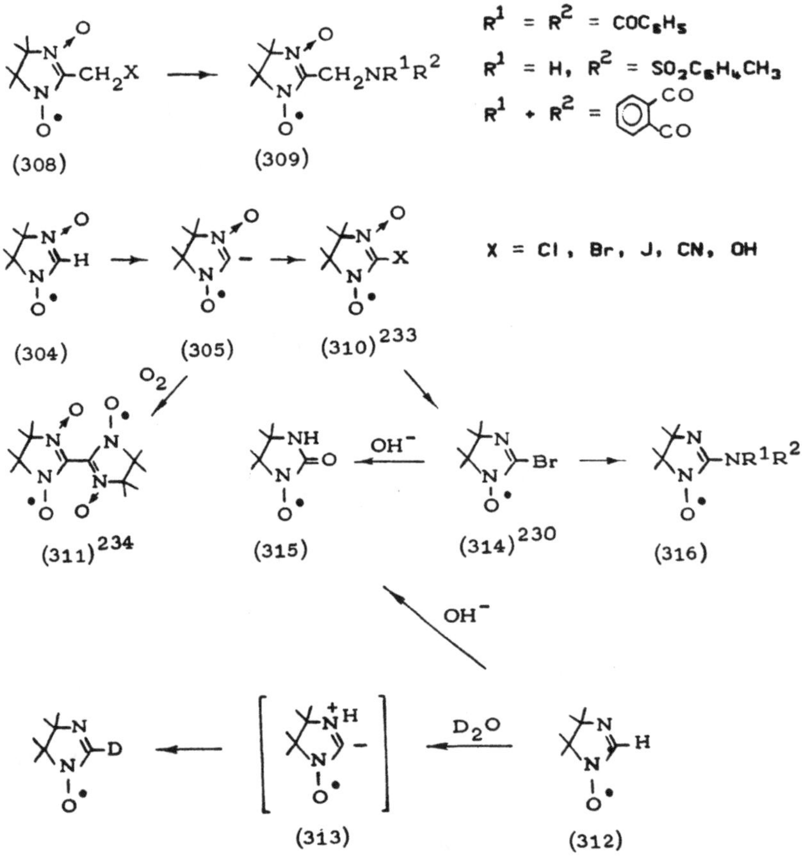

$$R^1 = R^2 = COC_6H_5$$
$$R^1 = H, \quad R^2 = SO_2C_6H_4CH_3$$
$$R^1 + R^2 = \text{(phthaloyl)}$$

(308) → (309)

(304) → (305) → (310)[233]

$$X = Cl, \ Br, \ J, \ CN, \ OH$$

(311)[234]   (315)   (314)[230]   (316)

(313)   (312)

SCHEME 33.

The possibility of varying substituents in the 4- and 5-positions of the heterocycle is restricted by the limited availability of different 1,2-bishydroxylamines with hydroxylamino groups at the tertiary carbon atoms. Recently, another method has been developed to generate 2-imidazoline imino nitroxides which allows variation of substituents in the 2- and 5-positions of the heterocycle. This method is based on the reaction of oxidative methoxylation of $4H$-imidazole-1(3)-oxides **317**, **318**.[236,237] The reaction with ethanol seems to be much slower, which allows it to be used as a solvent in the oxidative amination reaction leading to imino nitroxides **319**, **320** with the amino group in heterocycle, including positioned at the nitroxyl α-carbon.[226,238]

Imino and nitronyl nitroxides are formed in some oxidative reactions of imidazole-*N*-oxides together with 3-imidazoline and 3-imidazoline-3-oxide nitroxides (Schemes 35 and 37). This allows variation of substituents in the 4- and 5-positions of the heterocycle, and the introduction of functional groups in these positions,

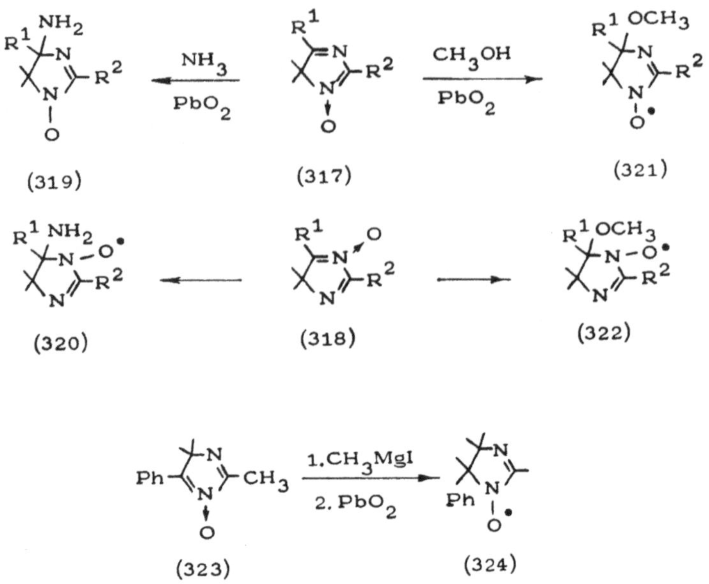

(319)   (317)   (321)

(320)   (318)   (322)

(323)   (324)

SCHEME 34.

expanding the synthetic possibilities of stable imino nitroxides. Another method to generate imino nitroxides consists of the reaction of 4*H*-imidazole-1-oxide **323** with methylmagnesium iodide with subsequent oxidation with lead dioxide leading to compound **324**.[239]

d. 3-Imidazoline-3-Oxides and 3-Imidazolines

3-Imidazoline-3-oxide derivatives **325** containing the sterically hindered hydroxylamino group in heterocycle were initially obtained on the heating of *anti*-1,2-hydroxylaminooximes **326** in excess acetone at 140 to 150°C in sealed tubes.[240] Substitution of acetone by its diethylketal allows the reaction to be conducted in a boiling solution of excess reagent and to be extended to various hydroxylaminooximes **326**.[241,242] It was shown later that condensation with acetone is catalyzed by acids, which made it possible to conduct the process in boiling acetone.[243] Hydroxylaminooximes **326** are quite readily available compounds, which may be obtained from the nitrosochlorides of olefins (R = alkyl) or α-haloketones.[244-248]

Subsequent oxidation of 1-hydroxy derivatives with lead or manganese dioxides or nitrogen oxides (R = C_6H_5) smoothly leads to nitroxides **335**.[254] It is interesting to note that nitroxides **335** are also formed in the oxidation of hydroxylaminooximes **334**.[72]

Apart from acetone, cycloalkylketones may participate in the condensation reaction: cyclohexanone and cyclopentanone; and some α-diketones: diacetyl and cyclohexandione-1,2.[251,252] In the latter case, 1-hydroxy derivatives **327** are formed, where oxidation under similar conditions leads to 4*H*-imidazole-1,3-dioxide

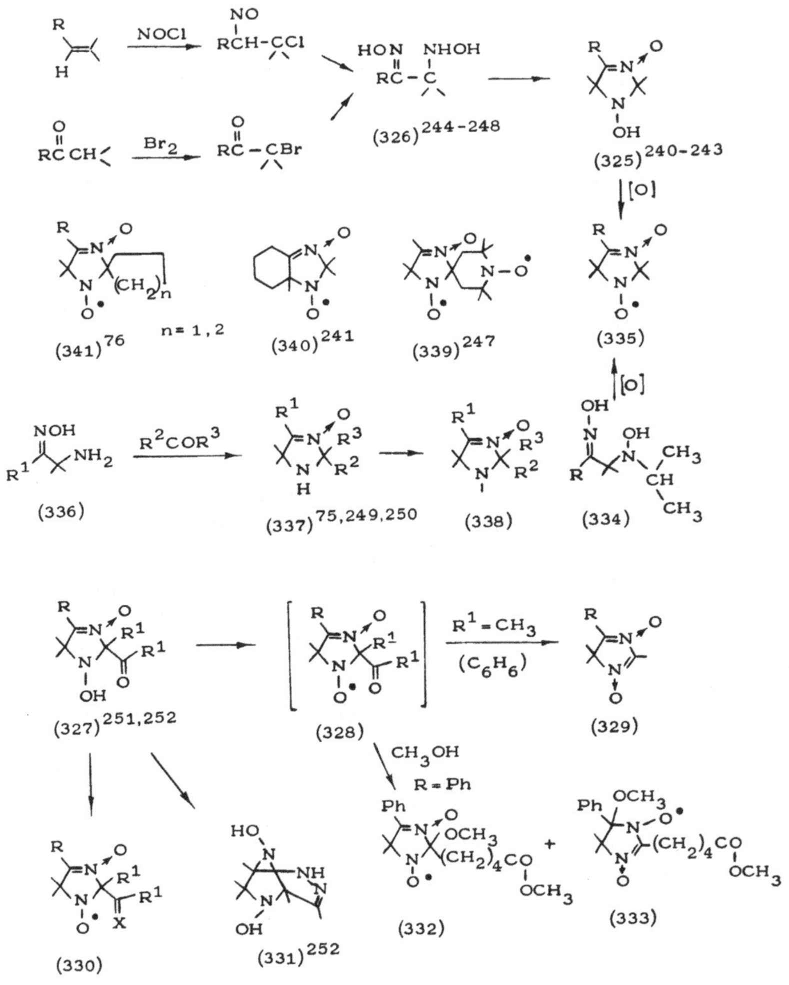

$(326)^{244-248}$ $(325)^{240-243}$

$(341)^{76}$ n = 1, 2 $(340)^{241}$ $(339)^{247}$ (335)

(336) $(337)^{75,249,250}$ (338) (334)

$(327)^{251,252}$ (328) (329)

(330) $(331)^{252}$ (332) (333)

X = NOH, NNHCH$_3$, NNHCONH$_2$, NNHCSNH$_2$, HOH

SCHEME 35.

derivatives **329**.[251,253] Upon oxidation by lead dioxide in methanol, further oxidation takes place to form a mixture of nitroxide **332** and imino nitroxide **333** (compare to Scheme 37).[253] The intermediate imidazole **329** is supposed to be oxidized to radical cation, and subsequent nucleophilic attack by methanol at positions 2 and 5 of its heterocycle leads to formation of the reaction products.[253] Subsitution of ketones **327** by their derivatives or reduction of the keto group to the alcoholic one and subsequent oxidation leads to stable nitroxides **330** containing various functional groups at the nitroxyl α-carbon.[251,252] At the same time, the hydrazone group formed in the reaction of compounds **327** (R^1 = CH$_3$) with hydrazine hydrate intramolecularly and attacks the nitrone group in heterocycle to form the bicyclic bishydroxylamino derivatives **331**. Compounds **331** may be oxidized to unstable monoradicals which may not be isolated individually because of the lability to disproportionation on solution concentration.[252]

An alternative method of synthesizing 3-imidazoline-3-oxide nitroxides is based on the condensation reaction of α-aminooximes **336** with ketones in the presence of catalytic amounts of diluted hydrochloric acid.[249,250] This method is advantageous compared to that based on the condensation reaction of hydroxylaminooximes **326** in that the reaction may be performed with various methylketones and cycloalkylketones including triacetonamine. Subsequent oxidation of the amino derivatives **337** under usual conditions, i.e., with hydrogen peroxide in the presence of tungstate affords nitroxides **335**. The spirobiradical **339** was obtained by this scheme.[247] The rigid conditions for generation of nitroxides from the amino derivatives **337** and the greater stability of the latter compared to the respective nitroxides makes it possible to use compounds **337** as tracers to examine the route of oil-field fluids in oil extraction using EPR spectroscopy. The compounds which are still more prospective in this respect are the 1-methyl derivatives **338** obtained from amines **337** by their reaction with formaldehyde and formic acid.[75,250] Nitroxides are generated after sampling and concentration of the sample directly in a spectrometer tube.[30]

Recently, dimerization of α,β-unsaturated oxime **342** has been found to lead to bicyclic compound **343** existing as a mixture of two tautomers **A** and **B**.[255,256] Under radical polymerization conditions in the presence of compound **343**, the nitroxide **344** was formed which was detected by EPR, though it was not obtained preparatively.[255] It is interesting to note that the reaction of compound **343** with methylmagnesium iodide leads to the piperidine derivative **345** containing the sterically hindered amino group and the oxime group at α-position to it.[256] Oxidation of compound **345** to nitroxide **346** was not reported.

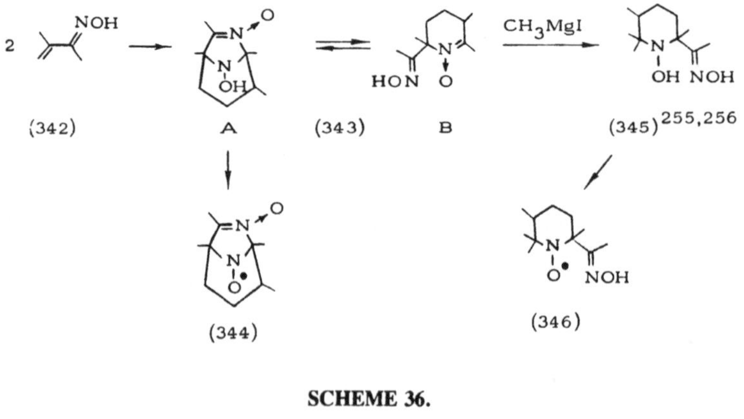

SCHEME 36.

The oxidation of 3-imidazoline-3-oxide derivatives **347** existing as a tautomeric mixture with acyclic hydroxylaminonitrones **348**, with lead dioxide in methanol leads to a mixture of nitronyl nitroxides **349** and 3-imidazoline-3-oxide nitroxides **350** containing the methoxy group at the nitroxyl α-carbon atom (Scheme 37). The product ratio depends on the character of substituents R^1 and R^2: on increased electron-donating character of R^1 and electron-accepting character of R^2, the content of nitroxide **350** increases. On the contrary, the donating substituents R^2 favor formation of nitronyl nitroxides **349**.[236,237] The oxidation proceeds via the stage of formation of 4*H*-imidazole-1,3-dioxides **351** which may be isolated by performing the reaction in ethanol. Upon oxidation of compounds **351** in methanol saturated with ammonia or methylamine, a mixture of radicals **349** and **350** is formed with an amino group at the nitroxyl α-carbon.[238] The reaction of methoxy derivatives of nitroxides **350** (Y = OCH_3) and nitronyl nitroxides **349** (Y = OCH_3) with ammonia leads to α-aminonitronyl nitroxides **352**.[238] The oxidation reaction of 3-imidazoline-3-oxides **347** (R^2 = H) may be performed not only in methanol but also in other alcohols to obtain 2,2-dialkoxy-3-imidazoline-3-oxides **353**.[236,237]

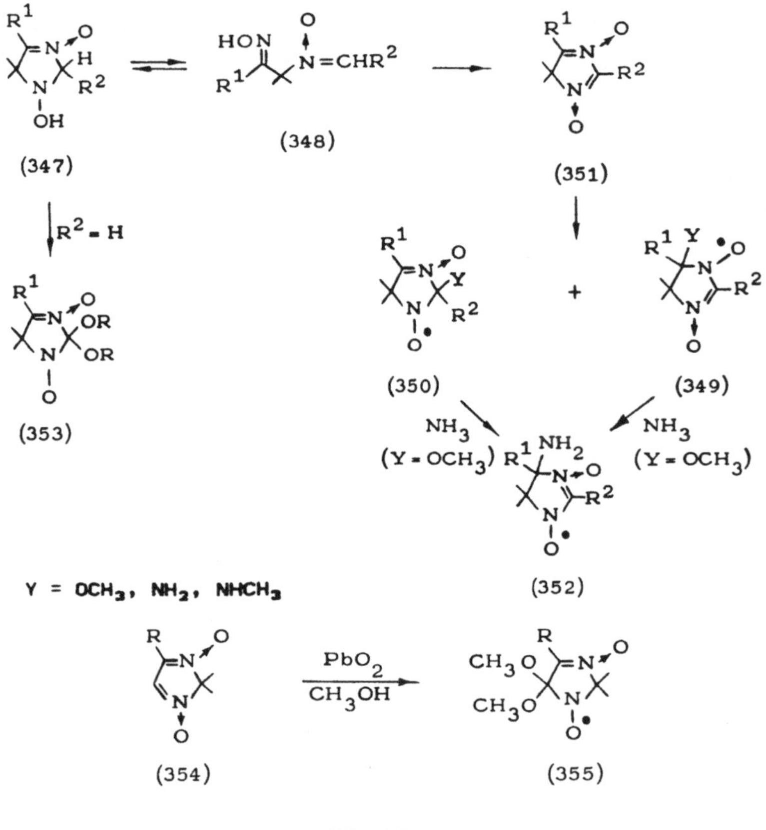

SCHEME 37.

Oxidation of 2*H*-imidazole-1,3-dioxides **354** in methanol by lead dioxide leads to 5,5-dimethoxy-3-imidazoline-3-oxide-1-oxyls **355**. In view of relatively mild oxidation conditions, the R substituent may include different functional groups.[257]

This set of chemical transformations (Schemes 35 and 37) affords stable nitroxides with a nontraditional environment of the nitroxyl group. Considering that the alkoxy-substituted derivatives are nearly as stable as the respective alkyl-substituted nitroxides and that substitution of alkyl groups by the alkoxy ones markedly changes the magnetic resonance properties of the nitroxyl group, it seems appropriate to use these compounds as synthons of new spin labels, probes, and chelating reagents.

It has been shown that the nitroxyl group of 3-imidazoline-3-oxide nitroxides is stable over a wide pH range and against oxidants. The synthetic possibilities of these compounds are determined by the presence of the nitrone group in heterocycle, due to which four reaction types are possible for them: (1) electrophilic substitution at the nitrone α-carbon; (2) nucleophilic 1,3-addition at the nitrone group; (3) dipolar cycloadduction reactions; and (4) electrophilic substitution at the arylnitrone group.

In addition, a lot of chemical transformations may be performed with functional derivatives obtained by one of the four reaction types. Schemes 38 and 39 represent some synthetically useful derivatives synthesized from 3-imidazoline-3-oxide, 1-hydroxy derivatives, and the respective nitroxides.

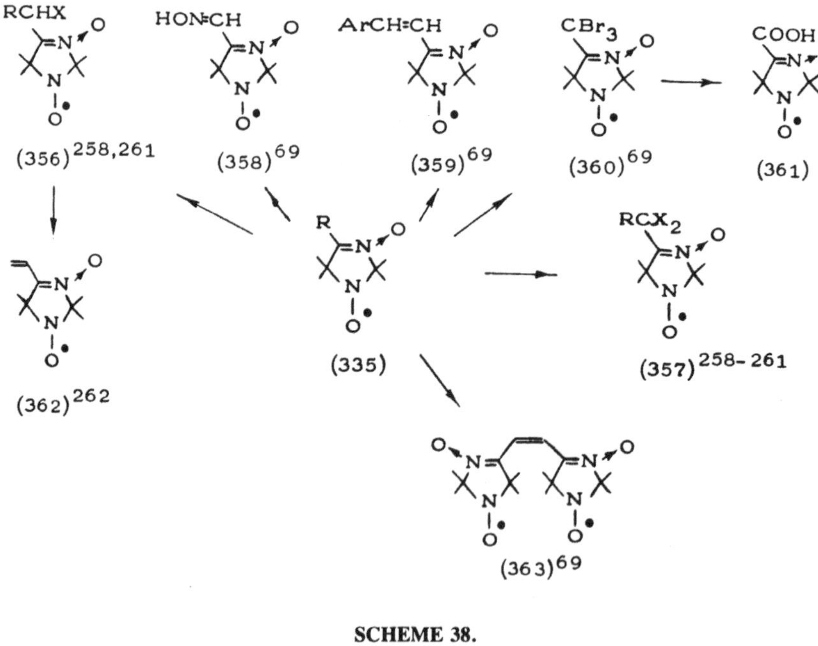

SCHEME 38.

The reactions of alkyl nitrones with electrophilic reagents may be catalyzed by both bases and acids. In the former case, the resonance-stabilized anion **F** is formed, which is capable of reacting with electrophilic reagents. In acid media, enolization of the alkylnitrone group is supposed to take place, producing the reactive enhydroxylamine **G**. This, as well as stability of the nitroxyl group of this class of nitroxides to oxidants, allows direct bromination of alkylnitrones **335** with bromine in methanol to form mono- (**356**, X = Br) and dibromides (**357**, X = Br).[260,261]

It is worthwhile to note the transformation of aldonitrone **335** (X = H) to the carboxylic acid salt 3-imidazoline derivative **364** proceeding by reaction with KCN. The cyanide ion initially adds at the nitrone group. Further dehydration and partial hydrolysis of the nitrile group lead to 3-imidazoline amide **365** in which alkaline hydrolysis gives an acid isolated in the form of its sodium salt **364**.[264]

Interesting synthetic possibilities open up when aldonitrone **335** (R = H) is used as the 1,3-dipole in 1,3-dipolar cycloadduction reactions. In particular, this reaction affords some enaminoketones **366** and amidines **367**, as well as thioamide **368** which oxidation with iodine leads to disulfide **369**. Compounds **368, 369** were suggested for use as spin labels for disulfide and sulfhydryl groups.[265] Amidines **367** have a pK in the range of 4 to 7 and are capable of reversible protonation leading to a change in the EPR spectrum. Due to this, they were suggested for use as pH indicators for biological microobjects.[263] In view of the easy decomposition of the initially formed cycloadducts **370** to amidines **367**, the possibility of using these compounds as precursors of amidines was investigated. In this connection, compounds **371, 372** obtained as a result of cycloaddition reaction of aldonitrone **335** (R = H) with diisocyanates may be regarded as spin labels, which, after binding with a biological substrate, may be easily modified to the amidine-containing, pH-sensitive spin labels. In particular, this approach may be used to obtain the macromolecular pH-sensitive spin probes. Another field of application of diisocyanate adducts is the reaction with alcohols (for example, with cholesterol) and amines (amino acids or their esters) as nucleophilic reagents leading to urethanes and ureas, respectively. Subsequent cleavage of the diaoxazolidine heterocycle leads to different pH-sensitive spin probes **373**.

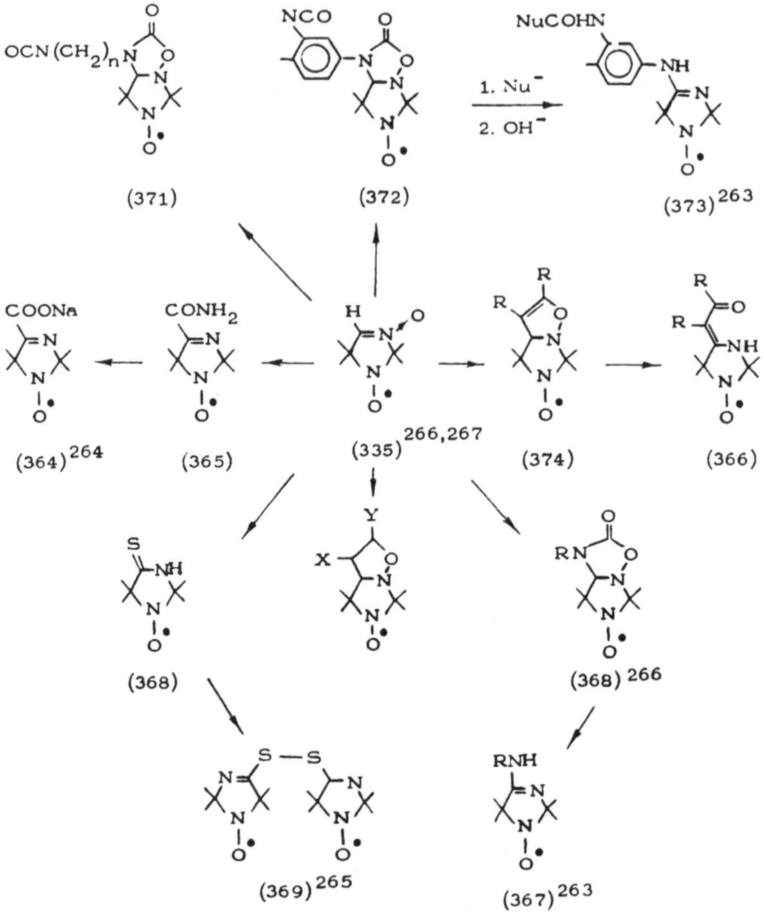

SCHEME 39.

Aldonitrone **335** (R = H) itself can be used as a spin label which allows a nitroxyl-containing fragment to be introduced into the substrates containing the C=C bonds. The reaction conditions are rather rigid, but this method of spin label introduction was used, for example, to obtain spin-labeled rubbers.[268,269] An advantage of this method is tight binding of the nitroxyl-containing fragment with the substrate.

The haloalkyl derivatives of 3-imidazoline-3-oxide **356, 357** are the key compounds in the synthesis of various functional derivatives of 3-imidazoline-3-oxide and 3-imidazoline nitroxides. Some transformation of these compounds are shown in the following scheme.

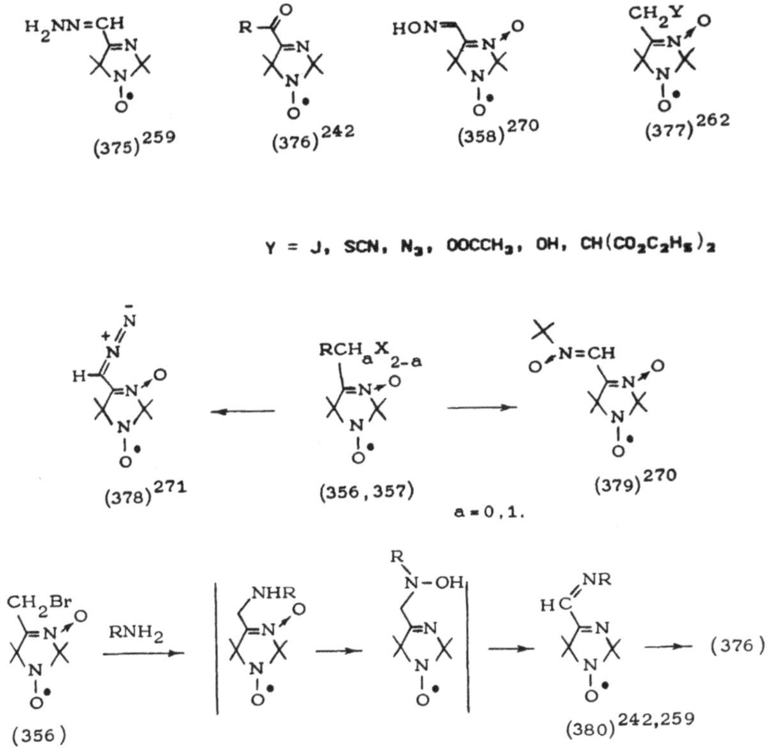

$Y = J, SCN, N_3, OOCCH_3, OH, CH(CO_2C_2H_5)_2$

$a = 0, 1.$

SCHEME 40.

The possibilities of modification of haloalkyl derivatives of 3-imidazoline-3-oxide arises, on the one hand, from the high nucleophilic mobility of halogen atoms and, on the other hand, from possible participation of the nitrone group in the intramolecular redox processes leading to 3-imidazoline derivatives.[242,259] The dihaloalkyl derivatives react with nucleophilic reagents generally with preservation of the nitrone group to form 3-imidazoline-3-oxide derivatives. It has been shown that the nucleophilic substitution reaction may proceed with preservation of the *N*-oxide oxygen atom also for the monobromo derivative **356**. Due to this, compound **356** is suggested to be an alkylating spin label. It should be noted that in the series of 3-imidazoline-3-oxides, the only stable paramagnetic diazonitrone **378** was obtained.[271]

In the reactions of dihalomethyl derivatives of 3-imidazoline-3-oxide with nucleophilic reagents, the nitroxyl group affects the transformations at the reaction center which does not bond directly to it.[272] This may be illustrated in the case of alkaline hydrolysis of dihalomethyl derivatives **357, 381**.

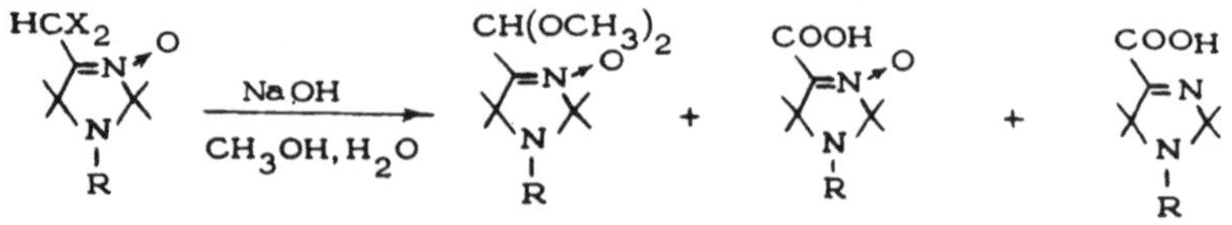

(381) (382) (383) (384)

Reaction time, h	X			
240	Cl (a)	60%	1–2%	–
120	Br (b)	90%	1–2%	–
70	Br (c)	20%	20%	15%

R = CH_3(a,b), NO (c)

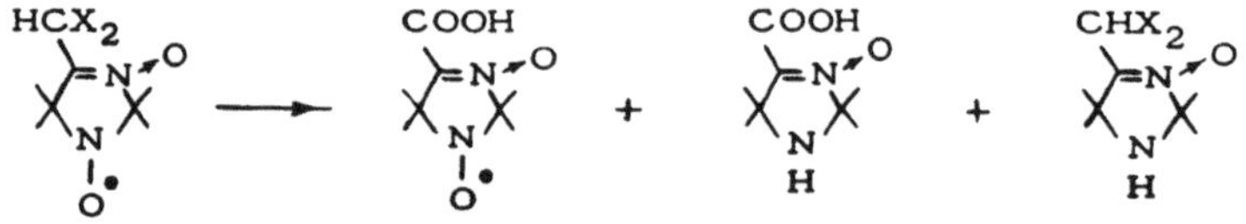

(357) (361) (385) (386)

SCHEME 41.

In the reaction of diamagnetic derivatives **381a,b** with NaOH in aqueous methanol, acetals **382** and carboxylic acids **383** are formed. The reaction with 1-nitroso derivative **381c** under similar conditions leads to acetal **382** and the carboxylic acids — the derivatives of 3-imidazoline **384** and 3-imidazoline-3-oxide **383c**. At the same time, the paramagnetic derivatives **357** react much more readily under similar conditions: the dibromo derivative **357** (X = Br) vanishes during 100 min, the dichloro derivative **357** (X = Cl), during 1 min. As a result of the reaction, the subsequent acetal is not formed even as an intermediate, and the reaction products are the diamagnetic **385** and paramagnetic **361** acids. Also, formation of the diamagnetic dihalomethyl derivatives **386**, which are unstable in these conditions, has been reported.[273] Thus, the presence of the radical center in a molecule leads to a changed set of products, sharply increasing the rate of reaction even when compared with the 1-nitroso derivative **381c** (the *N*-nitroso group is a close analog of the nitroxyl group in its electron-accepting effect), and changing the reactivity order of the chloro and bromo derivatives. These factors may be brought about by the specific effect of the nitroxyl group due to its paramagnetism on the reaction center not bound directly with it.[272]

The specific effect of the nitroxyl group was also reported in the reactions of 3-imidazoline-3-oxide dihalo derivatives with amines.[272] It shows itself in the effect on the oxidative ability of the nitrone group and in some other transformations.[272]

Thus, while conducting even nonradical reactions with 3-imidazoline-3-oxide derivatives, one must bear in mind the possible electronic spin effect of the nitroxyl group on the direction and rate of the transformation.

While investigating the properties of 4-aryl-substituted derivatives of 3-imidazoline-3-oxide **335** (R = aryl), it was shown that in solutions of concentrated sulfuric, chloro-, and fluorosulfonic acids, they completely disproportionate to form the protonated oxammonium salt **387** and the diprotonated hydroxylamino derivative **388**. Upon diluting the solutions with water, the starting nitroxides are quantitatively regenerated.[29] This makes it possible to perform the electrophilic substitution reactions with 4-aryl-3-imidazoline-3-oxide nitroxides in the above conditions. Thus, in nitration of the aryl derivatives with HNO_3 in H_2SO_4 (concentrated), the respective nitroaryl derivatives of nitroxides are formed, that is, the transformation proceeds with preservation of the nitroxyl group in the reaction products, though in the process the nitroxyl group disappears.[248,275]

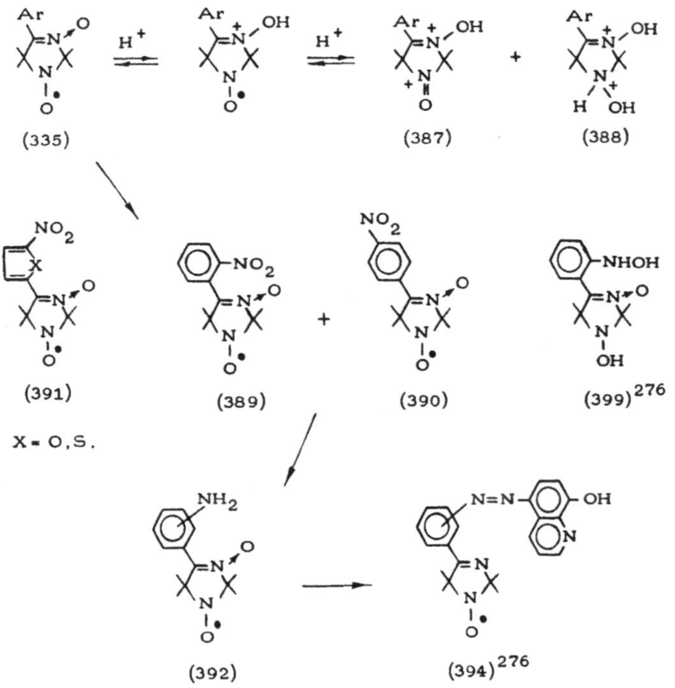

SCHEME 42.

The protonated nitrone group acts in the nitration reaction as a weak *ortho-para*-orientant, because the reaction forms a mixture of *ortho-* and *para*-nitro derivatives **389, 390**. The nitration reaction can also be performed with hetaryl derivatives **335**, in the case of 2-thienyl and 2-furyl derivatives, the 5-substituted isomers **391** are formed. Reduction of nitro derivatives with hydrazine hydrate in the presence of Raney nickel and subsequent oxidation lead to amines **392**. In the case of the *ortho*-isomer, the reduction may be stopped at the stage of the hydroxylamino derivative **393**. Diazotization of amines **392** and subsequent interaction with active arenes leads to azo compounds. In this way, spin-labeled analogs of 8-hydroxyquinoline **394** were obtained, which were suggested for use as paramagnetic analytical reagents.[276]

Thus, making use of the reactivity of the nitrone group in the 3-imidazoline-3-oxide heterocycle affords various functionally substituted 3-imidazoline-3-oxide, 3-imidazoline and imidazolidine nitroxides, and acyclic nitroxides. The presence of an additional nitrogen or *N*-oxide group in heterocycle makes the functional derivatives of nitroxides useful starting compounds in the synthesis of spin-labeled chelating reagents. Such compounds are of interest as spin-labeled analytical reagents for the EPR determination of both paramagnetic and diamagnetic metal ions.[19] Besides, paramagnetic ligands may serve as a basis for organic materials with macromolecular ferromagnetism (compare to Chapter 4).[24] Complexes of paramagnetic ligands with high-spin metal ions can be used as contrasting agents in the NMR tomography (compare to Reference 11). Scheme 43 represents some paramagnetic chelating reagents — the derivatives of 3-imidazoline and 3-imidazoline-3-oxide.

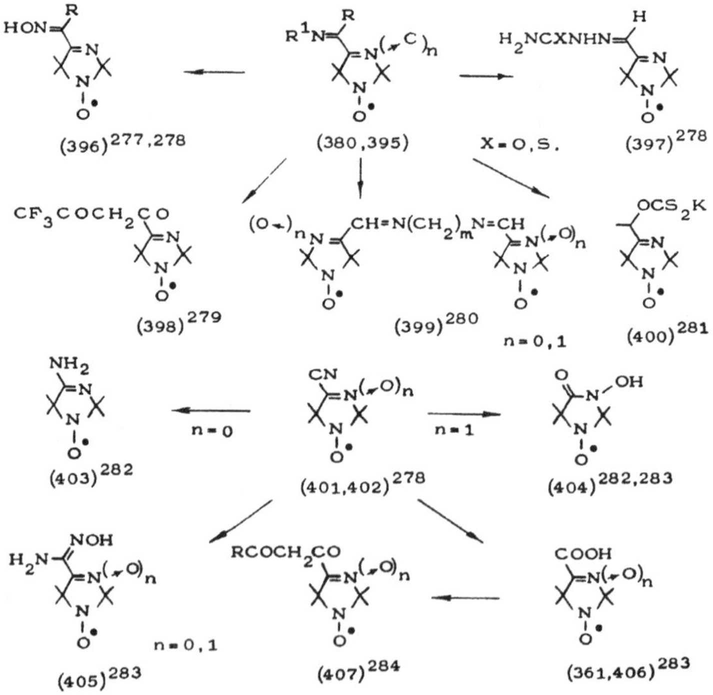

SCHEME 43.

As seen from this scheme, the paramagnetic complexing reagents obtained have different chelating groups in a molecule and different dentate character: coordination with metal may proceed either with or without participation of a heterocyclic nitrogen or *N*-oxide group. Topology of the molecule in the former case is such that nitroxyl is not involved in chelate formation. On the other hand, the derivatives **330** (Scheme 35) with a functional group at nitroxyl α-carbon are capable of forming coordination chelate compounds only with participation of the nitroxyl group. Thus, the chemistry of 3-imidazoline-3-oxide nitroxides permits the design of paramagnetic ligands of such a structure where participation of the nitroxyl group in chelate formation is unambiguously suggested.

Another approach to the synthesis of functional derivatives of 3-imidazoline nitroxides is to construct at first a 3-imidazoline heterocycle containing the sterically hindered hydroxylamino group. In this case, possibilities for functionalization are determined by the presence of the imino group and other functional groups in the side chain of the heterocycle which may be introduced into a molecule at the stage of heterocycle construction. This approach uses the condensation reaction of 1,2-hydroxylamino ketones **408** obtained by hydrolysis of 1,2-hydroxylamino oximes **326**, with ketones in the presence of ammonia or ammonium acetate.

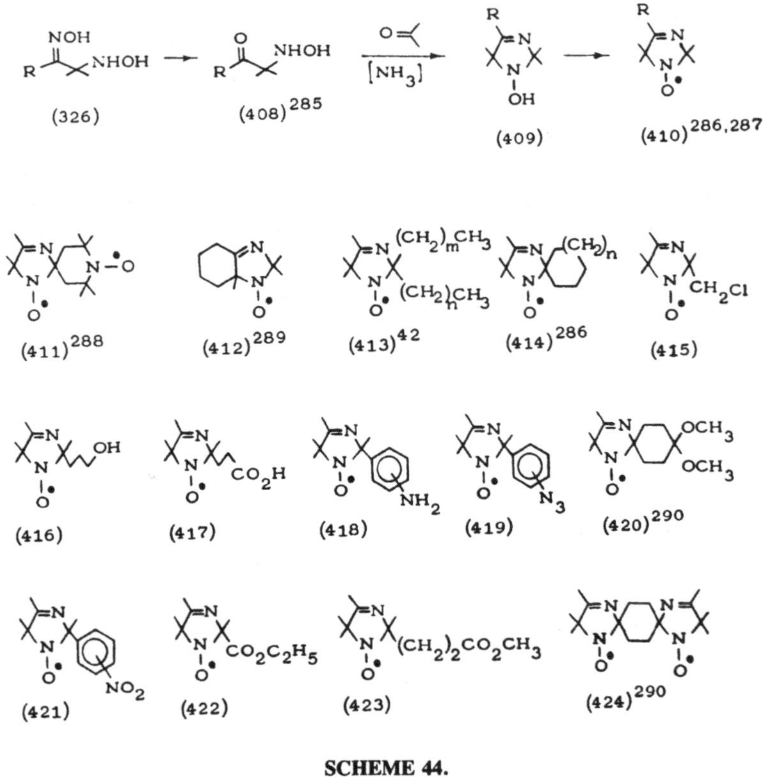

SCHEME 44.

The reaction is general enough and allows synthesis of various 3-imidazoline derivatives, first, due to availability of different α-hydroxylamino ketones **408**, and second, because it proceeds much more easily than in the case of α-hydroxylamino oximes **326**. This essentially expands the range of ketones that can undergo the condensation reaction. Thus, 3-imidazoline derivatives are formed in the reaction with various methylalkylketones, dialkylketones, cycloalkylketones, including tri-acetonamine, methylaryl-, and hetarylketones. The possibility of obtaining the spin-labeled sterine **6** by the reaction of hydroxylamino ketone **13** with 5α-cholestanone, and spin-labeled stearic acids **20** (Scheme 2) has been mentioned previously (Scheme 1). The functionally substituted ketones-esters of levulinic and pyruvic acids, chloroacetone, cyclohexanedione-1,4, etc., can easily undergo the condensation reaction. Thus, along with the imino group in heterocycle, one more functional group is introduced into the molecule, which can also be used in further transformations. Thus, the derivatives **416-419**, which may not be directly obtained as a result of the condensation reaction, become accessible as a result of further modification of substituents in the 2-position of the heterocycle using conventional techniques of organic synthesis.

The 1-hydroxy derivatives obtained in this manner are oxidized to the respective nitroxides in very mild conditions: by lead or manganese dioxides or by atmospheric oxygen without a catalyst.[254] Difficulties may arise in isolating the 1-hydroxy derivatives without radical admixture, since they may be oxidized by atmospheric oxygen even in solid state, while in solution the process occurs much more quickly.

Recently it has been shown that recyclization of bicyclic derivatives **425** obtained by condensation of hydroxylaminooximes **426** with acetylacetone[291] in alkaline media leads to compounds **427** whose oxidation gives bicyclic nitroxides **428** with imidazoline cycle in a molecule.[292] Formation of the 3-imidazoline derivative **429** was reported in studies on the reaction of dibromoketone **430** with aqueous ammonia. Subsequent oxidation with hydrogen peroxide in the presence of sodium tungstate leads to nitroxide **431**.[293]

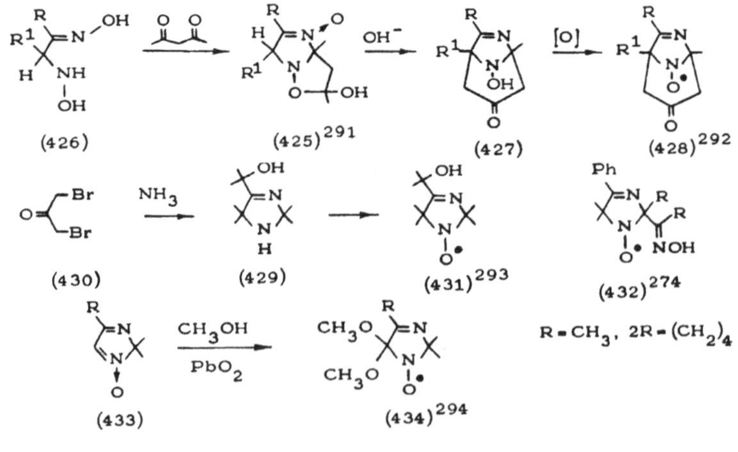

SCHEME 45.

Recently, an interesting method has been developed for the preparation of 3-imidazoline nitroxides with 5,5-dimethoxy groups **434** based on the oxidative methoxylation of 2*H*-imidazole-1-oxide derivatives **433**. The R substituent may also contain the functional group, since the oxidation proceeds under sufficiently mild conditions.[294]

In the reaction of 3-imidazoline-3-oxide nitroxides with hydrazine, not only the nitroxyl group is reduced but also the nitrone group is deoxygenated, leading to 3-imidazoline derivatives, but this reaction is, generally, of no preparative value. An exception is synthesis of oximes **432** whose preparation in some other way is difficult.[274]

As shown above, the possibilities for functionalization of 3-imidazoline derivatives are determined by the presence of the imino group in heterocycle and the presence of functional groups in the substituent in the 2-position of the heterocycle. Making use of the reactivity of the imino group involves certain difficulties, because in the case of paramagnetic derivatives its reactivity is substantially reduced by the electron-accepting effect of the nitroxyl group. This may be illustrated by the decrease of the pK value by 2 to 3 orders on passing from diamagnetic derivatives to the respective nitroxides.[295] On the other hand, the diamagnetic 1-hydroxy derivatives of 3-imidazoline, as well as imidazolidine, are rather labile hydrolytically: the heterocycle cleaves in 5% hydrochloric acid solutions in several minutes. Besides, the nitroxyl group in the 3-imidazoline heterocycle easily undergoes oxidation to the oxammonium group and further hydrolytic decomposition of the heterocycle.[296] In this connection, for functionalization of 3-imidazoline derivatives, special approaches are used to raise the reactivity of the imino group, which involve the transition to 3-imidazolinium salts and the respective enamines, *N*-formylenamines, and BH$_3$ and BF$_3$ complexes. The use of metallated 3-imidazoline derivatives in the synthesis of conjugated imidazolidine-1-oxyl enamines is rather fruitful. Schemes 46 and 47 represent some functional derivatives of imidazolidine nitroxides which may be obtained from imidazolinium salts **435** and enamines **436**.

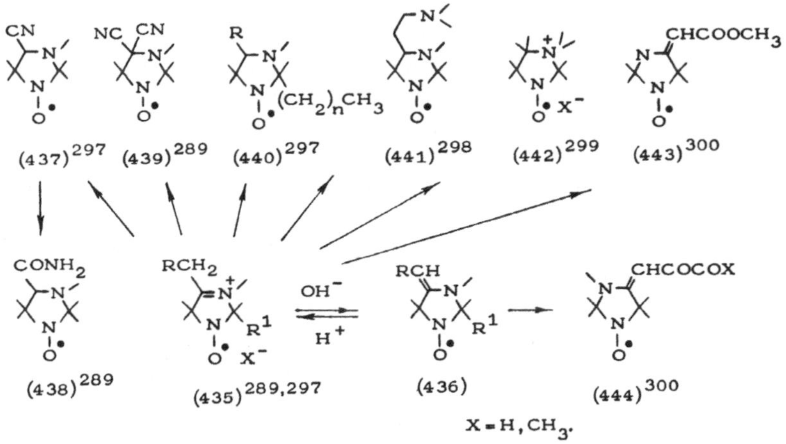

SCHEME 46.

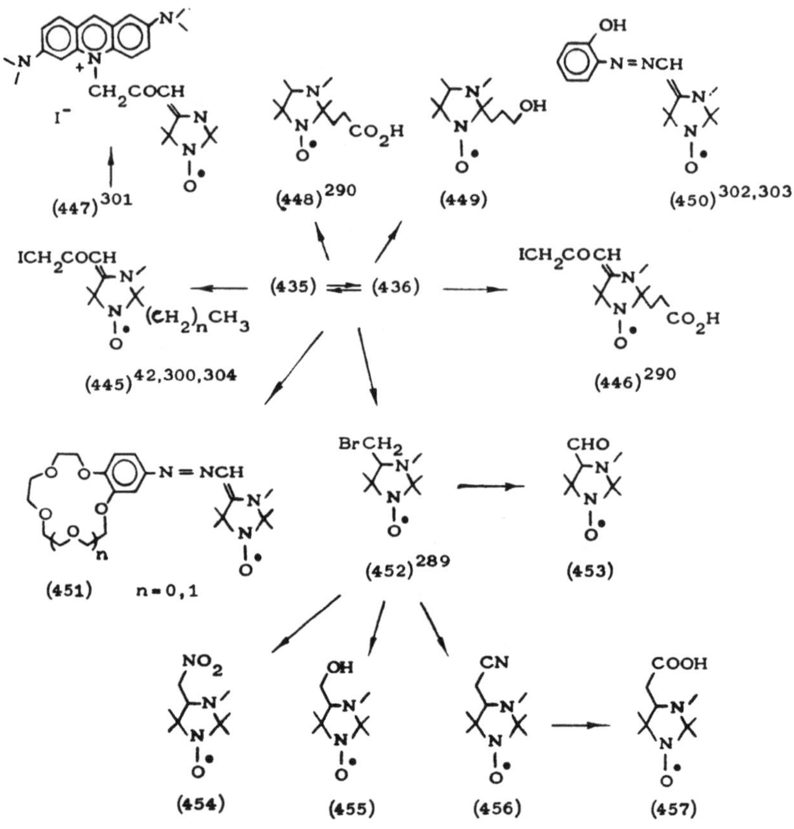

SCHEME 47.

The feasibility of transformations represented in Schemes 46 and 47 is determined by the ease of nucleophilic addition at the polarized C=N bond and the high nucleophilicity of the enamine group (compare to Reference 181). Using this approach one can obtain spin labels and probes, chelating compounds, and spin-labeled crown-esters. It should be noted that imidazolidine derivatives synthesized containing no exocyclic double bond have a pK in the range of 4 to 4.5, are capable of reversible protonation leading to pronounced change of the EPR spectrum, and may, therefore, be used as pH indicators for biochemical research (compare to Reference 8). In the presence of an additional functional group, for example, in compounds **448, 449, 452, 453, 455, and 457**, the imidazolidine derivatives synthesized may be used for creation of macromolecular pH-sensitive spin probes. Thus, it has been shown that the bromine atom in the imidazolidine molecule **452** possesses a high nucleophilic mobility owing to the anchimeric effect of the N-3 atom. This allowed compound **452** to be used in the synthesis of spin-labeled pH-sensitive human albumin.[295]

Scheme 48 shows some functionally substituted derivatives of 3-imidazoline and imidazolidine nitroxides. In their synthesis enamides **91** are used, formed in the formylation of 3-imidazolines **410, 421, and 423**, and the adducts of 3-imidazoline derivatives with borane **458** and BF_3 **459**. An advantage of these methods of activating the imino group to electrophilic substitution at the α-carbon atom and nucleophilic addition at the C=N bond is the fact that the nitrogen atom N-3 remains unsubstituted; after the transformations have been complete, the activating groups are removed directly in the course of reactions (Scheme 48).

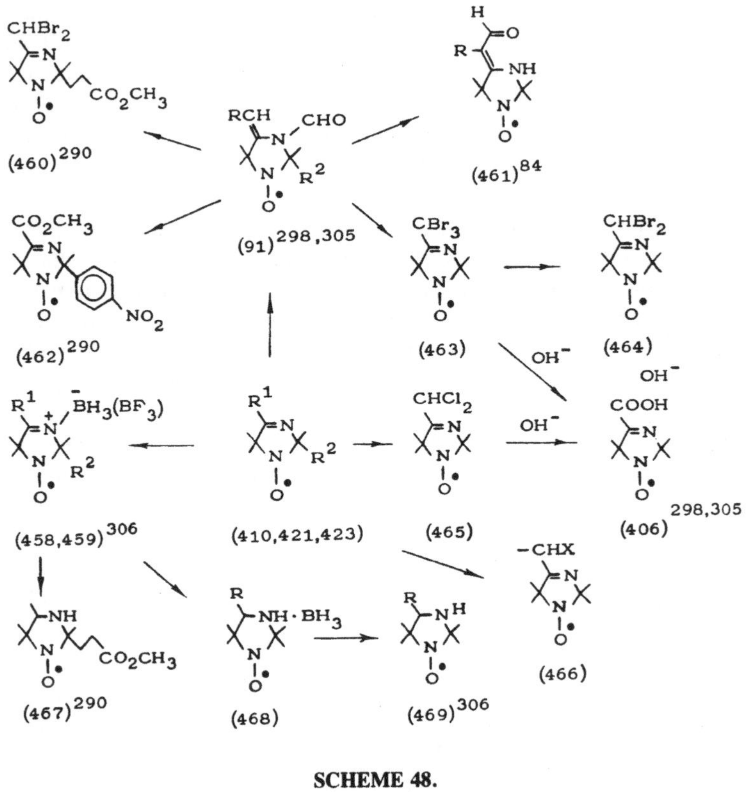

SCHEME 48.

The use of BH_3 and BF_3 adducts permits the reduction of the C=N bond in the 3-imidazoline heterocycle with preservation of the nitroxyl group. The ester group in these conditions also remains intact. On the contrary, in the 3-imidazoline molecule the ester group may be reduced by treatment with lithium aluminum hydride with preservation of the imino and nitroxyl groups is reduced to the hydroxylamino group. The catalytic hydration of nitro derivatives **421** (5% Pd/C) to the amino group also occurs with preservation of the C=N bond (Scheme 44).

It is worthwhile to note an unusual reaction of halogenation of enamide **91** (R = H) with excess sodium hypobromite in an aqueous-alcohol solution. Initially, the tribromo derivative **463** is formed in the reaction, which is quantitatively transformed to the dibromo derivative **464** upon heating of the reaction mixture at 30 to 40°C for a short time.[305] The haloalkyl derivatives **463** to **466** may also be obtained by direct halogenation of 3-imidazoline derivatives with halosuccinimides, but the yield in this case is much lower than with the use of *N*-formylenamines **91** as intermediates.[305] The alkaline hydrolysis of dichloro- **465**, dibromo- **464**, and tribromomethyl derivatives **463** gives carboxylic acid **406** in a high yield,[298,305] being a convenient method for the preparation of this interesting paramagnetic chelating reagent. The nucleophilic mobility of the halogen atoms in haloalkyl derivatives of 3-imidazoline is much lower than for the similar 3-imidazoline-3-oxide derivatives. Thus, the monochloro derivative **470** (Scheme 50) which is available with difficulty remains unchanged during prolonged time in the methanol solution of sodium methoxide, and the monobromomethyl derivative of 3-imidazoline was not obtained. Thus, the preparation of the carboxylic acid **406** is the only transformation of haloalkyl derivatives of 3-imidazoline which is of preparative interest.

Another group of transformations (Schemes 49 and 50) involves the reactions of metallated 3-imidazolines with electrophilic reagents. In view of the fact that the nitroxyl group in 3-imidazoline molecule (R = alkyl) is quantitatively reduced to the hydroxylamino group by treatment with organometal compounds, the reaction is equally feasible both with nitroxides **410** and the hydroxylamines **409**. The metallating reagents may include phenyllithium, lithium diisopropyl amide (LDA) and in some cases, sodium amide in liquid ammonia, methylmagnesium iodide, and sodium hydride in DMFA. In the latter case, the nitroxyl group remains intact, but applicability of these conditions to 3-imidazoline derivatives is rather limited.

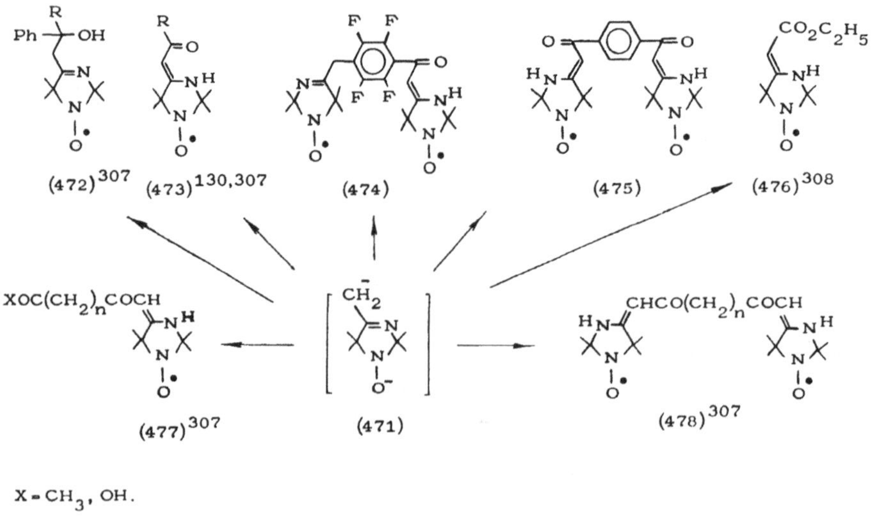

X = CH$_3$, OH.

SCHEME 49.

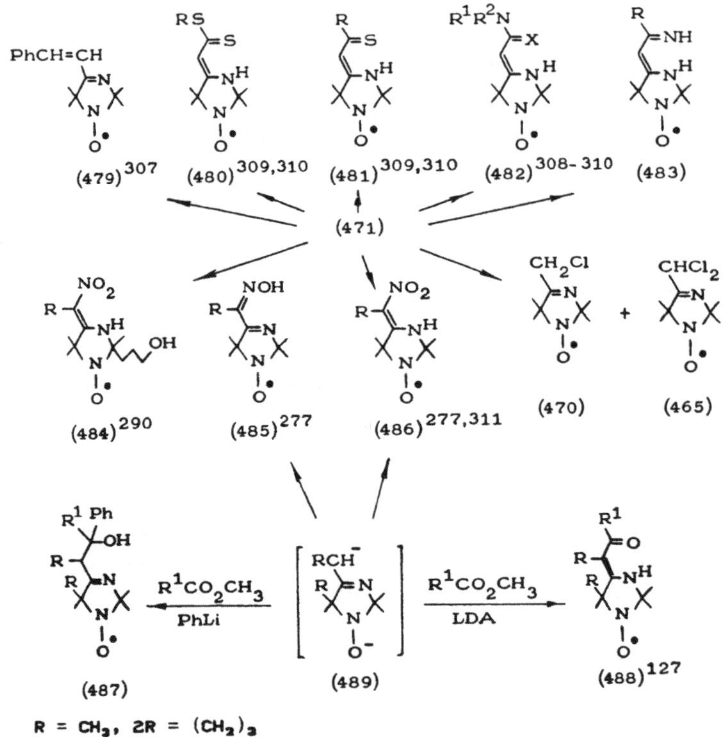

SCHEME 50.

The compounds that can react with the metallated derivative **471** include aldehydes and ketones, esters of mono- and dicarboxylic acids, nitriles, carbon disulfide, esters of mono- and dithiocarboxylic acids, isocyanates and isothiocyanates, methyl nitrate, amyl nitrite, etc. This reaction is used to synthesize various functional derivatives of imidazoline and imidazolidine nitroxides, in particular, conjugated heterocyclic enamines which are of interest as paramagnetic chelating reagents (compare to References 18 and 19) and have wide synthetic possibilities. Because in the reaction of metallated derivative **471** with diesters the condensation reaction may be performed at one or both ester groups, which affords esters of enaminoketocarboxylic acids **477** (that is molecules with one more functional group) the chloro-substituted enaminone **473** (R = CH$_2$Cl) is formed in the reaction of the metallated derivative **471** with ethyl chloroacetate.

It seems unusual that the mono- **470** and dichloromethyl derivatives **465** should be formed in the reaction of the metallated derivative **471** with *p*-toluenesulfochloride.

In the reaction of the metallated derivative **471** with esters, the yields of enaminones **473** considerably increases (from 30 to 90%) from phenyllithium to LDA. In the case of the derivatives **489**, substitution of phenyllithium by LDA leads to a changed structure of reaction products. Thus, in the presence of phenyllithium in the reaction with esters, alcohols **487** are formed, but when LDA is used as a metallating reagent, enaminones **488** are formed.[127] Such a difference in the structure of reaction products is due to the possible reaction of esters with excess phenyllithium, which is not observed to any pronounced degree when the low nucleophilic LDA is used.

In addition to having various synthetic possibilities and being a paramagnetic chelating reagent, nitroenamine **486** (R = H) is a rather strong C–H acid (pK ≈ 7). The EPR spectrum of the deprotonated form differs from that of the starting compound, which permits this compound to be used as a pH-sensitive spin probe. Due to this, compound **484** has been synthesized, which contains in its molecule the pH-sensitive nitroenamino group and the hydroxy group which one can use to modify the macromolecule to design the macromolecular pH-sensitive spin probe.

The 4-phenyl-substituted derivative **410** (R = C₆H₅) may be metallated with butyllithium. The addition of the organometal reagent at the C=N bond does not occur in this reaction (compare to Reference 312), the nitroxyl group is reduced in these conditions to form nearly equal amounts of the hydroxylamino group and butyl ether **490**. Formation of the 1-butoxy derivative may be avoided by introducing the 1-hydroxy derivative **409** into the reaction. Metallation by excess butyllithium occurs at the *ortho*-position, (compare to Reference 313) as indicated by the fact that the products of condensation of the metallated derivative **491** with benzaldehyde and benzophenone **492** exist in the form of the tautomeric mixture of the ''acyclic'' (**A**) and spirobicyclic (**B**) tautomeric forms.

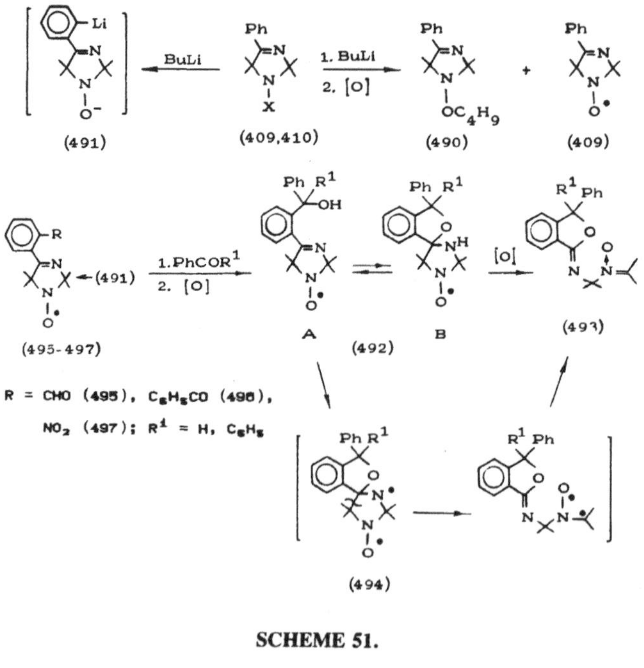

SCHEME 51.

Under the similar conditions as in oxidation of the hydroxylamino group to nitroxyl group (treatment with manganese dioxide in organic solvent), compounds **492** are easily oxidized to the diamagnetic derivatives **493**. This reaction is the only example of such an easy oxidative destruction of stable imidazoline and imidazolidine nitroxides. Judging by the structure of oxidation products, the reaction proceeds via the stage of aminyl radicals **494** formed due to the presence of the spirobicyclic tautomeric form (**B**), rather than via oxammonium salts, as it usually observed in processes of oxidative destruction of stable nitroxides (compare to Reference 29). Subsequent homolytical decomposition of the C^4–C^5 bond and recombination lead to compounds **493**.

The reaction of the metallated derivative **491** with ethyl formiate and ethyl benzoate leads to ketones **495, 496**. The preparation of nitrophenyl 3-imidazolines by nitration of imidazoline **409** in normal conditions seems to be difficult because of instability of oxammonium salt formed in these conditions (compare to References 296 and 314). The use of the reaction of the metallated derivative **491** with methyl nitrate affords the *ortho*-nitro derivative **497**. Due to regioselectivity of the metallation reaction, in this condition only one *ortho*-substituted product is formed.

Another example of the electrophilic substitution reaction which proceeds at the *ortho*-position of the 4-phenyl group of 3-imidazoline **410** (R = C_6H_5) is the reaction with the salts of Hg^{2+} which leads to the organomercuric radical **498**. In similar conditions, the organomercuric derivative **499** was obtained from 3-imidazoline-3-oxide **335** (R = C_6H_5).

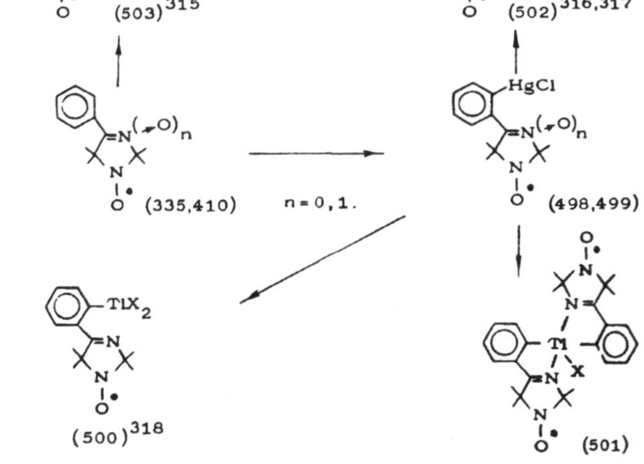

SCHEME 52.

In the reaction of compound **498** with thallium halides, the organothallium radical **500** is formed, and the reaction with thallium trifluoroacetate leads to biradical **501**. The organo-mercury biradicals **502** are formed in the reaction of chloromercuroates **498, 499** with sodium stannite and sodium iodide, respectively. Compounds **498–500** were suggested for use as spin labels on the SH-groups of protein molecules.[316-318] It should be noted that in the reaction of imidazoline **410** with palladium halides, biradicals **503** are formed.[315]

It has been shown in studies on the properties of conjugated enamines of imidazolidine derivatives that the reactions with electrophilic agents generally occur at the enamine carbon atom, though reactions can also occur as a result of electrophilic attack on the endocyclic nitrogen atom N-3, the carbonyl or nitro group oxygen in the case of enaminoketones and nitroenamines, respectively, and at the sulfur atom in the case of enaminothiocarbonyl compounds. The reactions at two nucleophilic centers have also been reported. The reactivity of conjugated enamines in nucleophilic reactions is rather low considering the enamino group but not other groups that may be present in a molecule. The hydrolytic stability of these compounds essentially depends on the presence of the nitroxyl group in a molecule. Thus, the paramagnetic derivatives are rather stable to hydrolysis, while in the case of 1-hydroxy derivatives, the imidazoline heterocycle easily undergoes cleavage in aqueous acids to form acyclic derivatives of hydroxylamine **504**. In the presence of suitable functional groups, the cyclization occurs with participation of the hydroxylamino group released as a result of cycle cleavage to form other heterocyclic systems.[126-128,130,318,319] Schemes 53 and 54 show some synthetically and practically useful functional derivatives, which may be obtained from the conjugated enamines of imidazolidine nitroxides.

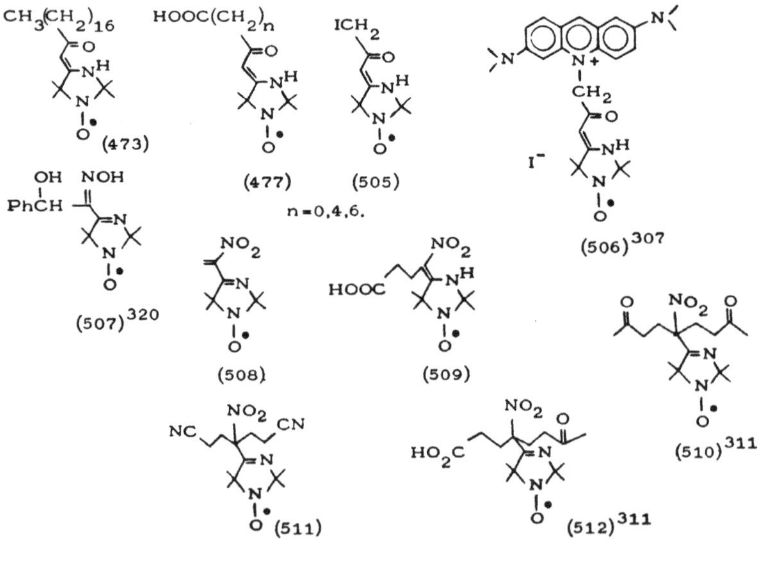

SCHEME 53.

Enaminoketones **477** are of interest as spin labels and probes (compare to Reference 14) which at the same time can form chelates.[18] In the presence of Zn^{2+} salts, the acridinium salt **506** has been reported to show changed fluorescence, which is attributed to complexation at the enaminoketone group leading to changed interaction of the lone electron of the nitroxyl group with the fluorescent acridinium group.[19] The hydroxyoxime **507** shows a high selectivity in the extraction of Cu^{2+} ions from aqueous into organic phase, which may underlie its application as a spin-labeled analytical reagent in copper determination.[19] A number of polyfunctional derivatives of nitroxides **508-514** were obtained by the reaction of nitroenamine **486** (R = H) with electrophilic reagents. It is also worthwhile to note formation of *O*-acylhydroximoyl chlorides **513** in the reaction of nitroenamine **486** with carboxyl chlorides. The chlorine atom in the molecule of these compounds possesses a high nucleophilic mobility, due to which chlorides **513** may be used as the highly active acylating spin labels. On the other hand, compounds **513** are the synthetic equivalents of paramagnetic nitrile oxide **515** in the reactions with nucleophiles. Scheme 54 represents some compounds obtained in the reaction of chloride **513** (R = C_6H_5) with nucleophilic reagents.[321] All the derivatives obtained in this way are the spin-labeled chelating reagents, because chlorine atom substitution is generally accompanied by elimination of the benzoyl group to form substituted α-hydroxyimines (Scheme 54).

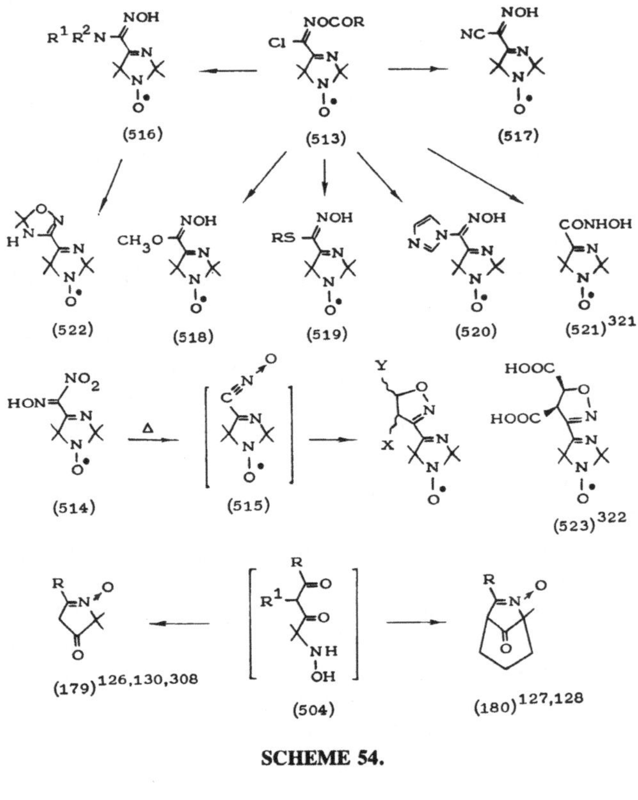

SCHEME 54.

In the thermolysis of nitrooxime **514**, the paramagnetic nitrile oxide **515** is formed which can easily form the 1,3-dipolar cycloaddition products both with activated and nonconjugated olefins. This in its turn may be used both for spin-label introduction at the C=C bonds of various molecules and for the preparation of bi- and polyfunctional derivatives of various nitroxides, for example, the diacid **523**. It should be noted that the cycloaddition reaction proceeds regio- and stereoselectively. Despite the easy cycloaddition process, the range of applications of the suggested method of spin-label introduction at the C=C bonds is presently confined to thermally stable substrates, as formation of nitrile oxide **515** proceeds to a pronounced degree at 90 to 100°C. Nitrogen oxide evolving in thermolysis does not interfere with the process, as it may be easily removed from the reaction mixture by a flow of an inert gas through the solution.

Scheme 21 presents examples of the transformations of enaminoketones **473**, **488** which proceed with cleavage of the imidazoline cycle but, nevertheless, lead to heterocyclic compounds **179**, **180**. These are the starting compounds for the synthesis of nitroxides of other classes: the derivatives of pyrrolidine-4-one **182** and 6-azabicyclo[3,2,1]octane **187**.

To sum up the discussion on 3-imidazoline and 3-imidazoline-3-oxide nitroxides, it is necessary to note very wide synthetic possibilities of these nitroxide classes, ease of their generation and functionalization, and the possibility to go over to other classes of stable nitroxides — heterocyclic and acyclic. The presence of an extra nitrogen atom in the 3-imidazoline and imidazolidine heterocycle makes these compounds prospective for the design of spin-labeled chelating compounds that are characterized either by practical impossibility for the nitroxyl group to participate in chelate formation or, vice versa, the requirement of its participation in chelate cycle formation. The chemistry of nitroxides of this class is currently developed to such an extent that numerous functional derivatives have been created, which seem to be more numerous and varied compared to other classes of stable nitroxides, but their practical application in molecular biology and biochemistry is restricted despite all the advantages of these compounds. This seems to be due to the fact that these functional derivatives are mostly the synthons of paramagnetic models of diamagnetic biogenic models rather than the paramagnetic models themselves. However, in view of the wide synthetic possibilities of these synthons, one can suppose that 3-imidazoline and 3-imidazoline-3-oxide nitroxides will still occupy the place they deserve among the practically useful spin labels and probes.

2. Six-Membered Heterocyclic Nitroxides
a. Tetrahydropyrimidines
The scheme for constructing the six-membered heterocyclic nitroxides with the 1,3-position of heteroatoms is similar to the scheme of constructing the respective five-membered heterocycles and consists of condensation of 1,3-bifunctional derivatives with carbonyl compounds: aldehydes, ketones, and more rarely, carboxylic acids. Thus, the condensation of 1,3-bishydroxylamine **524** with aldehydes and subsequent oxidation lead to six-membered nitronyl nitroxides; tetrahydropyrimidine derivatives **525** that are unstable and decompose in solution in several hours. Compounds **525** have not been isolated individually.[323]

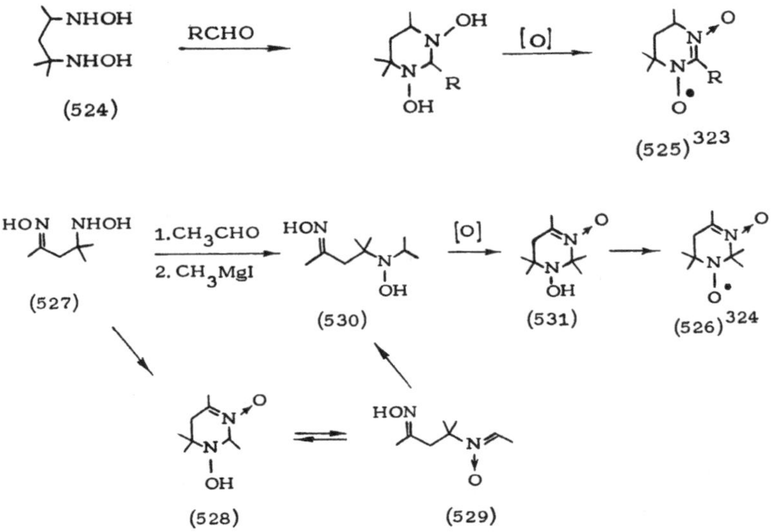

SCHEME 55.

Nitroxide **526** is somewhat more stable. It was isolated individually, but quickly decomposes at room temperature. The method for preparing the compound **526** is based on the condensation reaction of 1,3-hydroxylaminooxime **527** with acetaldehyde and subsequent interaction of the tetrahydropyrimidine-derivative formed **528**, which exists in tautomeric equilibrium with acyclic hydroximinonitrone **529** with methylmagnesium iodide. Subsequent oxidation of hydroxylaminooxime **530** by lead dioxide leads to compound **526**.[324] The condensation of hydroxylaminooxime **527** with acetone to obtain the precursor of nitroxide **526** does not occur.

The low stability of tetrahydropyrimidine nitroxides and their limited availability restrict the application of compounds of this class in further functionalization or as spin labels and probes.

b. Piperazines

Piperazine nitroxides are characterized by the 1,4-position of heteroatoms and, consequently, the corresponding diamagnetic derivatives: the amino and hydroxylamino derivatives and oxammonium salts must have a high hydrolytic stability.

One of the ways to obtain sterically hindered amino derivatives of the piperazine series is similar to the method for the preparation of imidazolidinones (Scheme 30) and consists of self-condensation of aminonitriles **285** in alkaline media.[325,329,330] The compounds **532** formed in this process contain the amino and imido groups in heterocycle, due to which the imido group, labile under the oxidation reaction conditions, must be protected by acylation or alkylation. The acyl group is removed after the oxidation stage, and this leads to nitroxides **534** containing the reactive imido group in heterocycle in addition to the nitroxyl group.[325] The reduction of the imido group of tetrahydropyrazine **532** with lithium aluminium hydride,

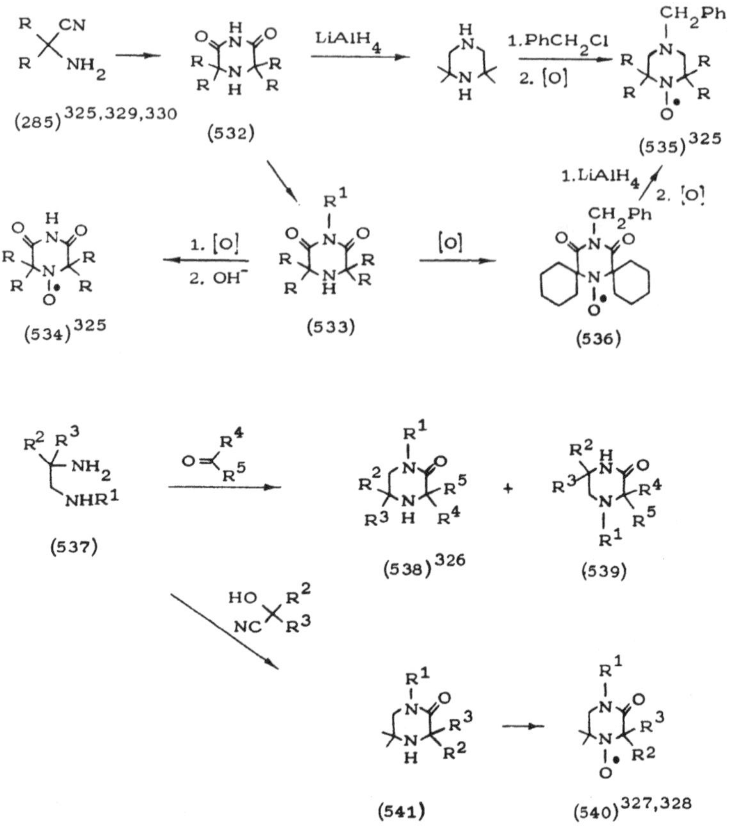

SCHEME 56.

subsequent alkylation, and oxidation of the secondary amino group with *m*-chloroperbenzoic acid lead to the nitroxide **535** (R = CH₃). Another route to a similar compound [2R = (CH₂)₅] consists of the reduction of the benzyl derivative **536** with lithium aluminium hydride and subsequent oxidation of the resulting hydroxylamino derivative by lead dioxide.[325]

An alternative way to prepare piperazine nitroxides is condensation of 1,2-diamines **537** with ketones and chloroform under the phase transfer catalysis conditions.[326] This results in a mixture of compounds **538** and **539** which may be avoided by performing the condensation of diamine **537** with ketone cyanohydrines under similar conditions. Subsequent oxidation of compounds **541** with *m*-chloroperbenzoic acid leads to nitroxides **540**.[327,328]

Analysis of the existing methods for the preparation of piperazine nitroxides shows that their advantage is the possibility of varying the alkyl substituents in the vicinity of the nitroxyl group. Functionalization of these compounds is possible due to the presence of the amide or imide groups in heterocycle, but currently there are no data on the use of these compounds in the synthesis of other functionally substituted derivatives of nitroxides.

c. Tetrahydrooxazines

The existing methods for the synthesis of tetrahydrooxazine derivatives are similar to methods for the synthesis of oxazolidine derivatives and differ from them in the use of 1,3-aminocarbinols as the starting compounds. The latter may be obtained from α,β-unsaturated ketones by interaction with ammonia and subsequent transformation of the keto group to the alcohol group by the reduction or interaction with organomagnesium reagents.[331] The latter can expand the range of the starting aminocarbinols, but in view of the necessity of subsequent oxidation of the heterocyclic amino group at the stage of radical center generation, the R′ substituent cannot contain functional groups labile in these conditions.

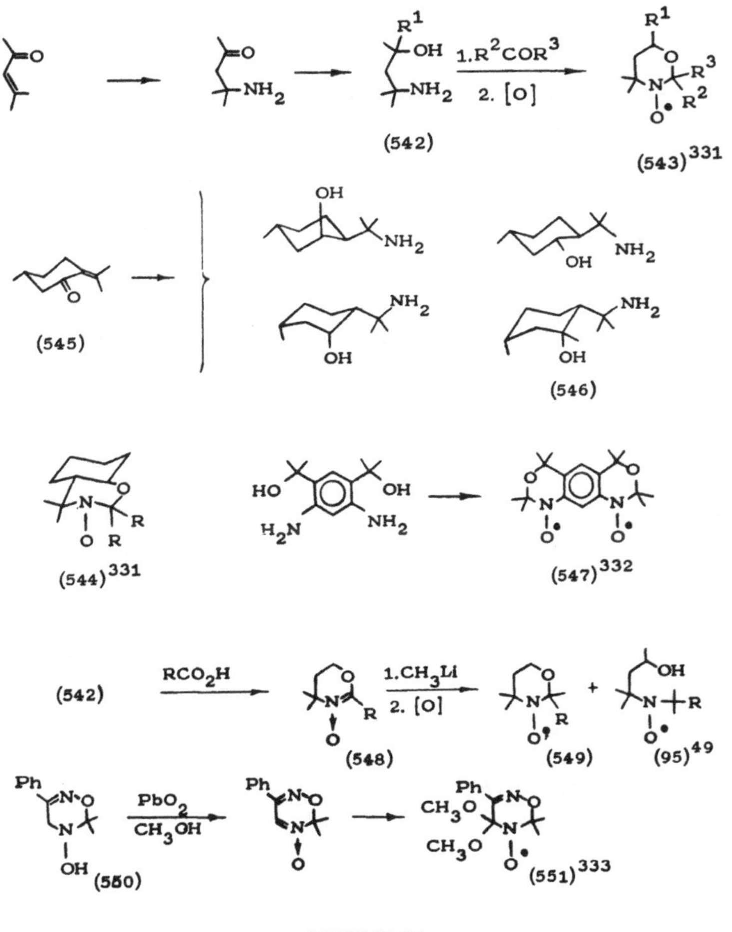

SCHEME 57.

The condensation of aminocarbinols with ketones and subsequent oxidation with *m*-chloroperbenzoic acid leads to nitroxides **543**. This approach may be used for the preparation of condensed tetrahydrooxazine nitroxides **536**. Thus, (+)-pulegone was converted to 1,3-aminocarbinols **546** in which the reaction with ketones and subsequent oxidation leads to nitroxides, one of which is shown in Scheme 57.[331] This approach was used for the synthesis of biradical **547**.[332]

As it has been noted earlier, the condensation of 1,2-aminocarbinols with carboxylic acids leads to oxazolines **32** which may be transformed to nitrones **34**. Further interaction with organometallic reagents and oxidation lead to oxazolidine nitroxides (Scheme 4). Similarly, the use of 1,3-aminocarbinol **542** gave oxazines **548** where interaction with methyllithium and subsequent oxidation lead to tetra-hydrooxazine nitroxide **549** and acyclic nitroxide **95**.[49] The possibility of oxidative methoxylation of compound **548** to prepare α-methoxy-substituted nitroxides (compare to Scheme 29) has not been investigated. At the same time, the oxidative methoxylation of the 1,2,5-oxadiazine derivative **550** has been shown to give nitroxide **551**.[333] This approach is suggested to be applicable to the synthesis of tetrahydrooxazine nitroxides.

The synthetic possibilities of tetrahydrooxazine nitroxides seem to be reduced to modifications of functional groups in substituents, but no data regarding this are currently available.

3. Seven-Membered Heterocyclic Nitroxides — Homopiperazines

There are only a few representatives of homopiperazine nitroxides, first, because of unavailability of simple and universal methods for the synthesis of this hetero-cyclic system and, second, because of the limited possibilities for functionalization of these compounds. The latter limitation is a consequence of the former, since, in addition to the sterically hindered amino or nitroxyl group, the available hom-opiperazine derivatives contain the amide or secondary amino group having rather limited synthetic possibilities, especially in view of the limited stability of the nitroxyl group.

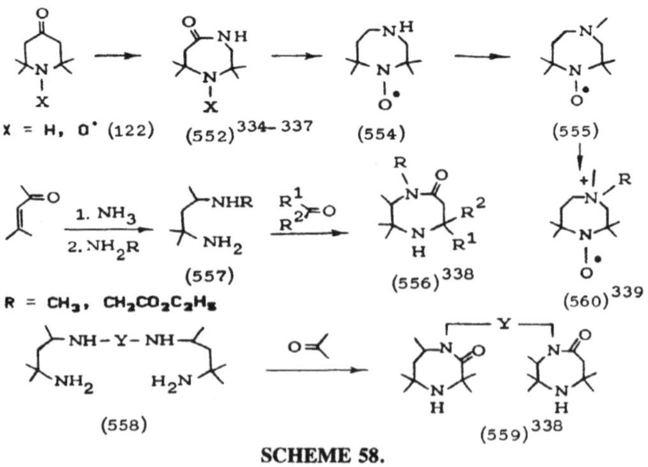

SCHEME 58.

The method for the synthesis of homopiperazines containing the sterically hindered amino or nitroxyl group consists of the piperidine heterocycle expansion using the Schmidt reaction in the case of triacetoneamine or the Beckmann rearrangements of diamagnetic (X = H) or paramagnetic (X = O) oximes.[334-336] An improved method has been developed for the preparation of nitroxide **552** consisting of the sequential treatment of nitroxide **122** with hydroxylamine hydrochloride and potassium carbonate.[337]

An original method for the preparation of 2,2-disubstituted sterically hindered amines **556** has been suggested in Reference 338. This method is similar to that for the synthesis of piperazine derivatives and consists of the reaction of easily available 1,3-diamines **557** with ketones and chloroform in phase transfer catalysis conditions. It should be noted that the reaction with triamine **558** leads to bisdiazepines **559**. The possibility of oxidizing the derivatives **556, 559** to the respective nitroxides has not been investigated, but in view of the structural similarity of these compounds to homopiperazine **552**, such a transformation seems to be quite possible.

The reduction of amide **553** with lithium aluminium hydride and subsequent oxidation with lead dioxide leads to nitroxide **554** containing the secondary amino group in heterocycle. It should be noted that the possibilities for functionalization of compounds **554** are reduced to the reactions at the amino group, and presently, only the alkylation reactions of compounds **554, 555** are known.[339]

E. ISOTOPE-SUBSTITUTED STABLE NITROXIDES

Synthesis of isotope-substituted stable nitroxides is of great interest in view of the fact that the intensity of the EPR spectra of nitroxides dramatically increases on hydrogen substitution by deuterium, due to the disappearance of a lone electron splitting on them. As a matter of fact, the smaller the number of hydrogen atoms that surround the nitroxyl group, the more the resolved spectrum is observed for the spin label. In this respect, the use of the alkylating spin label — the imidazoline derivative **445** — is more preferable than that of the widespread iodoacetamide spin label, the piperidine derivative **253**.

Substitution of the ^{14}N atom (s = 1) of the nitroxyl group by the ^{15}N atom (s = 0.5) leads to the decreased multiplicity of the EPR spectrum (doublet), resulting also in the increased spin label or probe sensitivity at the same integral intensity and facilitating interpretation of the immobilized EPR spectra. Synthesis of isotope-substituted nitroxides also opens up new prospects of their application, e.g., in studies on macromolecular tumbling with characteristic times in the range of 10^{-7} to 10^{-3} s. (For details on application of deuterated spin labels and probes, see Reference 3).

Synthesis of isotope-containing perdeutero and ^{15}N nitroxides seems to involve certain difficulties associated with the limited availability and high price of the starting isotope-containing compounds, the multistep character of the synthesis and a comparatively low yield of the end products. Due to this, simplified approaches have been worked out to the synthesis of partially isotope-substituted nitroxides: the hydrogen-containing derivatives with the ^{15}N isotope of the nitroxyl group, or, vice versa, the completely or partially deuterated derivatives with the ^{14}N isotope of the nitroxyl group. The choice between these approaches depends in every particular case on the problem being solved and the necessary sensitivity of a label or probe. Schemes 59 and 60 show examples of synthesis of some isotope-containing nitroxides.

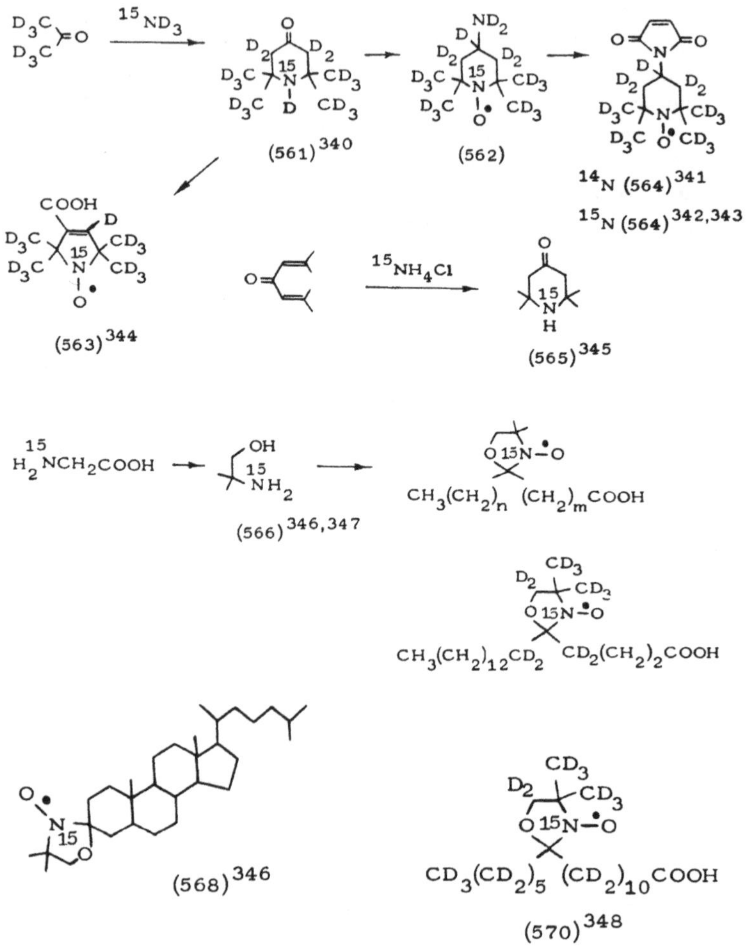

SCHEME 59.

Synthesis of the perdeutero-^{15}N derivative of piperidine **562** is complicated by the low yield of the final product from the not readily available $^{15}ND_3$, which is taken in large excess for the condensation at the stage of synthesis of isotope-substituted triacetoneamine **561**. Due to this, the acid **563** obtained from triaceto-neamine **561** is also not readily available. A simple and economical way for the synthesis of the nondeuterated ^{15}N triacetoneamine **565** consists of the reaction of foron with $^{15}NH_4Cl$.[345]

The isotope-containing oxazolidine nitroxides are obtained from perdeutero-^{15}N-1,2-aminocarbinol **566** synthesized from ^{15}N-glycine.[346,347] The procedure is essentially simplified when the nondeuterated ^{15}N-aminocarbinol **566** is used. Aminocarbinols **566** were used to obtain the protic ^{15}N-spin-labeled derivatives of stearic acid **567** and sterine **568**, partially-**569**, and perdeutero-^{15}N-stearic acids **570**.[346-348]

Recently, a method has been developed for the synthesis of spin-labeled stearic acid **571** containing the octadeutero-^{15}N-pyrrolidine-1-oxyl fragment in the molecule. The key compound in this synthesis is nanodeutero-^{15}N-pyrroline-1-oxide **572** which was obtained starting from perdeuteroacetone and $^{15}NH_2OH$.[11]

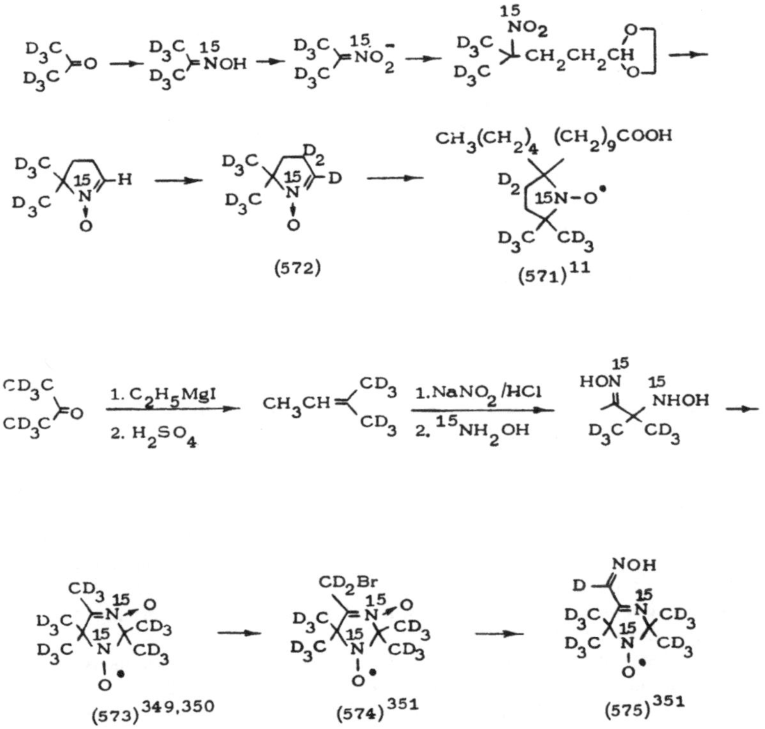
(572)
(571)11
1. C2H5MgI
2. H2SO4
1.NaNO2/HCl
2. 15NH2OH
(573)349,350
(574)351
(575)351

SCHEME 60.

There are wide possibilities for the synthesis of deutero- and ^{15}N-labeled derivatives of 3-imidazoline and 3-imidazoline-3-oxide nitroxides. The isotopes can be selectively introduced into different positions of the molecule, and the perdeutero-^{15}N derivatives **573** may be synthesized. Among the isotope-substituted nitroxides of this class that are of practical interest, it is worthwhile to mention the alkylating spin label **574** and the spin-labeled chelating reagent α-iminooxime **575**. The data on many isotope-substituted nitroxides of the 3-imidazoline and 3-imidazoline-3-oxide series, including compounds **573** to **575**, are reported in Reference 351.

To sum up the discussion on isotope-containing nitroxides, we will note that there are methods of introducing the ^{17}O-isotope into the nitroxyl group, generally based on the reaction of the aminyl radicals, generated by any method, with $^{17}O_2$.[352]

REFERENCES

1. **McConnel, H. M. and McFarland, B. C.**, Physics and chemistry of spin labels. *Q. Rev. Biophys.*, 3, 91, 1970.
2. **Likhtenstein, G. I.**, *Spin Labeling Method in Molecular Biology*, Wiley Interscience, New York, 1976.
3. **Berliner, L. J. and Reuben, J., Eds.**, *Spin Labeling: Theory and Applications*, Vols. 1, 2, and 8, Academic Press, New York, 1976, 1979, and 1989.
4. **Kuznetsov, A. N.**, *Spin Labeling Method*, Nauka, Moscow, 1976; *Chem. Abstr.*, 86, 148698e, 1977.
5. **Kevan, L. and Baglioni, P.**, Electron spin-echo modulation studies of ionic and nonionic micelle via stearic acid nitroxide probes, *J. Pure Appl. Chem.*, 62(2), 275, 1990.
6. **Salikhov, K. M., Semenov, A. G., and Tsvetkov, Yu. D.**, *Electron Spin Echo and its Applications*, Nauka, Novosibirsk, U.S.S.R., 1976.
7. **Kevan, L. and Bowman, M. K.**, *Modern Pulsed Continuous-Wave Electron Spin Resonance*, John Wiley & Sons, Chichester, England, 1990.
8. **Khramtsov, V. V. and Weiner, L. M.**, Proton exchange in stable nitroxyl radicals: pH-sensitive spin probes, in *Imidazoline Nitroxides*, Vol. 2, Volodarsky, L. B., Ed., CRC Press, Boca Raton, FL, 1988, chap. 2.
9. **Hyde, J. S. and Subczynski, W. K.**, Spin-label oximetry, in *Spin Labeling: Theory and Applications*, Vol. 8, Berliner, L. J. and Reuben, J., Eds., Plenum Press, New York, 1989, chap. 8.
10. **Hyde, J. S., Yin, J.-J., Feix, J. B., and Hubbel, W. L.**, Advances in spin label oximetry, *J. Pure Appl. Chem.*, 62(2), 255, 1990.
11. **Keana, J. F. W., Lex, L., Mann, J. S., May, J. M., Park, J. H., Pou, S., Prabhu, V. S., Rosen, G. M., Sweetman, B. J., and Wu, Y.**, Novel nitroxides for spin-labelling, -trapping, and magnetic resonance imaging applications, *J. Pure Appl. Chem.*, 62(2), 201, 1990.
12. **Sueki, M., Eaton, G. R., and Eaton, S. A.**, Multidimensional EPR imaging of nitroxides, *J. Pure Appl. Chem.*, 62(2), 229, 1990.
13. **Swarts, H. M.**, The use of nitroxides in viable biological systems: an opportunity and challenger for chemists and biochemists, *J. Pure Appl. Chem.*, 62(2), 235, 1990.
14. **Lyakhovich, V. V. and Weiner, L. M.**, Application of stable nitroxyl radicals for studying metabolizing systems of xenobiotics, in *Imidazoline Nitroxides*, Vol. 2, Volodarsky, L. B., Ed., CRC Press, Boca Raton, FL, 1988, chap. 1.
15. **Bowry, V., Lustyk, J., and Ingold, K. U.**, Calibration of very fast alkyl radical "clock" rearrangements using nitroxides, *J. Pure Appl. Chem.*, 62(2), 213, 1990.
16. **Park, J. H. and Trommer, W. E.**, Advantages of ^{15}N and deuterium spin probes for biomedical electron paramagnetic resonance investigations, in *Spin Labeling: Theory and Applications*, Vol. 8, Berliner, L. J. and Reuben, J., Eds., Plenum Press, New York, 1989, chap. 11.
17. **Bowman, M. K., Michalski, T. J., Peric, M., and Halpern, H. J.**, Fourier-transform EPR and low-frequency EPR studies of nitroxides, *J. Pure Appl. Chem.*, 62(2), 271, 1990.
18. **Larionov, S. V.**, Imidazoline nitroxides in coordination chemistry, in *Imidazoline Nitroxides*, Vol. 2, Volodarsky, L. B., Ed., CRC Press, Boca Raton, FL, 1988, chap. 3.

19. **Nagy, V. Yu.**, Imidazoline nitroxides in analytical chemistry, in *Imidazoline Nitroxides*, Vol. 2, Volodarsky, L. B., Ed., CRC Press, Boca Raton, FL, 1988, chap. 4.

20. **Sosnovsky, J., Lukszo, J., Guttierez, P. L., and Scheffler, K.**, EPR and ENDOR of sodium complexes of spin labelled monoazacrown ethers, *Z. Naturforsch., B.*, 42b, 376, 1986.

21. **More, J. K., More, K. M., Eaton, G. R., and Eaton, S. S.**, Metal-nitroxyl interactions. 55. Manganese (III)-nitroxyl electron-electron spin-spin interaction, *J. Pure Appl. Chem.*, 62, 241, 1990.

22. **Rassat, A.**, Magnetic properties of nitroxide multiradicals, *J. Pure Appl. Chem.*, 62, 223, 1990.

23. **Rey, P., Laugier, J., Caneschi, A., and Gatteschi, D.**, High-spin metal-notroxyl molecular species, in *Int. Conf. Nitroxide Radicals* (Abstr.), Novosibirsk, Russia, 1989, 2C.

24. **Ovcharenko, V. I., Ikorskii, V. N., Vostrikova, K. E., Burdukov, A. B., Romanenko, G. V., Pervukhina, N. V., and Podberezskaja, N. V.**, Low temperature ferromagnetics based on imidazoline nitroxide complexes, *Int. Conf. Nitroxide Radicals (Abstr.)*, Novosibirsk, U.S.S.R., 1989, 18C.

25. **Yamaguchi, M., Miyazawa, T., Takata, T., and Endo, T.**, Application of redox system based on nitroxides to organic synthesis, *J. Pure Appl. Chem.*, 62, 217, 1990.

26. **Abakumov, G. A., Muraev, V. A., Razuvaev, G. A., Tikhonov, V. D., Chechet, Yu. V., and Nechuev, A. I.**, Electrochemical aspects of single-electron transfer in organic reactions, *Dokl. Akad. Nauk S.S.S.R.*, 230, 589, 1976; *Chem. Abstr.*, 86, 10065q, 1977.

27. **Rozantsev, E. G.**, *Free Nitroxyl Radicals*, Plenum Press, New York, 1970.

28. **Forrester, A. R., Hay, J. M., and Thomson, R. V.**, *Organic Chemistry of Stable Free Radicals*, Academic Press, London, 1968.

29. **Shchukin, G. I. and Grigir'ev, I. A.**, Oxidation-reduction properties of nitroxides, in *Imidazoline Nitroxides*, Vol. 1, Volodarsky, L. B., Ed., CRC Press, Boca Raton, FL, 1989, chap. 6.

30. **Bukin, I. I.**, Nitroxides in oil-field development, in *Int. Conf. Nitroxide Radicals (Abstr.)*, Novosibirsk, U.S.S.R., 1988, 69P.

31. **Hubbel, W. L. and McConnel, H. M.**, Motion of steroid spin labels in membranes, *Proc. Natl. Acad. Sci. U.S.A.*, 63, 16, 1969.

32. **Keana, J. F. W., Keana, S. B., and Beetham, D.**, A new versatile ketone spin label, *J. Am. Chem. Soc.*, 89, 3055, 1967.

33. **Keana, J. F. W., Norton, R. S., Morello, M., Van Engen, D., and Clardy, J.**, Mononitroxides and proximate dinitroxides derived by oxidation of 2,2,4,4,5,5-hexasubstituted imidazolidines. A new series of nitroxide and dinitroxide spin labels, *J. Am. Chem. Soc.*, 100, 934, 1978.

34. **Keana, J. F. W. and Dinerstein, R. J.**, A new highly anisotropic dinitroxide spin label. A sensitive probe for membrane structure, *J. Am. Chem. Soc.*, 93, 2808, 1971.

35. **Metzner, E. K., Libertini, L. J., and Calvin, M.**, Electron spin exchange in rigid biradicals, *J. Am. Chem. Soc.*, 99, 4500, 1977.

36. **Gaffney, B. G. and Mich, R. G.**, A new measurement of surface charge in model and biological lipid membranes, *J. Am. Chem. Soc.*, 98, 3044, 1976.

37. **Dvolaitzeky, M., Billard, J., and Poldy, F.**, Smectic, E, C and A free radicals, *Tetrahedron*, 32, 1835, 1976.

38. **Keana, J. F. W., Bernard, E. M., and Roman, R. B.**, Selective reduction of doxyl and proxyl nitroxide carboxylic acids to the corresponding alcohols with borane methyl sulfide, *Synth. Commun.*, 8, 169, 1978.

39. **Borin, M. L., Kedik, S. A., Volodarskii, L. B., and Shvets, V. I.**, Lipid spin probes comprising an imidazoline nitroxyl fragment, *Bioorg. Khim.*, 10, 251, 1984; *Chem. Abstr.*, 100, 153303j, 1984.

40. **Borin, M. L., Kedik, S. A., Volodarskii, L. B., and Shvets, V. I.**, Steric acid derivatives comprising an imidazoline nitroxyl fragment, *Bioorg. Khim.*, 10, 1558, 1984; *Chem. Abstr.*, 102, 131595s, 1985.

41. **Volodarskii, L. B., Reznikov, V. A., and Grigor'ev, I. A.**, Chemical properties of heterocyclic nitroxides, in *Imidazoline Nitroxides*, Vol. 1, Volodarsky, L. B., Ed., CRC Press, Boca Raton, FL, 1989, chap. 3.

42. **Subkhankulova, T. N., Lyakhovich, V. V., Weiner, L. M., and Reznikov, V. A.,** Affinity modification of the membrane-bound microsomal cytochrome P-450 by substrate lipophylic analogs, *Biochimia,* 54, 17, 1989; *Chem. Abstr.,* 111, 19927q, 1989.

43. **Eastman, M. P., Patterson, D. E., Bartsch, R. A., Liu, Y., and Eller, P. G.,** Three novel spin labeled crown ethers, *J. Phys. Chem.,* 86, 2052, 1982.

44. **Ullman, E. F., Osiecki, J. H., Boocock, D. G. B., and Darcy, R.,** Studies of stable free radicals. X. Nitronyl nitroxide monoradicals and biradicals as possible small molecule spin labels, *J. Am. Chem. Soc.,* 94, 2049, 1972.

45. **Weinkam, R. J. and Jorgensen, E. C.,** Free radical analogues of histidine, *J. Am. Chem. Soc.,* 93, 7028, 1971.

46. **Abdallah, M. A., Andre, J.-J., and Biellmann, J.-F.,** A new spin-labelled analogue of nicotinamide adenine dinucleotide, *Bioorg. Chem.,* 6, 157, 1977.

47. **Tronchet, J. M. J., Winter-Mihaly, E., Habashi, F., Schwarzenbach, D., Likic, U., and Geoffroy, M.,** Deoxyhydroxylaminosugar derivatives and corresponding diglycosylnitroxides free radicals. Preliminary communication, *Helv. Chim. Acta,* 64, 610, 1981.

48. **Keana, J. F. W. and Lee, T. D.,** A versatile synthesis of doxyl spin labels bypassing the usual ketone precursors, *J. Am. Chem. Soc.,* 97, 1273, 1975.

49. **Lee, T. D. and Keana, J. F. W.,** Nitrones and nitroxides derived from oxazolines and dihydrooxazines, *J. Org. Chem.,* 41, 3237, 1976.

50. **Williams, J. C., Mehlhorn, R. J., and Keith, A. D.,** Synthesis and novel uses of nitroxide motion probes, *Chem. Phys. Lipids,* 7, 207, 1971.

51. **Terekhina, A. I., Vesela, I. V., Kamernitskii, A. V., and Lisitsa, L. I.,** Biological activity of transformed steroids. VII. Synthesis and hormonal effect of some 16α, 17α-oxazolines of the pregnane group, *Khim. Farm. Zh.,* 11, 97, 1977; *Chem. Abstr.,* 88, 51079g, 1978.

52. **Ramasseul, R. and Rassat, A.,** Nitroxides. XLIX. Steroidal nitroxides, *Tetrahedron Lett.,* 1623, 1971.

53. **Shechter, H., Ley, D. E., and Zeldin, L.,** Addition reactions of nitroalkanes with acrolein and methyl vinyl ketone. Selective reduction of nitrocarbonyl compounds to nitrocarbinols, *J. Am. Chem. Soc.,* 74, 3664, 1952.

54. **Rundel, M.,** Methoden zur herstellung und umwandlung von Nitronen, in *Houben-Weyl: Methoden der Organishen Chemie,* B 10/4, Muller, E., Ed., Georg Thieme Verlag, Stuttgart, 1968, 311.

55. **Keana, J. F. W.,** Synthesis and chemistry of nitroxide spin labels, in *Spin Labeling in Pharmacology,* Holtzman, J. L., Ed., Academic Press, Orlando, FL, 1984, 1.

56. **Hideg, K. and Hankovszky, H. O.,** Chemistry of spin-labelled amino acids and peptides: some new mono- and bifunctional nitroxide free radicals, in *Spin Labeling: Theory and Applications,* Vol. 8, Berliner, L. J. and Reuben, J., Eds., Plenum Press, New York, 1989, chap. 9.

57. **Hideg, K.,** Novel, potentially useful spin-label reagents, *J. Pure Appl. Chem.,* 62, 207, 1990.

58. **Keana, J. F. W.,** New aspects of nitroxide chemistry, in *Spin Labeling: Theory and Applications,* Vol. 2, Berliner, L. J. and Reuben, J., Eds., Academic Press, New York, 1979, 115.

59. **Lee, T. D. and Keana, J. F. W.,** Nitroxides derived from 3,1-dehydro-2,5-dimethyl-2H-pyrrol-1-oxide: a new series of minimum steric perturbation lipid spin labels, *J. Org. Chem.,* 43, 4226, 1978.

60. **Nazarcki, R. B. and Skowroncki, R.,** Sterically crowded five-membered heterocyclic systems. Part 3. Unexpected formation of stable flexible pyrrolidinoxyl biradicals via nitrone aldol dimers: a spectroscopic and mechanistic study, *J. Chem. Soc., Perkin Trans. 1,* 1603, 1989.

61. **Meyers, A. I., Temple, D. L., Nolen, R. L., and Mihelich, E. D.,** Oxazolidines. IX. Synthesis of homologated acetic acids and esters, *J. Org. Chem.,* 39, 2778, 1974.

62. **Hideg, K. and Lex, L.,** Synthesis of various new nitroxide free radical fatty acids, *J. Chem. Soc., Perkin. Trans. 1,* 1431, 1986.

63. **Hideg, K. and Lex, L.,** Synthesis of new 2-mono- and 2,5-difunctionalized pyrrolin-1-oxyl spin labels, *J. Chem. Soc., Perkin Trans. 1,* 1117, 1987.

64. **Lee, T. D., Birrel, G. B., and Keana, J. F. W.,** A new series of minimum steric perturbation nitroxide spin labels, *J. Am. Chem. Soc.,* 100, 1618, 1978.

65. **Lee, T. D., Birrel, G. B., Bjorman, P., and Keana, J. F. W.**, ESR studies of azethoxyl nitroxides incorporated into thiourea crystals, membrane model systems and chromatophores from *Rhodopseudomones sphaeroides* Ga., *Biochim. Biophys. Acta*, 550, 369, 1979.

66. **Keana, J. F. W., Cuomo, J., Lex, L., and Seyedrezai, S. E.**, Azethoxyl nitroxide spin-labeled crown ethers and cryptands with N–O group positioned near the cavity, *J. Org. Chem.*, 18, 2617, 1983.

67. **Keana, J. F. W., Tamura, T., McMillen, D. A., and Jost, P. C.**, Synthesis and characterization of a novel cholesterol nitroxide spin label. Application to the molecular organization of human high density lipoprotein, *J. Am. Chem. Soc.*, 103, 1901, 1981.

68. **Keana, J. F. W., Heo, G. S., and Ganghan, G.**, Stereospecific synthesis of difunctionalized 2,5-disubstituted *cis*-2,5-dimethylpyrrolidine (azethoxyl) nitroxides by oxidative cleavage of protected 8-azabicyclo[3,2,1]octane precursors, *J. Org. Chem.*, 50, 2346, 1985.

69. **Volodarskii, L. B., Grigor'ev, I. A., and Kutikova, G. A.**, Reaction of 1-hydroxy-2,2,4,5,5-pentamethyl-Δ³-imidazoline with aldehydes, bromine, amyl nitrite and nitrosobenzene in the presence of bases, *Zh. Org. Khim.*, 9, 1974, 1973; *Chem. Abstr.*, 79, 146457q, 1973.

70. **Schlude, H.**, Oxidation of hydroxylamines to nitroxyl radicals with Fremy's salt, *Tetrahedron*, 29, 4007, 1973.

71. **Kutikova, G. A. and Volodarskii, L. B.**, Reaction of stable hydroxyimino radicals of 3-imidazoline-3-oxide with methylmagnesium iodide, *Zh. Org. Khim.*, 6, 1505, 1970; *Chem. Abstr.*, 73, 87851g, 1970.

72. **Martin, V. V., Kobrin, V. S., and Volodarskii, L. B.**, Synthesis and properties of N-alkylhydroxylamino oximes. Stable nitroxyl radicals with α-ketoxime group, *Izv. Sib. Otd. Akad. Nauk S.S.S.R., Ser. Khim. Nauk*, 3, 153, 1977; *Chem. Abstr.*, 87, 117517m, 1977.

73. **Volodarskii, L. B., Martin, V. V., and Kobrin, V. S.**, Formation of oximes of N-alkylhydroxylamino ketones during the reaction of 1-hydroxy-3-imidazoline-3-oxides with Grignard reagent, *Zh. Org. Khim.*, 12, 2267, 1976; *Chem. Abstr.*, 86, 72520t, 1977.

74. **Martin, V. V., Volodarskii, L. B., and Vishnivetskaya, L. A.**, Interaction of sterically hindered imidazoline oxides which are precursors of stable nitroxyl radicals with organometallic reagents, *Izv. Sib. Otd. Akad. Nauk S.S.S.R., Ser. Khim. Nauk*, 4, 94, 1981; *Chem. Abstr.*, 96, 6648, 1982.

75. **Ivanovskaya, L. Yu., Derendyaev, B. G., Martin, V. V., and Kobrin, V. S.**, Mass spectra of imidazoline derivatives. I. sterically hindered 3-imidazoline-3-oxides, *Izv. Sib. Otd. Akad. Nauk S.S.S.R., Ser. Khim. Nauk*, 5, 136, 1977; *Chem. Abstr.*, 21561r, 1978.

76. **Martin, V. V.**, Synthesis and Reactions of Sterically Hindered 3-Imidazoline 3-Oxides, Ph.D. dissertation, Institution of Organic Chemistry, Novosibirsk, Russia, 1984.

77. **Wieland, H. and Roth, K.**, Weitere unter Suchungen uber Derivate des Vierwertigen Stickstoffs, *Chem. Ber.*, 53, 210, 1920.

78. **Martinie-Hanbrouck, J. and Rassat, A.**, Nitroxides. LX. Isolement et autodecomposition du N-methyl N-tri-*t*-butyl-2,4,6-phenyl nitroxyde, *Tetrahedron*, 30, 433, 1974.

79. **Berti, C.**, A new route to symmetric diarylnitroxides, *Synthesis*, 793, 1983.

80. **Calder, A. and Forrester, A. R.**, Nitroxide radicals. Part VI. Stability of meta- and para-alkyl substituted phenyl-*t*-butylnitroxides, *J. Chem. Soc. C.*, 1459, 1969.

81. **Forrester, A. R. and Hepburn, S. P.**, Nitroxide radicals. Part XV. *p*-Methoxy- and *p*-phenoxyphenyl *t*-butyl nitroxides, *J. Chem. Soc., Perkin Trans. 1*, 2208, 1974.

82. **Forrester, A. R. and Hepburn, S. P.**, Nitroxide radicals. Part VIII. Stability of *ortho*-alkyl substituted phenyl *t*-butyl nitroxides, *J. Chem. Soc. C.*, 1277, 1970.

83. **Keana, J. F. W., Prabhu, V. S., Ohmia, S., and Klopfenstein, C. E.**, 2-(9,10-Dimethoxyantracenyl)-*tert*-butyl-nitroxide. An ESR spectroscopic indicator for singlet oxygen, *J. Am. Chem. Soc.*, 107, 5020, 1985.

84. **Reznikov, V. A., Vishnivetskaya, L. A., and Volodarskii, L. B.**, Reactions of the heterocyclic enamides of imidazolidine nitroxyl radicals with electrophilic and nucleophilic reagents, *Khim. Geterotsikl. Soedin.*, p. 620, 1988; *Chem. Abstr.*, 110, 135142h, 1989.

85. **Kedik, S. A., Rozantsev, E. G., and Usvyatsov, A. A.**, Oxidizing cleavage of a triacetonamine heterocycle in an aqueous solution of hydrogen peroxide, *Dokl. Akad. Nauk S.S.S.R.*, 257, 1382, 1981; *Chem. Abstr.*, 95, 96975g, 1981.

86. **Kedik, S. A. and Usvyatsov, A. A.**, Oxidative splitting of a triacetonamine heterocycle by potassium permanganate in aqueous medium, *Izv. Akad. Nauk S.S.S.R., Ser. Khim.*, p. 1932, 1981; *Chem. Abstr.*, 95, 219742p, 1981.

87. **Keana, J. F. W. and Pou, S.**, Synthesis and properties of some nitroxide α-carboxylate salts, *J. Org. Chem.*, 54, 2417, 1989.

88. **Hussain, S. A., Jenkins, T. C., Perkins, M. J., and King, T. J.**, X-ray crystallographic studies of two acyl tert-butyl nitroxides, *J. Chem. Soc. Pakistan*, 8, 159, 1986; *Chem. Abstr.*, 105, 82027w, 1986.

89. **Perkins, M. J., Berti, C., Brooks, D. J., Grierson, L., Grime, J. A.-M., Jenkins, T. C., and Smith, S. L.**, Acyl nitroxides: reactions and reactivity, *J. Pure Appl. Chem.*, 62, 195, 1990.

90. **Dmitriev, P. I.**, Stable *N*-benzoyl- and *N*-silyl-nitroxides, in *Int. Conf. Nitroxide Radicals*, (Abstr.), Novosibirsk, Russia 1989, 4P.

91. **Aurich, H. G.**, The chemistry of vinyl nitroxides, *J. Pure Appl. Chem.*, 62, 183, 1990.

92. **Luckhurst, G. R. and Sundholm, F.**, The synthesis and electron resonance spectrum of the nitroxide, 2,2,3,3-tetramethylaziridine-1-oxyl, *Tetrahedron Lett.*, 675, 1971.

93. **Keana, J. F. W., Dinerstain, R. J., and Dolata, D. P.**, On the recently reported synthesis of the nitroxide, 2,2,3,3-tetramethylaziridine-1-oxyl, *Tetrahedron Lett.*, 119, 1972.

94. **Espie, J.-C. and Rassat, A.**, Nitroxides. XLV. Radicaux libres azetidiniques (note preliminaire), *Bull. Soc. Chim. France*, 4385, 1971.

95. **Espie, J.-C., Ramasseul, R., and Rassat, A.**, Nitroxides. LXXXV. Nitroxides azetidiniques, *Tetrahedron Lett.*, 795, 1978.

96. **Zhdanov, R. I.**, *Paramagnetue Modeli Biologieckij Aktivnukh Soedinenii (Paramagnetic Models of Biologically Active Compounds)*, Nauka, Moskow, 1981.

97. **Alcock, N. W., Goldin, B. T., Ioannou, P. V., and Sawyor, J. F.**, Preparation, crystal structure and reactions of a new spin-labelling reagent, *cis*-3,5-dibromo-1-oxo-2,2,6,6-tetramethylpiperidine-1-yloxy, *Tetrahedron*, 33, 2969, 1977.

98. **Krinitokaya, L. A.**, Synthesis and application of 2,2,5,5-tetramethylpyrrolidine-3-carboxylic acid or amide 1-oxyl, U.S.S.R. Patent, 688497; *Chem. Abstr.*, 92, 41754s, 1980.

99. **Hideg, K., Hankovczky, O. H., Lex, L., and Kulczar, G.**, Nitroxyls. VI. Synthesis and reactions of 3-hydroxymethyl-2,2,5,5-tetramethyl-2,5-dihydropyrrole-1-oxyl and formyl derivatives, *Synthesis*, 911, 1986.

100. **Hankovczky, H. O., Hidog, K., Sar, P. C., Lovac, M. J., and Jorkovich, G.**, Synthesis and dehydrobromination of α-bromo aldehyde and ketone nitroxyl radical spin labels, *Synthesis*, 59, 1990.

101. **Hankovszky, H. O., Hideg, K., and Lex, L.**, Nitroxyls. VIII. Synthesis of nitroxylphosphinimines; a convenient route to amine, isothiocyanate, aminocarbonylaziridine and carbodiimide nitroxyls, *Synthesis*, 147, 1981.

102. **Hideg, K., Lex, L., Hankovczky, H. O., and Tigyi, J.**, Nitroxyls. III. Synthesis of spin-labelled amino acids and their reactive derivatives, *Synthesis*, 914, 1978.

103. **Stone, T. J., Buckman, T., Nordic, R. L., and McConnell, H. M.**, Spin labeled biomolecules, *Proc. Natl. Acad. Sci. U.S.A.*, 54, 1010, 1965.

104. **Shapiro, A. B., Pavlikov, V. V., and Rozantsev, E. G.**, Stable radicals in diene synthesis, *Dokl. Akad. Nauk S.S.S.R.*, 232, 398, 1977; *Chem. Abstr.*, 86, 171293c, 1977.

105. **Shapiro, A. B., Skripnichenko, L. N., Chumakov, V. M., Pavlikov, V. V., and Rozantsev, E. G.**, Interaction of paramagnetic Mannich-bases with cysteine and biopolymer molecules, *Bioorg. Khim.*, 3, 707, 1977; *Chem. Abstr.*, 87, 34765q, 1977.

106. **Rozantsev, E. G. and Krinitskaya, L. A.**, Free iminoxyl radicals in the hydrogenated pyrrole series, *Tetrahedron*, 21, 491, 1965.

107. **Keana, J. F. W., Hideg, K., Birrell, G. B., Hankovszky, H. O., Ferguson, G., and Parvoz, M.**, New mono- and difunctionalized 2,2,5,5-tetramethylpyrrolidine- and Δ³-pyrroline-1-oxyl nitroxide spin labels, *Can. J. Chem.*, 60, 1439, 1982.

108. **Hankovszky, H. O., Hideg, K., Lex, L., Kulcsar, G., and Halasz, H. A.**, Methods for preparation of heterobifunctional nitroxides: α,β-unsaturated ketones, β-ketoesters, cyano-nitro-derivatives, *Can. J. Chem.*, 60, 1432, 1982.

109. **Wenzel, H. R., Becker, G., and von Goldammer, E.**, Synthese eines Bifunktionellen spin-labels zur quervernetzung von Proteinen, *Chem. Ber.*, 111, 2453, 1978.

110. **Raccat, A. and Rey, P.**, Nitroxides. XXIII. Preparation diaminoacides radicalaires et de lear sels complexes, *Bull. Soc. Chim. Fr.*, 815, 1967.

111. **Hideg, K., Hankovszky, H. O., and Halasz, H. A.**, Conjugate addition with organometallic and nitration of nitroxide (aminoxyl) free radicals. Synthesis of potentially useful cross-linking spin label reagents, *J. Chem. Soc., Perkin Trans.*, 1 2905, 1988.

112. **Hankovszky, H. O., Hideg, K., and Jerkovich, G.**, Synthesis of 3-substituted 2,5-dihydro-2,2,5,5-tetramethyl-1*H*-pyrrol-1-yloxyl radicals, useful for spin-labelling of biomolecules, *Synthesis*, 526, 1989.

113. **Crouch, R. K., Ebreg, T. G., and Govindjee, R.**, A bacteriorhodopsin analogue containing the retinal nitroxide free radical, *J. Am. Chem. Soc.*, 103, 7364, 1981.

114. **Grigor'ev, I. A., Shchukin, G. I., and Volodarskii, L. B.**, ESR, ^{1}H and ^{13}C NMR studies of nitroxide transformations in strong acids, *Izv. Akad. Nauk S.S.S.R., Ser. Khim.*, p. 2277, 1986; *Chem. Abstr.*, 107, 197363c, 1987.

115. **Varand, V. L., Grigor'ev, I. A., Vacil'eva, L. I., and Volodarskii, L. B.**, Polarographic reduction of nitroxyl radicals of imidazolidine and imidazoline derivatives, *Izv. Sib. Otd. Akad. Nauk S.S.S.R., Ser. Khim. Nauk*, (6), 99, 1982; *Chem. Abstr.*, 98, 88611g, 1983.

116. **Dikanov, S. A., Grigor'ev, I. A., Volodarskii, L. B., and Tsvetkov, Yu. D.**, Oxidative properties of nitroxyl radicals in reactions between them and sterically hindered hydroxylamines, *Zh. Fis. Khim.*, 56, 2762, 1982; *Chem. Abstr.*, 98, 125151e, 1983.

117. **Belkin, S., Mehlhorn, R. J., Hideg, K., Hakovczky, H. O., and Packer, L.**, Reduction and destruction rates of nitroxide spin probes, *Arch. Biochem. Biophys.*, 256, 232, 1987.

118. **Hankovczky, H. O., Hideg, K., and Lex, L.**, Nitroxyls. VII. Synthesis and reactions of highly reactive 1-oxyl-2,2,5,5-tetramethyl-2,5-dihydro-3-ylmethyl sulfonates, *Synthesis*, 914, 1980.

119. **Keana, J. F. W., Lee, T. D., and Bernard, E. M.**, Side-chain substituted 2,2,5,5-tetramethylpyrrolidine-*N*-oxyl (proxyl) nitroxides. A new series of lipid spin labels showing improved properties for the study of biological membranes, *J. Am. Chem. Soc.*, 98, 3052, 1976.

120. **Birrell, G. B., Lee, T. D., Griffith, O. H., and Keana, J. F. W.**, Synthesis and properties of chlorophyll-derived nitroxide spin labels, *Bioorg. Chem.*, 7, 409, 1978.

121. **Keana, J. F. W., Boyd, S. A., McMillen, D. A., Bernard, E. M., and Jact, P. C.**, Synthesis of charged amphipathic nitroxide lipid spin labels and an example of their application in membrane studies, *Chem. Phys. Lipids*, 31, 339, 1982.

122. **Roman, R. B. and Keana, J. F. W.**, Nitroxide spin labeled analogs of the non-ionic detergent Triton X-100, *Chem. Phys. Lipids*, 31, 161, 1982.

123. **Keana, J. F. W., Seyedrezai, S. E., and Gaughan, G.**, Difunctionalized *trans*-2,5-disubstituted pyrrolidine (azethoxyl) nitroxide spin labels, *J. Org. Chem.*, 48, 2644, 1983.

124. **Tse-Tang, M. W., Gaffney, B. J., and Kelly, R. E.**, Synthesis of bifunctional spin label molecules and their orientations in membranes, *Heterocycles*, 15, 965, 1981.

125. **Reznikov, V. A. and Volodarskii, L. B.**, Interaction of 2-substituted 5,5-dimethyl-4-oxo-1-pyrroline-1-oxides with nucleophilic reagents and synthesis of the pyrrolidinone-derived nitroxides, *Izv. Akad. Nauk S.S.S.R., Ser. Khim.*, 390, 1990.

126. **Reznikov, V. A. and Volodarskii, L. B.**, Synthesis of *N*-yloxy 5,5-disubstituted pyrrolidine-3-ones, U.S.S.R. Patent 1244145, 1986.

127. **Reznikov, V. A., Urzhuntseva, I. A., and Volodarskii, L. B.**, Synthesis of bicyclic β-oxonitrones, derivatives of 6-azabicyclo[3,2,1]octen and their interaction with nucleophilic reagents, *Izv. Akad. Nauk S.S.S.R., Ser. Khim.*, 682, 1991.

128. **Reznikov, V. A., Urzhuntseva, I. A., and Volodarsky, L. B.**, Recyclization of imidazolidine enaminoketones: a new route to nitroxide derivatives of bicyclo[3,2,1]azooctane, in *Int. Conf. Nitroxide Radicals* (Abstr.), Novosibirsk, Russia, 1988, 13P.

129. **Reznikov, V. A. and Volodarskii, L. B.**, Method of producing 2-substituted 5,5-dimethyl-4-oxopyrroline-1-oxides, U.S.S.R. Patent 1356400; *Chem. Abstr.*, 110, 8037c, 1989.

130. **Reznikov, V. A. and Volodarskii, L. B.**, Recyclization of enaminoketones of imidazolidine to 1-pyrroline-4-on-1-oxides, *Khim. Geterotsikl. Soedin.*, 921, 1990.

131. **Reznikov, V. A., Vishnivetskaya, L. A., and Volodarskii, L. B.**, Interaction of 2-substituted 5,5-dimethyl-4-oxo-1-pyrroline-1-oxides with electrophilic reagents, *Izv. Akad. Nauk. S.S.S.R., Ser. Khim.*, 395, 1990.

132. **Balaban, A. T., Negoita, N., and Baican, R.**, A new stable spiropyranic aminyloxide (nitroxide), *Org. Magn. Res.*, 9, 553, 1977.

133. **Berti, C., Colonna, M., Greci, L., and Marchetti, L.**, Stable nitroxide radicals from phenilisatogen and arylimino-derivatives with organo-metallic compounds, *Tetrahedron*, 31, 1745, 1975.

134. **Greci, L.**, Unexpected bromination of an indolinonic nitroxide radical, *J. Chem. Res. Miniprint*, 204, 1979.

135. **Berti, C., Greci, L., and Poloni, M.**, Stereoselective reduction of indoline nitroxide radicals, *J. Chem. Soc., Perkin Trans. 2*, 710, 1980.

136. **Dopp, D., Greci, L., and Nour-el-Din, A. M.**, Reaction of 5,7-di-*tert*-butyl-3,3-dimethyl-3*H*-indole 1-oxide with Grignard reagents. A new stable aminyl oxide (nitroxide), *Chem. Ber.*, 116, 2049, 1983.

137. **Griffiths, P. G., Rizzardo, E., and Solomon, D. H.**, Quantitative studies on free radical reactions with the scavenger 1,2,3,3-tetramethylisoindolinyl-2-oxy, *Tetrahedron Lett.*, 23, 1309, 1982.

138. **Benassi, R., Taddei, F., Greci, L., Marchetti, L., Andreetti, G. B., Bocelli, G., and Sgarabotto, P.**, Crystal and molecular structure of the *N*-oxyl radicals 1,2-dihydro-3-oxo-2,2-diphenyl-3*H*indole-1-oxyl and 1,2-dihydro-2,2-diphenylquinoline-1-oxyl. Attempted calculation of hyperfine coupling constants by the INDO-SCF-MO method, *J. Chem. Soc., Perkin Trans. 2*, 786, 1980.

139. **Ramasseul, R. and Rassat, A.**, Nitroxides. XXXIII. Radicaux: nitroxides pyrroliques encombres. Un pyrryloxyle stable, *Bull. Soc. Chim. Fr.*, 4330, 1970.

140. **Giroud, A. M. and Rassat, A.**, Synthesis of mononitroxide and dinitroxide radicals derived from isoindoline, *Bull. Soc. Chim. Fr.*, 48, 1979.

141. **Sholle, V. D., Golubev, V. A., and Rozantsev, E. G.**, Reduction products of isoindoline nitroxyl radicals, *Izv. Akad. Nauk S.S.S.R., Ser. Khim.*, 1204, 1972; *Chem. Abstr.*, 77, 114169g, 1972.

142. **Berti, C. and Greci, L.**, Nucleophilic substitutions on 1,2-dihydro-2,2-disubstituted-3-oxo-3*H*-indole-1-oxyl radicals. Direct acyloxylation and methoxylation, *J. Org. Chem.*, 46, 3060, 1981.

143. **Berti, C., Colonna, M., Greci, L., and Marshetti, L.**, Homolytic substitutions in indoline nitroxide radicals. I. Reaction with aroyloxy radicals, *Tetrahedron*, 33, 3149, 1977.

144. **Kaliska, V., Toma, S., and Lesho, J.**, Synthesis and mass spectra of piperidine and pirazine *N*-oxyl radicals, *J. Coll. Czech. Commun.*, 52, 2266, 1987.

145. **Murray, R. W. and Singh, M.**, A high yield one step synthesis of hydroxylamines, *Synth. Commun.*, 19, 3509, 1989.

146. **Cseko, J., Hankovzky, H. O., and Hideg, K.**, Synthesis of novel highly reactive 1-oxyl-2,2,6,6-tetramethyl-1,2,5,-6,-tetrahydropyridine derivatives, *Can. J. Chem.*, 63, 940, 1985.

147. **Dvolaitzki, M., Taupin, C., and Poldy, F.**, Nitroxides piperidiniques — synthese de nouvelles "sondes paramagnetiques", *Tetrahedron Lett.*, 1469, 1975.

148. **Rosen, G. M. and Rauckman, E. J.**, A new route to the synthesis of nitroxide carboxylic acids, *Org. Prep. Proced. Int.*, 10, 17, 1978.

149. **Pavlikov, V. V., Rozantsev, E. G., Shapiro, A. B., and Sholle, V. D.**, Halogen-containing acetylenic nitroxyls, *Izv. Akad. Nauk S.S.S.R., Ser. Khim.*, 197, 1980; *Chem. Abstr.*, 93, 7967z, 1980.

150. **Fioshin, M. Ya., Avrutskaya, I. A., Bogdanova, N. P., Kedik, S. A., and Surov, I. I.**, New electrochemical method for the preparation of 1-oxyl-2,2,6,6-tetramethyl-4-aminopiperidine, *Izv. Akad. Nauk S.S.S.R., Ser. Khim.*, 1691, 1983; *Chem. Abstr.*, 99, 175552c, 1983.

151. **Schlude, H.**, A new reagent for the spin labelling of aldehydes and ketones, *Tetrahedron Lett.*, 2179, 1976.

152. **Shapiro, A. B. and Dmitriev, P. I.**, Organometallic nitroxyl radicals of piperidine, *Dokl. Akad. Nauk S.S.S.R.*, 257, 898, 1981; *Chem. Abstr.*, 95, 115692f, 1981.

153. **Myshkina, L. A., Rozynov, B. V., and Rozantsev, E. G.**, New stable paramagnets in a series of nitropiperidines, *Izv. Akad. Nauk S.S.S.R., Ser. Khim.*, 1416, 1980; *Chem. Abstr.*, 93, 186117s, 1980.

154. **Rauckman, E. J., Rosen, G. M., and Abou-Donia, M. B.**, The use of trimethyl sulfonium iodide in the synthesis of biologically active nitroxides, *Org. Prep. Proceed. Int.*, 8, 159, 1976.

155. **Rassat, A. and Rey, P.**, Nitroxides. XXIII. Preparation d'aminoacides radicalaires et de leurs sels complexes, *Bull. Soc. Chim. Fr.*, 815, 1967.

156. **Wong, L. T. L., Schwenk, R., and Hsia, J. C.**, New synthesis of nitroxyl radicals of the piperidine and tetrahydropyridine series, *Can. J. Chem.*, 54, 3381, 1974.

157. **Rauckman, E. J., Rosen, G. M., and Abou-Donia, M. B.**, Synthesis of a useful spin labeled probe, 1-oxyl-4-carboxyl-2,2,6,6-tetramethylpiperidine, *J. Org. Chem.*, 41, 564, 1976.

158. **Pavlikov, V. V., Murav'ev, V. V., Shapiro, A. B., Taits, S. Z., and Rozantsev, E. G.**, Ethynylation of 4-oxo-2,2,6,6-tetramethylpiperidin-1-oxyl, *Izv. Akad. Nauk S.S.S.R., Ser. Khim.*, 1200, 1980; *Chem. Abstr.*, 93, 186107p, 1980.

159. **Shapiro, A. B., Skripnichenko, L. N., Pavlikov, V. V., and Rozantsev, E. G.**, Synthesis of nitroxyl radicals based on 4-ethynyl-4-hydroxy-2,2,6,6-tetramethylpiperidine, *Izv. Akad. Nauk S.S.S.R., Ser. Khim.*, 151, 1979; *Chem. Abstr.*, 90, 203838k, 1979.

160. **Whitlock, H. W. and Adams, S. P.**, 2,2,6,6-Tetramethylpiperidine-*N*-oxyl-4-carboxylic anhydride, *J. Org. Chem.*, 44, 3433, 1979.

161. **Gala, D., Schultz, R., and Kreilick, R.**, Synthesis and properties of some spin labeled long-chain aliphatic acids, *Can. J. Chem.*, 60, 710, 1982.

162. **Rauckman, E. J. and Rosen, G. M.**, Synthesis of spin labeled probes: esterification and reduction, *Synth. Commun.*, 6, 325, 1976.

163. **Smith, P. H., Eaton, G. R., and Eaton, S. S.**, Metal nitroxyl interactions. 38. Effect of the nickel-nitroxyl linkage on the electron-electron spin-spin interactions in fluid and frozen solution, *J. Am. Chem. Soc.*, 106, 1986, 1984.

164. **Pavlikov, V. V., Shapiro, A. B., and Rozantsev, E. G.**, New polyyne derivatives of 2,2,6,6-tetramethylpiperidine-1-oxyl, *Izv. Akad. Nauk S.S.S.R., Ser. Khim.*, 128, 1980; *Chem. Abstr.*, 92, 214424q, 1980.

165. **Rosen, G. M.**, Use of sodium cyanoborohydride in the preparation of biologically active nitroxides, *J. Med. Chem.*, 17, 358, 1974.

166. **Sinha, B. and Chignell, C. F.**, Synthesis and biological activity of spin-labeled analogues of biotin, hexamethonium, decamethonium, dichloroisoproterenol and propranolol, *J. Med. Chem.*, 18, 669, 1975.

167. **Feix, J. B. and Butterfield, D. A.**, Selective spin labeling of sialic acid residues of glycoproteins and glycolipids in erythrocyte membranes. A novel method to study cell surface interaction, *FEBS Lett.*, 115, 185, 1980.

168. **Korshak, Yu. V., Medvedeva, T. V., Ovchinnikov, A. A., and Spector, V. N.**, Organic polymer ferromagnet, *Nature*, 326, 370, 1987.

169. **Cao, Y., Wang, P., Hu, Z., Li, S., Zhang, L., and Zhao, J.**, Magnetic characterization of organic ferromagnet/poly-BIPO and its analog, *Solid State Commun.*, 68, 817, 1988.

170. **Wiley, D. W., Calabreze, J. C., and Miller, J. S.**, Structural and magnetic characterization of α- and β-2,4-hexadiyne-1,6-diyl *bis*(2,2,5,5-tetramethyl-1-oxyl-3-pyrroline-3-carboxylate) and its termal degradation product, *J. Chem. Soc., Chem. Commun.*, 1523, 1989.

171. **Miller, J. S., Glatzhofer, D. T., Laversanne, R., Brill, T. B., Timken, M. D., O'Connor, C. J., Zhang, J. H., Calabrese, J. C., Epstein, A. J., Chittipeddi, S., and Vaca, P.**, Structural and magnetic characterization of α- and β-4,4'-(butadiene-1,4-diyl)*bis*(2,2,6,6-tetramethyl-4-hydroxypiperidine-1-oxyl) and characterization of its thermal degradation product, *Chem. Mat.*, 2, 60, 1990.

172. **Krinitskaya, L. A. and Volodarskii, L. B.**, Synthesis of 3-functionally-substituted 4-oximinopiperidine nitroxyls, *Izv. Akad. Nauk S.S.S.R., Ser. Khim.*, 1618, 1984; *Chem. Abstr.*, 102, 6132r, 1985.

173. **Sen', V. D.**, Synthesis of 3,4-diamino-2,2,6,6-tetramethylpiperidine-1-oxyl, *Izv. Akad. Nauk S.S.S.R., Ser. Khim.*, 2094, 1989.

174. **Mori, H., Ohara, M., and Kwan, T.,** Carboxylation of the nitroxide radical of 2,2,6,6-tetramethylpiperidone-1-oxyl with carbon dioxide and potassium phenoxide, and the physical properties of the products, *Chem. Pharm. Bull.*, 28, 3178, 1980.

175. **Briere, R., Espie, J.-C., Ramasseul, R., Rassat, A., and Rey, P.,** Nitroxides. 91. β-Cetoenolates nitroxides, *Tetrahedron Lett.*, 941, 1979.

176. **Volodarsky, L. B.,** Advances in the chemistry of stable nitroxides, *J. Pure Appl. Chem.*, 62, 177, 1990.

177. **Sholle, V. D., Kagan, E. Sh., Mikhailov, V. I., Rozantsev, E. G., Frangopol, P. T., Frangopol, M., Pop, V. I., and Bonga, Gh.,** A new spin label for SH groups in proteins: the synthesis and some applications in labeling of albumin and erythrocyte membranes, *Rev. Roum. Biochim.*, 17, 291, 1980.

178. **Rozantsev, E. G., Dagonneau, M., Kagan, E. S., Mikhailov, V. I., and Sholle, V. D.,** Synthesis of 3-substituted derivatives of 2,2,6,6-tetramethylpiperidine: potential new spin labels, *J. Chem. Res. Miniprint*, 2901, 1979.

179. **Sen', V. D.,** Synthesis of nitroxyl and oxammonium derivatives of anticancer agents, in *Int. Conf. Nitroxide Radicals* (Abstr.), Novosibirsk, Russia, 1989, 8P.

180. **Volodarskii, L. B., Grigor'eva, L. N., Dulepova, N. V., and Tikhonov, A. Ya.,** Preparation of α-hydroxylaminooximes of triacetonamine derivatives and their reactions with carbonyl compounds, *Izv. Akad. Nauk S.S.S.R., Ser. Khim.*, 406, 1988; *Chem. Abstr.*, 110, 95160d, 1989.

181. **Cook, A. G., Ed.,** *Enamines: Synthesis, Structure and Reactions,* Marcel Dekker, New York, 1969.

182. **Gagnaire, G., Jennet, A., and Pierre, J.-L.,** Regulation by potassium ions of spin exchange and dipolar splitting in a biradical. A simple allosteric system, *Tetrahedron Lett.*, 30, 6507, 1989.

183. **Bordeaux, D., Gagnaire, G., Lajzerowich, J., and Commandeur, G.,** The synthesis of nitroxide free radicals of 2,2,6,6-tetramethyl-1,2,3,6-tetrahydropyridine. The structure of 3,4-epoxy-2,2,6,6-tetramethylpiperidine oxyl, $C_9H_{16}NO_2$, *Acta Crystallog.*, C 39, 1656, 1983.

184. **Kirichenko, L. N. and Medzhidov, A. A.,** Complexes of metals with paramagnetic xanthate ligands, *Izv. Akad. Nauk S.S.S.R., Ser. Khim.*, 2849, 1969; *Chem. Abstr.*, 72, 96151k, 1970.

185. **Solozhenkin, P. M., Shvengler, F. A., Kopitsya, N. I., and Semikopnyi, A. I.,** Complexes of metals with a paramagnetic dithiocarbamate ligand, *Dokl. Akad. Nauk S.S.S.R.*, 262, 904, 1982; *Chem. Abstr.*, 96, 209775h, 1982.

186. **Gaffney, B. J.,** The chemistry of spin labels, in *Spin Labeling: Theory and Applications,* Vol. 1, Berliner, L. J., Ed., Academic Press, New York, 1976, chap. 5.

187. **Griffith, H. O. and McConnell, H. M.,** A nitroxide-maleimide spin label, *Proc. Natl. Acad. Sci. U.S.A.*, 55, 8, 1966.

188. **Azzi, A., Bragadin, M. A., Tamburro, A. M., and Santato, M.,** Site-directed spin labeling of mithochondrial membrane, *J. Biol. Chem.*, 248, 2520, 1973.

189. **Kornberg, R. D. and McConnell, H. M.,** Inside-outside transitions of phospholipids in vesicle membranes, *Biochemistry,* 10, 1111, 1971.

190. **Lepock, J. R., Morse, P. D., Mehlhorn, R. J., Hammerstedt, R. H., Shiper, W., and Keith, A. D.,** Spin labels for cell surface, *FEBS Lett.*, 60, 185, 1975.

191. **Annaev, B., Ivanov, V. P., Roichman, L. M., and Rozantsev, E. G.,** Spin label for heme-containing compounds, *Izv. Akad. Nauk S.S.S.R., Ser. Khim.*, 2814, 1971; *Chem. Abstr.*, 77, 71621d, 1972.

192. **Vlictotra, E. J., Nolte, R. J. M., Zwikker, J. W., Drenth, W., and Meijer, E. W.,** Synthesis and magnetic properties of a rigid high spin density polymer with piperidine-N-oxyl pending groups, *Macromolecules,* 23, 946, 1990.

193. **McConnell, H. M., Deal, W., and Ogata, R. T.,** Spin labelled hemoglobin derivatives in solution, polycrystalline suspensions and single crystals, *Biochemistry,* 8, 2580, 1969.

194. **Berti, C., Colonna, M., Greci, L., and Marchetti, L.,** Stable nitroxide radicals from acridine N-oxides with Grignard reagents, *Gazz. Chim. Ital.*, 108, 659, 1978.

195. **Colonna, M., Greci, L., and Poloni, M.,** Stable nitroxide radicals. Reaction between 2-cyanobenzoquinone and 4-cyanobenzoquinoline *N*-oxides and the Grignard reagent, *J. Heterocycl. Chem.*, 17, 1437, 1980.

196. **Lin, J. S. and Olcott, H. S.,** Ethoxyquin nitroxide, *J. Agric. Food Chem.*, 23, 798, 1975.

197. **Berti, C., Colonna, M., and Greci, L.,** Stable nitroxide radicals from 2-substituted quinoline-*N*-oxides with organometallic compounds, *Tetrahedron*, 32, 2147, 1976.

198. **Colonna, M., Greci, L., and Poloni, M.,** Quinoline nitroxide radicals. IPSO attack in the reaction between 2-methoxy and 2-cyanoquinoline *N*-oxide and phenylmagnesium bromide, *J. Heterocycl. Chem.*, 17, 293, 1980.

199. **Shibaeva, R. P., Rozenberg, L. P., Shapiro, A. B., and Povarov, L. S.,** Crystal and molecular structure of 4-(spirotetrahydro-2'-furyl)-2-spirocyclohexyl-1,2,3,4-tetrahydroquinidine and a stable nitroxyl radical formed during its catalytic oxidation, *Zh. Strukt. Khim.*, 22, 140, 1981; *Chem. Abstr.*, 96, 121894k 1982.

200. **Shapiro, A. B., Ivanov, V. P., Khvostach, O. M., and Rozantsev, E. G.,** Products of 2,2,6,6-tetramethyl-4-oxopiperidine-1-oxy condensation with formaldehyde, *Izv. Akad. Nauk S.S.S.R., Ser. Khim.*, 1688, 1973; *Chem. Abstr.*, 79, 105106y, 1973.

201. **Rassat, A. and Ronzand, J.,** Nitroxydes. LXXI. Synthese de derives nitroxydes du tropane. Etude par RPE et RMN des interactions hyperfines a longue dans ces radicaux, *Tetrahedron*, 32, 239, 1976.

202. **Rassat, A. and Rey, P.,** Nitroxydes. XLVII. Nitroxydes isoquinuclidiniques, *Tetrahedron*, 28, 741, 1972.

203. **Rassat, A. and Rey, P.,** Nitroxides: photochemical synthesis of trimethylisoquinuclidine-*N*-oxyl, *Chem. Commun.*, 1161, 1971.

204. **Dupeyre, R. M., Rassat, A., and Ronzand, J.,** Nitroxides. LII. Synthesis and electron spin resonance studies of *N*,*N*'-dioxy-2,6-diazaadamantane, a symmetrical ground state triplet, *J. Am. Chem. Soc.*, 96, 6559, 1974.

205. **Nelson, J. A., Chou, S., and Spenser, T. A.,** Reaction of 4,4-dimethyloxazolidine-*N*-oxyl (doxyl) derivatives with nitrogen dioxide: a novel and efficient reconversion into the parent ketone, *Chem. Commun*, 1, 1580, 1971.

206. **Chou, S., Nelson, J. A., and Spenser, T. A.,** Oxidation and mass spectra of 4,4-dimethyloxazolidine-*N*-oxyl (doxyl) derivatives of ketones, *J. Org. Chem.*, 39, 2356, 1974.

207. **Cella, J. A., Kelley, J. A., and Kenehan, E. F.,** Oxidation of nitroxides by *m*-chloroperbenzoic acid, *Tetrahedron Lett.*, 2869, 1975.

208. **Nelson, J. A., Chou, S., and Spenser, T. A.,** Oxidative demethylation at C-4 of a steroid via nitroxide photolysis, *J. Am. Chem. Soc.*, 97, 648, 1975.

209. **Rohde, O., Van, S. P., Kester, W. R., and Griffith, O. H.,** Spin labels as molecular rules. The conformational analysis of a model system. 1,4-Didoxylcyclohexane oriented in a crystalline host, *J. Am. Chem. Soc.*, 96, 5311, 1974.

210. **Michon, P. and Rassat, A.,** Nitroxides. LXIX. 1,4-Bis (4',4'-dimethyloxazolidine-3'-oxyl) cyclohexane structure determination by electron spin resonance, *J. Am. Chem. Soc.*, 97, 696, 1975.

211. **Morat, C. and Rassat, A.,** Synthesis of oxazolidines substituted by adamantane and stable nitroxide oxazolidines derived from adamantane, *Tetrahedron Lett.*, 4561, 1979.

212. **Dvolaitzki, M. and Taupin, C.,** Synthesis and use of some spin-labeled long chain quaternary ammonium salts, *Nouv. J. Chim.*, 1, 355, 1977.

213. **Gaffney, B. J., Willingham, G. L., and Schepp, R. S.,** Synthesis and membrane interactions of spin-label bifunctional reagents, *Biochemistry*, 22, 881, 1983.

214. **Willingham, G. L. and Gaffney, B. J.,** Reactions of spin-label cross-linking reagents with red blood cell proteins, *Biochemistry*, 22, 892, 1983.

215. **Bakunova, S. M., Grigor'ev, I. A., Kirilyuk, I. A., Gatilov, Yu. V., Bagryanskaya, I. Yu., and Volodarskii, L. B.,** Synthesis of stable nitroxide radicals of the oxazolidine series with methoxy groups at α-carbons of radical centre, *Izv. Akad. Nauk. Russ., Ser. Khim.*, 966, 1992.

216. **Meyers, A. I., Mihelich, E. D., and Nolen, R. L.,** Oxazolines. X. Synthesis of α-butyrolactones, *J. Org. Chem.*, 39, 2783, 1974.

217. **Meyers, A. I., Temple, D. L., Haidukewych, D., and Mihelich, E. D.,** Oxazolines. XI. Synthesis of functionalized aromatic and aliphatic acids. A useful protecting group for carboxylic acids against Grignard and hydride reagents, *J. Org. Chem.,* 39, 2787, 1974.

218. **Noland, W. E., Sundberg, R. J., and Michaelson, M. I.,** Synthetic studies involving 1-aminocyclohexanecarbonitrile, *J. Org. Chem.,* 28, 3576, 1963.

219. **Toda, T., Morimura, S., and Murayama, K.,** Studies on stable free radicals. VII. The mechanism for cyclization reaction of α-amino nitriles with carbonyl compounds, *Bull. Chem. Soc. Jpn.,* 45, 557, 1972.

220. **Paskal, R., Taillades, J., and Commeyras, A.,** Strecker and related systems. X. Decomposition and hydration of secondary α-aminonitriles in aqueous basic media. Autocatalytic hydration process and catalysis by acetone, *Tetrahedron,* 34, 2275, 1978.

221. **Piloty, O. and Graf Schwerin, B.,** Ueber die existenz von Derivaten des vierwerthigen Stickstoffs, *Chem. Ber.,* 34, 2354, 1901.

222. **Toda, T., Morimura, S., Mori, E., Horiuchi, H., and Murayama, K.,** Studies of stable free radicals. VI. Synthesis of substituted 4-imidazolidinone-1-oxyls, *Bull. Chem. Soc., Jpn.,* 44, 3455, 1971.

223. **Murayma, K., Toda, T., and Mori, E.,** Japan Patent 7,615,001, 1976.

224. **Shchukin, G. I., Starichenko, V. F., Grigor'ev, I. A., Dikanov, S. A., Gulin, V. I., and Volodarskii, L. B.,** Formation of methoxy-nitrones and stable nitroxyl radicals with gem-dimethoxy groups at α-carbon in oxidation of aldonitrones in methanol, *Izv. Akad. Nauk S.S.S.R., Ser. Khim.,* 125, 1987; *Chem. Abstr.,* 107, 236596c.

225. **Grigor'ev, I. A., Starichenko, V. F., Kirilyuk, I. A., and Volodarskii, L. B.,** Formation of nitroxides with α-fluorine atoms in the reaction of nitrones with xenone difluoride, *Izv. Akad. Nauk S.S.S.R., Ser. Khim.,* 933, 1989; *Chem. Abstr.,* 111, 214051b, 1989.

226. **Grigor'ev, I. A., Volodarsky, L. B., Starichenko, V. F., and Kirilyuk, I. A.,** Synthesis of stable nitroxides with amino groups and fluorine atoms at α-carbon of the radical center, *Tetrahedron Lett.,* 30, 751, 1989.

227. **Reznikov, V. A. and Volodarskii, L. B.,** Interaction of β-oxo nitrones of imidazoline and pyrroline with nucleophilic reagents, *Khim. Geterotsikl. Soedin.,* 912, 1991.

228. **Osiecki, J. H. and Ullman, E. F.,** Studies of free radicals. I. α-Nitronyl nitroxides, a new class of stable radicals, *J. Am. Chem. Soc.,* 90, 1078, 1968.

229. **Boocock, D. G. B., Darcy, R., and Ullman, E. F.,** Studies of free radicals. II. Chemical properties of nitronyl nitroxides. A unique radical reaction, *J. Am. Chem. Soc.,* 90, 5945, 1968.

230. **Ullman, E. F., Call, L., and Osiecki, J. H.,** Stable free radicals. VIII. New imino amidino and carbamoyl nitroxides, *J. Org. Chem.,* 35, 3623, 1970.

231. **Hadeau, J. S. and Boocock, D. G. B.,** A system for a nitric oxide dozimeter: a free radical reagent on silica gel, *Anal. Chem.,* 49, 1672, 1977.

232. **Khechtel, J. R., Janzen, E. G., and Davie, E. R.,** Determination of chlorine dioxide in sewage effluents, *Anal. Chem.,* 50, 202, 1978.

233. **Boocock, D. G. B. and Ullman, E. F.,** Studies of stable radicals. III. A 1,3-dioxy-2-imidazolidone zwitterion and its stable nitronyl nitroxide radical anion, *J. Am. Chem. Soc.,* 90, 6873, 1968.

234. **Ullman, E. F. and Boocock, D. G. B.,** "Conjugated" nitronyl-nitroxide and imino-nitroxide biradicals, *J. Chem. Soc., Chem. Commun.,* 20, 1161, 1969.

235. **Beak, P. and Reitz, D. B.,** Dipole-stabilized carbanions: novel and useful intermediates, *Chem. Rev.,* 78, 275, 1978.

236. **Grigor'ev, I. A., Volodarsky, L. B., Starichenko, V. F., Shchukin, G. I., and Kirilyuk, I. A.,** Route to stable nitroxides with alkoxy groups at α-carbon — the derivatives of 2- and 3-imidazolines, *Tetrahedron Lett.,* 26, 5085, 1985.

237. **Grigor'ev, I. A., Kirilyuk, I. A., Starichenko, V. F., and Volodarskii, L. B.,** Oxidative alkoxylation of 4H-imidazole N-oxides as a new method of synthesis of stable 2- and 3-imidazoline nitroxyl radicals with alkoxy groups at the α-carbon atom of the radical center, *Izv. Akad. Nauk S.S.S.R., Ser. Khim.,* 1624, 1989; *Chem. Abstr.,* 112, 55708y, 1990.

238. **Grigor'ev, I. A., Starichenko, V. F., Kirilyuk, I. A., and Volodarskii, L. B.**, Synthesis of stable nitroxyl radicals with amino group at the carbon α atom to the radical center by oxidative amination of 4*H*-imidazole *N*-oxides, *Izv. Akad. Nauk. S.S.S.R., Ser. Khim.*, 661, 1989; *Chem. Abstr.*, 111, 232661e, 1989.

239. **Kobrin, V. S.**, Synthesis and Properties of 4*H*-Imidazol Derivatives, Ph.D. dissertation, Institution of Organic Chemistry, Novosibirsk, Russia, 1977.

240. **Volodarskii, L. B., Kutikova, G. A., Sagdeev, R. Z., and Molin, Yu. N.**, A route to stable nitroxide radicals of imidazoline *N*-oxide, *Tetrahedron Lett.*, 9, 1065, 1968.

241. **Volodarskii, L. B. and Kutikova, G. A.**, Preparation of stable iminoxyl radicals of 3-imidazoline-3-oxide, *Izv. Akad. Nauk S.S.S.R., Ser. Khim.*, 937, 1971; *Chem. Abstr.*, 75, 76685w, 1971.

242. **Grigor'ev, I. A. and Volodarskii, L. B.**, Intramolecular oxidation of the 4-aminoalkyl group of the nitrone grouping in stable iminoxyl radicals of 3-imidazoline 3-oxide derivatives, *Zh. Org. Khim.*, 11, 1328, 1975; *Chem. Abstr.*, 83, 206157d, 1975.

243. **Volodarskii, L. B., Martin, V. V., Shchukin, G. I., Grigor'ev, I. A., Koschevaya, E. N., and Rodionov, V. A.**, U.S.S.R. Patent 804637, 1980.

244. **Volodarskii, L. B. and Putsykin, Yu. G.**, Chemistry of α-hydroxylaminooximes. XI. Preparation and properties of acyclic α-hydroxylaminooximes, *Zh. Org. Khim.*, 3, 1686, 1967; *Chem. Abstr.*, 68, 39169r, 1967.

245. **Putsykin, Yu. G. and Volodarskii, L. B.**, Aliphatic α-hydroxylaminooximes from nitrosochlorides of olefins, *Izv. Sib. Otd. Akad. Nauk S.S.S.R., Ser. Khim. Nauk*, 4, 101, 1968; *Chem. Abstr.*, 70, 46764a, 1969.

246. **Volodarskii, L. B., Koptyug, V. A., and Lysak, A. N.**, Preparation of hydroxylamino oximes and establishment of their sin- and anti-configurations, *Zh. Org. Khim.*, 2, 114, 1966; *Chem. Abstr.*, 64, 14139h.

247. **Grigor'ev, I. A., Shchukin, G. I., and Volodarskii, L. B.**, Effect of a radical center on oxidative properties of the nitrone group in the reaction of nitroxyl radicals of 3-imidazoline 3-oxide with hydrazine, *Izv. Akad. Nauk S.S.S.R., Ser. Khim.*, 1140, 1983; *Chem. Abstr.*, 100, 6386k, 1983.

248. **Volodarskii, L. B., Grigor'ev, I. A., Kirilyuk, I. A., and Amitina, S. A.**, Nitration of 4-aryl- and 4-hetaryl-2,2,5,5-tetramethyl-3-imidazoline-3-oxydes and corresponding nitroxides, *Zh. Org. Khim.*, 21, 443, 1985; *Chem. Abstr.*, 102, 166657n, 1985.

249. **Volodarskii, L. B., Martin, V. V., and Kobrin, V. S.**, U.S.S.R. Patent 707914, 1980.

250. **Martin, V. V. and Volodarskii, L. B.**, Synthesis and some reactions of sterically-hindered 3-imidazoline 3-oxides, *Khim. Geterotsikl. Soedin.*, 103, 1979; *Chem. Abstr.*, 90, 186860s, 1979.

251. **Amitina, S. A. and Volodarskii, L. B.**, Synthesis of sterically hindered 1-hydroxy-2-acetyl-3-imidazoline-3-oxide and stable nitroxyl radicals based on them, *Izv. Akad. Nauk S.S.S.R., Ser. Khim.*, 2135, 1976; *Chem. Abstr.*, 86, 106467y, 1977.

252. **Grigor'ev, I. A., Shchukin, G. I., Dikanov, S. A., Kuznetsova, I. K., and Volodarskii, L. B.**, Formation of 2,3,6,8-tetraazabicyclo [3,2,1]octen-3 derivatives in the reaction of 2-acyl-3-imidazoline-3-oxides with hydrazine and their oxidation to the corresponding nitroxyl radicals and 1,2,4-triazine 4-oxides, *Izv. Akad. Nauk S.S.S.R., Ser. Khim.*, 1092, 1982; *Chem. Abstr.*, 109937y, 1982.

253. **Grigor'ev, I. A., Shchukin, G. I., Khramtsov, V. V., Weiner, L. M., Starichenko, V. F., and Volodarskii, L. B.**, Transformation of 3-imidazoline 3-oxide nitroxides into nitronyl nitroxides, *Izv. Akad. Nauk S.S.S.R., Ser. Khim.*, 2342, 1985.

254. **Volodarsky, L. B.**, Ed., *Imidazoline Nitroxides, Vol. 1, Synthesis and Properties*, CRC Press, Boca Raton, FL, 1988.

255. **Ota, T., Macuda, S., and Tanaka, H.**, Studies on the polymerization of acrolein. XV. Cyclodimerization of 3-methyl-3-buten-2-one oxime, *Chem. Lett.*, 111, 1981.

256. **Shulz, M., Mogel, L., and Rombach, J.**, Zur ring-kettentautomerie und Reactionen eines Bicyclischen 1-hydroxy-3-imidazolin-3-oxide. *J. Prakt. Chemie*, 328, 589, 1986.

257. **Kirilyuk, I. A., Grigor'ev, I. A., and Volodarskii, L. B.**, Synthesis of nitroxide radicals — derivatives of 5,5-dimethoxy-3-imidazoline-3-oxide-1-oxyl on the basis of 2*H*-imidazole-1,3-dioxides, *Izv. Akad. Nauk S.S.S.R., Ser. Khim.*, 2122, 1991.

258. **Kutikova, G. A. and Volodarskii, L. B.**, U.S.S.R. Patent 408948, 1973; *Chem. Abstr.*, 80, 120950h, 1974.

259. **Grigor'ev, I. A. and Volodarskii, L. B.**, Reaction of 4-bromomethyl- and 4-dibromomethyl derivatives of 2,2,5,5-tetramethyl-Δ^3-imidazoline 3-oxides with hydrazine and primary amines, *Zh. Org. Khim.*, 10, 118, 1974; *Chem. Abstr.*, 80, 108447s, 1974.

260. **Volodarsky, L. B., Martin, V. V., Grigor'ev, I. A., Shchukin, G. I., and Vishnivetskaya, L. A.**, Belgian Patent 896,094, 1983.

261. **Martin, V. V., Volodarskii, L. B., Shchukin, G. I., Vishnivetskaya, L. A., and Grigor'ev, I. A.**, The acid catalyzed reactions of α-alkylnitrones, derivatives of 3-imidazoline-3-oxides, *Izv. Akad. Nauk S.S.S.R., Ser. Khim.*, 161, 1985.

262. **Grigor'ev, I. A., Martin, V. V., Shchukin, G. I., and Volodarskii, L. B.**, Reaction of α-haloalkylnitrones — derivatives of nitroxyl radicals of 3-imidazoline-3-oxide with nucleophilic agents, *Izv. Akad. Nauk S.S.S.R., Ser. Khim.*, 2711, 1979; *Chem. Abstr.*, 92, 198318, 1980.

263. **Berezina, T. A., Martin, V. V., Volodarskii, L. B., Khramtsov, V. V., and Weiner, L. M.**, Synthesis of amidines — imidazoline nitroxide derivatives — a new series of pH-sensitive spin probes and spin labels, *Bioorg. Khim.*, 16, 262, 1990.

264. **Volodarskii, L. B., Kutikova, G. A., Kobrin, V. S., Sagdeev, R. Z., and Molin, Yu. N.**, Synthesis and spectral studies of stable iminoxyl radicals. Derivatives of 2,2,5,5-tetramethyl-3-imidazoline-1-oxyl-4-carboxylic acid and its amide, salt and copper (II) complex, *Izv. Sib. Otd. Akad. Nauk S.S.S.R., Ser. Khim. Nauk.* (3), 101, 1971; *Chem. Abstr.*, 76, 153673r, 1972.

265. **Khramtsov, V. V., Yelinova (Popova), V. I., Weiner, L. M., Berezina, T. A., Martin, V. V., and Volodarskii, L. B.**, Quantitative determination of SH-groups in low and high-molecular-weight compounds by an electron spin resonance method, *Anal. Biochem.*, 182, 58, 1989.

266. **Volodarsky, L. B., Martin, V. V., and Leluch, T. F.**, A stable nitroxide — 2,2,5,5-tetramethyl-3-imidazoline-3-oxide-1-oxyl — a reagent for spin labeling via 1,3-dipolar cycloaddition, *Tetrahedron Lett.*, 4801, 1985.

267. **Martin, V. V., Volodarskii, L. B., Voinov, M. A., Berezina, T. A., and Lelyukh, T. F.**, 1,3-Dipolar cycloaddition of 3-imidazoline-1-oxyl-3-oxide to dipolarophiles containing a carbon-carbon double bond, *Izv. Akad. Nauk S.S.S.R., Ser. Khim.*, 1875, 1988; *Chem. Abstr.*, 110, 231515c, 1989.

268. **Barachkova, I. I., Martin, V. V., Anfimov, B. N., Wasserman, A. M., and Volodarskii, L. B.**, Spin-labeled rubbers having a rigidly bound nitroxyl center, *Vysokomol. Soedin.*, 29, 354, 1987; *Chem. Abstr.*, 107, 79204d, 1987.

269. **Barashkova, I. I., Anfimov, B. N., and Wasserman, A. M.**, Adsorption of spin-labeled divinyl rubbers (SKD) on silicon dioxide, in *Int. Conf. Nitroxide Radicals* (Abstr.), Novosibirsk, Russia, 1989, 83P.

270. **Grigor'ev, I. A., Shchukin, G. I., and Volodarskii, L. B.**, Reaction of 1-hydroxy-1-dihalomethyl-Δ^3-imidazoline-3-oxides and their radicals with nucleophilic agents, *Zh. Org. Khim.*, 11, 1332, 1975; *Chem. Abstr.*, 83, 193171b, 1975.

271. **Shchukin, G. I. and Volodarskii, L. B.**, Oxidation of hydrazones of 4-oxoalkyl-3-imidazoline and 3-imidazoline-3-oxide with formation of imidazo[1,2,3]triazoles and diazocompounds with nitroxyl radical center, *Izv. Akad. Nauk S.S.S.R., Ser. Khim.*, 233, 1979; *Chem. Abstr.*, 90, 137741v, 1979.

272. **Grigor'ev, I. A., Shchukin, G. I., and Volodarskii, L. B.**, Studies of the effect of the radical center on chemical reactivity, in *Imidazoline Nitroxides, Vol. 1, Synthesis and Properties*, Volodarsky, L. B., Ed., CRC Press, Boca Raton, FL, 1988, chap. 5.

273. **Grigor'ev, I. A., Shchukin, G. I., and Volodarskii, L. B.**, Effect of radical center on the alkaline hydrolysis of 4-(dihalomethyl)-3-imidazoline-3-oxides, *Izv. Akad. Nauk S.S.S.R., Ser. Khim.*, 2787, 1983; *Chem. Abstr.*, 100, 208717q, 1981.

274. **Grigor'ev, I. A., Shchukin, G. I., and Volodarskii, L. B.**, Effect of a radical center on oxidative properties of the nitrone group in the reaction of nitroxyl radicals of 3-imidazoline-3-oxide with hydrazine, *Izv. Akad. Nauk S.S.S.R., Ser. Khim.*, 1140, 1983; *Chem. Abstr.*, 100, 6386k, 1984.

275. **Volodarsky, L. B., Grigor'ev, I. A., Grigor'eva, L. N., and Kirilyuk, I. A.,** Nitration of imidazoline-*N*-oxide nitroxides containing the aryl nitrone group, *Tetrahedron Lett.,* 25, 5809, 1984.

276. **Kirilyuk, I. A., Grigor'ev, I. A., and Volodarskii, L. B.,** Synthesis of 3-imidazoline-3-oxide nitroxides containing anilino group and their diazotation and diazocoupling, *Izv. Akad. Nauk S.S.S.R., Ser. Khim.,* 1588, 1987; *Chem. Abstr.,* 108, 167379r, 1988.

277. **Grigor'ev, I. A., Druganov, A. G., and Volodarskii, L. B.,** Nitrozation of sterically hindered 3-imidazolines. Formation of 4-nitromethylene-2,2,5,5-tetramethylimidazolidine-1-oxyl, *Izv. Sib. Otd. Akad. Nauk S.S.S.R., Ser. Khim. Nauk,* 4, 131, 1976; *Chem. Abstr.,* 85, 177316p, 1976.

278. **Grigor'ev, I. A., Shchukin, G. I., Druganov, A. G., and Volodarskii, L. B.,** Synthesis and properties of 3-imidazoline aldehyde and ketone derivatives containing nitroxyl radical center, *Izv. Sib. Otd. Akad. Nauk S.S.S.R., Ser. Khim. Nauk,* 1, 80, 1979; *Chem. Abstr.,* 91, 5159x, 1979.

279. **Skripnichenko, L. N., Schapiro, A. B., Rozantsev, E. G., and Volodarskii, L. B.,** Paramagnetic complexes of 4-trifluoroacetoacetyl-2,2,5,5-tetramethyl-3-imidazoline-1-oxyl, *Izv. Akad. Nauk S.S.S.R., Ser. Khim.,* 109, 1982; *Chem. Abstr.,* 96, 154365a, 1982.

280. **Dikanov, S. A., Shchukin, G. I., Grigor'ev, I. A., Rukin, S. I., and Volodarskii, L. B.,** Spin exchange in nitroxyl biradicals of imidazoline series with polymethylene bridges, *Izv. Akad. Nauk S.S.S.R., Ser. Khim.,* 565, 1985; *Chem. Abstr.,* 103, 70836w, 1985.

281. **Larionov, S. V., Ovcharenko, V. I., and Sagdeev, R. Z.,** Complexes of cobalt (III), nickel (II) and thallium (I) with paramagnetic xanthohenates, *Koord, Khim.,* 5, 1034, 1979; *Chem. Abstr.,* 91, 167620t, 1979.

282. **Shchukin, G. I. and Volodarskii, L. B.,** Reaction of amides of derivatives of 3-imidazoline and 3-imidazoline 3-oxide with sodium hypobromide, *Izv. Akad. Nauk S.S.S.R., Ser. Khim.,* 228, 1979; *Chem. Abstr.,* 90, 152078c, 1979.

283. **Shchukin, G. I., Grigor'ev, I. A., and Volodarskii, L. B.,** Reactions of nitriles of 3-imidazoline and 3-imidazoline 3-oxide derivatives with nucleophilic reagents, *Izv. Sib. Otd. Akad. Nauk S.S.S.R., Ser. Khim. Nauk,* 4, 81, 1984; *Chem. Abstr.,* 102, 113684u, 1985.

284. **Grigor'ev, I. A., Shchukin, G. I., and Volodarskii, L. B.,** 4-β-Dicarbonyl derivatives of stable nitroxyl radicals of 3-imidazoline, *Izv. Sib. Otd. Akad., Nauk S.S.S.R., Ser. Khim. Nauk,* 4, 135, 1977; *Chem. Abstr.,* 88, 6796c, 1978.

285. **Volodarskii, L. B. and Sevastyanova, T. K.,** Synthesis and properties of α-hydroxylamino ketones, *Zh. Org. Khim.,* 8, 1687, 1971; *Chem. Abstr.,* 75, 140425r, 1971.

286. **Sevastyanova, T. K. and Volodarskii, L. B.,** Preparation of stable iminoxyl radicals of 3-imidazolines, *Izv. Akad. Nauk S.S.S.R., Ser. Khim.,* 2339, 1972; *Chem. Abstr.,* 78, 58313f, 1973.

287. **Volodarskii, L. B. and Sevastyanova, T. K.,** U.S.S.R. Patent 389097, 1973.

288. **Zhdanov, R. I., Romashina, T. N., Volodarskii, L. B., and Rozantsev, E. G.,** Method for introducing a spin label into biologically active compounds, *Dokl. Akad. Nauk S.S.S.R.,* 236, 605, 1977; *Chem. Abstr.,* 88, 22737q, 1978.

289. **Roznikov, V. A. and Volodarskii, L. B.,** Halo derivatives of imidazolidine nitroxyl radicals and their properties, *Izv. Sib. Otd. Akad. Nauk S.S.S.R., Ser. Khim. Nauk,* 3, 89, 1984; *Chem. Abstr.,* 102, 6297y, 1985.

290. **Reznikov, V. A. and Volodarskii, L. B.,** Synthesis of bifunctional derivatives of imidazoline nitroxides, *Khim. Geterotsikl. Soedin.,* 772, 1990.

291. **Grigor'eva, L. N., Tikhonov, A. Ya., Martin, V. V., and Volodarskii, L. B.,** Condensation reaction of 1,2-hydroxylaminooximes with acetylacetone. Conversion of tetrahydroimidazo[1,2-b]isoxazoles to 2*H*-imidazole, 1-hydroxypyrrole and 4-oxo tetrahydropyridine derivatives, *Khim. Geteronsikl. Soedin.,* 765, 1990.

292. **Grigor'eva, L. N. and Tikhonov, A. Ya.,** Synthesis of nitroxide derivatives of 3-oxo-6,8-diazabicyclo[3,2,1]oct-6-ene-8-oxyl, in *Int. Conf. Nitroxide Radicals* (Abstr.), Novosibirsk, Russia, 1989, 15P.

293. **Rijk, E. A. V. and Tesser, G. I.,** An unexpected formation of a new 3-imidazoline free radical, *Tetrahedron,* 46, 2129, 1990.

294. **Kirilyuk, I. A., Grigor'ev, I. A., and Volodarskii, L. B.**, Synthesis of 2*H*-imidazole-1-oxides and of stable nitroxide radicals on their basis, *Izv. Akad. Nauk S.S.S.R., Ser. Khim.*, 2113, 1991.

295. **Khramtsov, V. V., Weiner, L. M., Eremenko, S. I., Belchenko, O. I., Schastnev, P. V., Grigor'ev, I. A., and Reznikov, V. A.**, Proton exchange in stable nitroxyl radicals of the imidazoline and imidazolidine series, *J. Magn. Res.*, 61, 397, 1985.

296. **Shchukin, G. I., Grigor'ev, I. A., and Volodarskii, L. B.**, Recyclization of the imidazoline ring in nitroxyl radicals of 3-imidazoline derivatives in oxidation by halogens and nitrous acid, *Izv. Akad. Nauk S.S.S.R., Ser. Khim.*, 1581, 1981; *Chem. Abstr.*, 95, 187150a, 1981.

297. **Volodarskii, L. B., Reznikov, V. A., and Kobrin, V. S.**, Preparation and properties of imidazolinium salts containing a nitroxyl radical center, *Zh. Org. Khim.*, 15, 415, 1979; *Chem. Abstr.*, 91, 5158w, 1979.

298. **Reznikov, V. A. and Volodarskii, L. B.**, Nitroxyl radicals of 3-imidazoline and 3-imidazolinium in a Vielsmeier reaction, *Izv. Akad. Nauk S.S.S.R., Ser. Khim.*, 1137, 1982; *Chem. Abstr.*, 97, 109929p, 1982.

299. **Reznikov, V. A.**, Synthesis and Properties of Functional Derivatives of 3-Imidazoline Nitroxides, Ph.D. dissertation, Institution of Organic Chemistry, Novosibirsk, Russia, 1982.

300. **Reznikov, V. A. and Volodarskii, L. B.**, Enaminoketones of imidazolidine nitroxide radicals, *Izv. Sib. Old. Akad. Nauk S.S.S.R., Ser. Khim. Nauk*, 5, 109, 1980; *Chem. Abstr.*, 94, 156817y, 1981.

301. **Reznikov, V. A. and Volodarskii, L. B.**, Imidazolidine enaminoketones — new spin labels, *Izv. Akad. Nauk S.S.S.R., Ser. Khim.*, 926, 1979; *Chem. Abstr.*, 91, 39389n, 1979.

302. **Kelareva, M. P., Zolotov, Yu. A., Bodnya, V. A., Volodarskii, L. B., and Reznikov, V. A.**, A spin labeled hydroxyazo compound as an extractant and potential reagent for the determination of metals by ESR, *Izv. Akad. Nauk S.S.S.R., Ser. Khim.*, 1434, 1981; *Chem. Abstr.*, 95, 107840u, 1981.

303. **Zolotov, Yu. A., Bodnya, V. A., Kelareva, M. P., Morosanova, E. I., Volodarskii, L. B., and Reznikov, V. A.**, A stable free-radical hydroxyazo compound as extractant and reagent for determination of metals by electron spin resonance, *Zh. Anal. Khim.*, 36, 981, 1982; *Chem. Abstr.*, 97, 229261m, 1982.

304. **Subkhankulova, T. N., Lyakhovich, V. V., Reznikov, V. A., Elinova, V. I., and Weiner, L. M.**, Study on interaction of a lipophilic spin-labeled substrate analogue of Cyt. P-450 with liposomes and microsomes, *Biol. Membr.*, 3, 720, 1986; *Chem. Abstr.*, 105, 167717n, 1986.

305. **Reznikov, V. A. and Volodarskii, L. B.**, Haloderivatives of nitroxyl radicals of imidazoline, *Izv. Akad. Nauk S.S.S.R., Ser. Khim.*, 2565, 1984; *Chem. Abstr.*, 102, 185008, 1985.

306. **Reznikov, V. A. and Volodarskii, L. B.**, Reaction of heterocyclic imines — derivatives of 3-imidazolines with diborane and sodium borohydride, *Izv. Akad. Nauk S.S.S.R., Ser. Khim.*, 2337, 1985; *Chem. Abstr.*, 105, 152993a, 1986.

307. **Reznikov, V. A., Reznikova, T. I., and Volodarskii, L. B.**, Reaction of 1-hydroxy-2,2,4,5,5-pentamethyl-3-imidazoline and the appropriate nitroxyl radical with aldehydes, ketones and esters, *Zh. Org. Khim.*, 18, 2135, 1982; *Chem. Abstr.*, 98, 53775m, 1983.

308. **Reznikov, V. A. and Volodarskii, L. B.**, Recyclization of enamino carbonyl and -thiocarbonyl derivatives of imidazolidine to pyrrolines, *Izv. Akad. Nauk S.S.S.R., Ser. Khim.*, 437, 1991.

309. **Reznikov, V. A. and Volodarsky**, Enaminothiones of imidazoline nitroxides, a new route to acetylenic compounds of 3-imidazolines, *Tetrahedron Lett.*, 1301, 1986.

310. **Reznikov, V. A. and Volodarskii, L. B.**, Synthesis of enaminothiones derived from imidazolidine nitroxides and their reaction with sodium hypobromate, *Izv. Akad. Nauk S.S.S.R., Ser. Khim.*, 2585, 1988; *Chem. Abstr.*, 110, 212689w, 1989.

311. **Reznikov, V. A. and Volodarskii, L. B.**, Nitroenamines — derivatives of imidazolidine nitroxyls, *Zh. Org. Khim.*, 23, 214, 1987; *Chem. Abstr.*, 107, 175938s, 1987.

312. **Layer, R. W.**, The chemistry of imines, *Chem. Rev.*, 63, 489, 1963.

313. **Bruce, M. I.**, Cyclometallation reactions, *Angew. Chem., Int. Ed. Engl.*, 16, 73, 1977.

314. **Grigor'ev, I. A., Shchukin, G. I., and Volodarskii, L. B.**, ESR, ^{1}H, and ^{13}C NMR studies of nitroxide transformations in strong and super strong acids, *Izv. Akad. Nauk S.S.S.R., Ser. Khim.*, 2277, 1986; *Chem Abstr.*, 107, 197363c, 1987.

315. **Ovcharenko, V. I., Larionov, S. V., Mironova, G. N., and Volodarskii, L. B.**, Dimeric complexes of palladium (II) with a nitroxyl radical of imidazoline, containing a palladium-carbon bond and bridge atoms of halogenes, *Izv. Akad. Nauk S.S.S.R., Ser. Khim.*, 645, 1979; *Chem. Abstr.*, 91, 5310, 1979.

316. **Shapiro, A. B., Volodarskii, L. B., Krasochka, O. N., Atovmyan, L. O., and Rozantsev, E. G.**, Synthesis and structure of an organomercury imidazoline nitroxyl biradical, *Dokl. Akad. Nauk. S.S.S.R.*, 248, 1135, 1979; *Chem. Abstr.*, 92, 94532z, 1980.

317. **Shapiro, A. B., Volodarskii, L. B., Krasochka, O. N., and Atovmyan, L. O.**, Cyclometallation of the nitroxyl radical of imidazoline with a nitrone function, *Dokl. Akad. Nauk S.S.S.R.*, 255, 1146, 1980; *Chem. Abstr.*, 95, 62326f, 1981.

318. **Shapiro, A. B., Volodarskii, L. B., Krasochka, O. N., and Atovmyan, L. O.**, Synthesis and structure of cyclothallated nitroxyl radicals of imidazoline, *Dokl. Akad. Nauk S.S.S.R.*, 251, 1140, 1980; *Chem. Abstr.*, 91, 102319u, 1981.

319. **Reznikov, V. A., Martin, V. V., and Volodarskii, L. B.**, Recyclization of α-hydroxyimino-β-oxo derivatives of 3-imidazoline and 3-imidazoline-3-oxide to pyrrolines, *Khim. Geterotsikl. Soedin.*, 1195, 1990.

320. **Reznikov, V. A., Reznikova, T. I., and Volodarskii, L. B.**, Reaction of enaminones of imidazolidine nitroxides with electrophiles, *Izv. Sib. Otd. Akad. Nauk S.S.S.R., Ser. Khim., Nauk*, 5, 128, 1982; *Chem. Abstr.*, 98, 215525j, 1983.

321. **Reznikov, V. A., Vishnivetskaya, L. A., and Volodarskii, L. B.**, 2,2,5,5-Tetramethyl-3-imidazoline-1-oxyl-1-*O*-acyl hydroximic acid chlorides, *Izv. Akad. Nauk S.S.S.R., Ser. Khim.*, 441, 1991.

322. **Reznikov, V. A., Vishnivetskaya, L. A., and Volodarsky, L. B.**, 4-Hydroxyiminonitromethyl-2,2,5,5-tetramethyl-3-imidazoline-1-oxyl as the spin-labeled nitrile oxide precursor, in *Int. Conf. Nitroxide Radicals* (Abstr.), Novosibirsk, Russia, 1989, 14P.

323. **Griofa, S. N., Darcy, R., and Conlon, M.**, Synthesis and electron resonance of 3-oxyl-1,3-diazacyclohexene-1-oxide (1,3-nitrone-nitroxide) radicals, *J. Chem. Soc., Perkin Trans. 1*, 651, 1977.

324. **Volodarskii, L. B. and Tikhonov, A. Ya.**, Preparation of 1-hydroxy-2,2,4,6,6-pentamethyl-1,2,5,6-tetrahydropyrimidine-3-oxide and a nitroxyl radical based on it, *Izv. Akad. Nauk S.S.S.R., Ser. Khim.*, 2619, 1977; *Chem. Abstr.*, 88, 74369r, 1978.

325. **Yoshioka, T., Mori, E., and Murayama, K.**, Studies on stable free radicals. XIII. Synthesis and ESR spectral properties of hindered piperazine *N*-oxyls, *Bull. Chem. Soc. Jpn.*, 45, 1885, 1972.

326. **Lai, J. T.**, Hindered amines. Novel synthesis of 1,3,3,5,5-pentasubstituted 2-piperazinones, *J. Org. Chem.*, 45, 754, 1980.

327. **Lai, J. T.**, Hindered amines. III. Highly regioselective synthesis of 1,3,3,5,5-pentasubstituted-2-piperazinones and their nitroxyl radicals, *Synthesis*, 40, 1981.

328. **Lai, J. T.**, Hindered amines. 4-Hydroxy-1,3,3,5,5-pentasubstituted 2-piperazinones, *Synthesis*, 124, 1984.

329. **Sudo, R. and Ichihara, S.**, Reactions of alicyclic aminonitrile, *Bull. Chem. Soc. Jpn.*, 36, 34, 1963.

330. **Ramey, C. E. and Luzzi, J. J.**, Certain dioxopiperazinyl alkylphenols, U.S. Patent 3,919,234, 1975.

331. **Rassat, A. and Rey, P.**, Nitroxydes. LXII. Nitroxydes oxaziniques synthese et etudde conformationelle, *Tetrahedron*, 30, 3315, 1974.

332. **Rassat, A. and Sieveking, H. U.**, Ein stabiles aromatisches diradical mit starker dipolar electronen-wechselwirkung, *Angew. Chem.*, 84, 353, 1972.

333. **Grigor'ev, I. A.**, private communication.

334. **Dickerman, S. C. and Lindwall, H. G.**, Studies on piperidone chemistry. I. A synthesis of 5-homopiperazines, *J. Org. Chem.*, 14, 530, 1949.

335. **Rozantsev, E. G., Chudinov, V. A., and Sholle, V. D.**, Beckman-Chapman rearrangement of 2,2,6,6-tetramethyl-4-hydroxyiminopiperidine 1-oxyl to 2,2,7,7-tetramethyl-5-oxohexahydro-1,4-diazepine 1-oxyl, *Izv. Akad. Nauk S.S.S.R., Ser. Khim.*, 2114, 1980; *Chem. Abstr.*, 94, 30720t, 1981.

336. **Rozantsev, E. G. and Papko, R. A.**, 2,2,7,7-Tetramethyl-5-oxohomopiperazine *N*-oxide, a stable free radical, *Izv. Akad. Nauk S.S.S.R., Ser. Khim.*, 2254, 1962; *Chem. Abstr.*, 58, 10203g, 1963.

337. **Kedik, S. A., Rozantsev, E. G., and Usvyatsov, A. A.**, 2,2,7,7-Tetramethyl-5-oxohexahydro-1,4-diazepine-1-oxyl, U.S.S.R. Patent 1047903, 1983.

338. **Son, P. and Lai, J. T.**, Synthesis of hexahydro-3,3,5,5,7-pentaalkyl-2*H*-1,4-diazepine-2-one from 1,3-diamines and ketones, *J. Org. Chem.*, 46x, 323, 1981.

339. **Ramasseul, R., Rassat, A., and Rey, P.**, Nitroxydes. LXXII. Preparation d'amines et d'ammonium quaternaire nitroxydes curarisants potentiels, *Tetrahedron Lett.*, 839 1975.

340. **Yost, Y., Polnasze, C. F., Mason, R. P., and Holtzman, J. L.**, Application of spin labeling to drag assays. 1. Synthesis of 2,2,6,6-tetramethylpiperidine-4-on-1-oxyl-[15]N-16-^{2}H, *J. Labelled Compd. Radiopharm.*, 18, 1089, 1981.

341. **Venkataramu, S. D., Pearson, D. E., Beth, A. H., Perkins, R. C., Park, C. R., and Park, J. H.**, Synthesis of perdeuterio-*N*-(1-oxyl-2,2,6,6-tetramethyl-4-piperidinyl)-maleimid, a highly sensitive spin probe, *J. Labelled Compd. Radiopharm.*, 18, 371, 1981.

342. **Beth, A. H., Venkataramu, S. D., Balasubramanian, K., Dalton, L. R., Robinson, B. H., Pearson, D. E., Park, C. R., and Park, J. H.**, ^{15}N-Substituted and [2]H-substituted maleimide spin labels. Improved sensitivity and resolution for biological EPR studies, *Proc. Natl. Acad. Sci. U.S.A.*, 78, 967, 1981.

343. **Bates, B. A., Johnson, M. E., and Currie, B. L.**, Stable isotope substituted spin labels. 2. An improved synthesis of perdeuterio-[15]N-(1-oxyl-2,2,6,6-tetramethyl-4-piperidenyl) maleimide, *J. Labelled Compd. Radiopharm.*, 20, 33, 1983.

344. **Yost, Y., Polnaszek, C. F., and Holtzman, J. L.**, Application of spin labeling to drug assays. 3. [15]N, [2]H[13]-2,2,5,5,-tetramethylpyrroline-1-oxyl-3-carboxylic acid coupled to phenyton, *J. Labelled Compd. Radiopharm.*, 20, 707, 1983.

345. **Dervan, P. B., Squillacote, M. E., Lahti, P. M., Sylvester, A. P., and Roberts, J. D.**, [15]NMR spectrum of a 1,1-diazene *N*-(2,2,6,6-tetramethylpiperiridyl)-nitrene, *J. Am. Chem. Soc.*, 103, 1120, 1981.

346. **Venkataramu, S. D., Pearson, D. E., Beth, A. H., Balasubramanian, K., Park, C. R., and Park, J. H.**, Synthesis of ^{15}N-5-doxylstearic acid for improved EPR characterization of lipid motion in biomembranes, *J. Labelled Compd. Radiopharm.*, 20, 433, 1983.

347. **Park, J. H. and Trommer, W. E.**, Approaches to the chemical synthesis of [15]N- and deuterium substituted spin labels, in *Biological Magnetic Resonance*, Berliner, L. J. and Reuben, J., Eds., Plenum Press, New York, 1989, appendix.

348. **Venkataramu, S. D., Pearson, D. E., Beth, A. H., Park, C. R., and Park, J. H.**, High resolution spin labeled fatty acid. Synthesis and EPR spectral characteristics, *Tetrahedron Lett.*, 26, 1403, 1985.

349. **Grigor'ev, I. A., Mitasov, M. M., Shchukin, G. I., and Volodarskii, L. B.**, Use of Raman spectroscopy for identifying the nitroxyl group of cyclic nitroxyl radicals, *Izv. Akad. Nauk S.S.S.R., Ser. Khim.*, 2606, 1979; *Chem. Abstr.*, 92, 120133c, 1980.

350. **Volodarskii, L. B. and Grigor'ev, I. A.**, Synthesis of heterocyclic nitroxides, in *Imidazoline Nitroxides, Vol. 1, Synthesis and Properties*, Volodarsky, L. B., Ed., CRC Press, Boca Raton, FL, 1988, chap. 2.

351. **Koptyug, V. A.**, Ed., *Atlas of Adsorption Spectra of Aromatic and Heterocyclic Compounds*, Vol. 27, Institute of Organic Chemistry, Novosibirsk, Russia, 1983.

352. **Aurich, H. G., Hahn, K., Stork, K., and Weiss, W.**, Aminyl-oxide (nitroxide). XXIV. Empirische Ermittlung der spin-dichteverteinlung in Aminyloxiden, *Tetrahedron*, 33, 969, 1977.

Chapter 4

COMPLEXES WITH METAL-NITROXYL GROUP
COORDINATION

I. INTRODUCTION

The chemistry of coordination compounds of transition metals with stable nitroxides is one of the more modern and actively developed directions in inorganic chemistry. Research results, syntheses, and applications of metal complexes with nitroxides are covered in reviews and monographs.[1-14,98] The number of new coordination compounds with nitroxides is ever growing, and new chemical and physical properties of these compounds are being increasingly identified. Along with the chemistry of these compounds, attention is being paid to future uses of these compounds in the design of new materials with technologically important physical properties.[8-14] Thus, solid-phase magnetism of such compounds has always been of great interest. Their molecules contain several paramagnetic centers of different electronic natures (p- and d- or f-electrons), making them convenient objects for investigation of fine aspects of exchange interactions in heterospin exchange clusters. Interest in the solid-phase magnetism of these compounds grew sharply after Anderson and Kuechler had shown that nitroxyl group (NG) coordination by a paramagnetic metal ion may not involve spin pairing of metal ion and NG paramagnetic centers.[15] This has stimulated syntheses of metal complexes with nitroxides in which their solid phase contains the $M-O\dot{-}N<$ bonds. Synthetic studies were expected to produce compounds with a ferromagnetic character of both intramolecular (intracluster) and intermolecular (intercluster) exchange interactions, implying synthesis of a new series of ferromagnetics. It is presently known that the antiferromagnetic interactions that may arise in NG coordination do not hinder the magnetic phase transition.[14] However, in this case also, the paramagnetic organic fragment must be coordinated by the metal ion.

In complexes without a metal-NG bond, where the paramagnetic ligand is coordinated by means of donor atoms of other groups, with the NG being non-coordinated, the $>N\dot{-}O$ fragment acts as a spin label in a molecule. Its presence practically does not affect the chemical behavior of the organic ligand. In terms of the general synthetic strategy for the nitroxyl-containing compounds discussed in the beginning of chapter 3, this situation may be attributed to approach 2, i.e., it falls within the general approach providing NG introduction into a molecule simultaneously with molecular frame construction (in this case, coordination frame construction). Literature on the synthesis and investigation of nitroxide coordination compounds without ion metal-NG coordination has been exhaustively reviewed in References 1, 6 through 9, 11, and 12 and is therefore not discussed here. Attention is focused on the complexes where the solid phase undergoes NG coordination by the paramagnetic metal ion. This has been done because of intensive work in this direction during the recent decade and the necessity to review these investigations. From the viewpoint of nitroxide chemistry, the metal complexes with NG coordination, i.e., the compounds where the $>N \dot{-} O$ fragment is functionally active, are of most interest.

It should be noted that by now synthetic approaches have been developed and various types of coordination compounds have been isolated in the individual form in molecules in which the paramagnetic fragment of organic ligand is situated in close vicinity to the metal ion. Along with metal complexes with nitroxides, these include complexes with paramagnetic diimines, phenoxyls, semiquinones, porphyrins and other ligands (e.g., References 20 to 31). However, the exceptionally high kinetic stability of many nitroxides both in solution and solid phase[32-36] has made them among the most widely used paramagnetic organic ligands, while investigations on physical properties of metal complexes with nitroxides have contributed to our knowledge of the paramagnetism of heterospin systems.[10,14,16,75,111]

The methodological approaches to the synthesis of nitroxide coordination compounds with metal-NG coordination essentially differ from those for nitroxides themselves. Isolation of these approaches as a separate group is associated with coordination, which is a radically new phenomenon in the synthesis of transition metal complexes. In turn, coordination of nitroxides depends not only on metal nature and donor environment in the starting metal-containing matrix but also on nitroxide structure. In the synthesis of complexes with metal-NG coordination, functionalization of a nitroxide or its generation is generally not needed, though recently examples of complexation have been found which also involve nitroxide generation in the presence of a metal and formation of the $M-O\dot{-}N<$ bond.[13,17] As a rule, the nitroxide chosen as a synthon remains unchanged on formation of a coordination compound. The aim here is to find favorable conditions for the formation of a coordination compound of definite stoichiometry, establish the stoichiometry, and most importantly, to determine the type of metal-paramagnetic ligand coordination.

In discussing the direct metal-NG coordination, we must point out one important circumstance. Presently, there are no other methods to prove the presence of the metal-NG bond in the solid phase of a complex except for X-ray analysis. There is a certain specific of this research: development of procedures for the synthesis of complexes with metal-NG coordination must proceed in parallel with methodological efforts to obtain these compounds as monocrystals suitable for the X-ray structure analysis aimed at proving the presence of the above coordination in the compound prepared. If a nitroxide molecule contains other donor functional groups in addition to NG, this being the most common and interesting situation, then we are to determine not only coordinations around the metal but also the structure dimension type of a solid-phase compound: molecular (O-D); chain (I-D); layer (2-D); or three-dimensional (3-D) type.

The latter circumstance, in turn, is an important part of intensive research to create the so-called "molecular ferromagnetics". This term occurs widely in recent publications, and first reviews on molecular ferromagnetics have appeared[14,18,19] in which titles contain this term as a key expression. Its usage needs a brief commentary. It may seem to be internally contradictory, because ferromagnetism is inherent in a phase (macroobject) rather than a separate molecule. Research on molecular ferromagnetics presently involves synthesis of organic, organometallic, or coordination compounds in which solid phases are formed by separate molecules or ions containing paramagnetic centers and capable of magnetic-phase transition to the ferromagnetic state. Another aim of research is the possibility of altering magnetic parameters of phases (spontaneous magnetization magnitude, phase-transition temperature, etc.) chemically, i.e., by altering the starting reagents, which generally have molecular structure. The tendency in the search for objects with the desired properties is to use experimentally established magnetic-structural correlations implying the relationship between the observed magnetic properties of a compound and its chemical/structural characteristics. Thus, in the term molecular ferromagnetic, the word ferromagnetic is used in its usual sense, while the word molecular reflects the methodology, actually, that of crystalline design: paramagnetic molecules/ions must have such a structure that on solid-phase formation they would give a favorable steric structure for magnetic-phase transition to the ferromagnetic state. Another term used in literature along with the term molecular ferromagnetic is "molecular ferromagnetism" implying the ferromagnetic character of exchange interactions between lone electrons of paramagnetic centers separated by molecular bond systems inside molecules. The sense of this term is usually clear from the context.

Section VI gives examples of solid-phase molecular ferromagnetic design on the basis of coordination compounds of paramagnetic transition metal ions with nitroxides, with metal-NG coordination. Still, before this it is necessary to consider synthetic approaches to molecular design of metal-NG coordination complexes based on the use of metal-containing matrices and nitroxides as synthons, since these studies underlie the molecular ferromagnetic design.

II. MONOFUNCTIONAL NITROXIDE

The monofunctional nitroxide ligands include, for example, di-*tert*-butyl-nitroxyl (L^1), 2,2,6,6-tetramethylpiperidine-1-oxyl (L^2), and 2,2,5,5-tetramethylpyrrolidine-1-oxyl (L^3). Metal complexes with these ligands 1 to 44 are well known. The results of their investigation may be summarized as follows:

1. Thermodynamic studies of complex formation in solutions[37-41] have shown that nitroxides having no other functional groups in addition to NGs show a very weak donating ability. Therefore, synthesis of their complexes requires the use of rather strong Lewis acids as metal-containing acceptors: anhydrous halides,[42-50,52] perchlorates,[53-56] hexafluoroacetylacetonates,[37,41,57,58,68-71,74] and halogenated metal carboxylates.[59-62,65-67] Schemes 1 through 3 show the structures of the complexes obtained.
2. Synthesis of complexes with L^1 to L^3 is preferably carried out in an inert atmosphere and anhydrous low polar solvents (solid phases of compounds

can also be sensitive to moisture). Most complexes are rather weak and markedly dissociate in solution. Therefore, it would be practical to have an excess of nitroxide in a reaction mixture to ensure isolation of the solid phase, which leads to a higher fraction of a mixed-ligand complex in the solution.

3. Mixed-ligand complex formation may be hindered by redox processes in the metal-NG system.

4. The exchange interactions in the solid phases of nitroxide complexes containing only an NG as a donor functional group are of antiferromagnetic character. Spin pairing of metal ion and the coordinated NG does not make any essential contribution to the enthalpy of the mixed-ligand complex formation.

Among the specific features found for the reactions of metals with monofunctional nitroxides, one should note the reaction with Pd^{2+} salts. For reasons presently unknown, the interaction of L^1 and L^2 with Pd^{2+} salts leads to the formation of diamagnetic dinuclear compounds **5, 6, and 8** (Schemes 1 and 2) containing bridge halogen atoms in high yields.[42,44,45] In complexes **5-8**, the hydroxylamine anions L^1 and L^2 are coordinated by the Pd^{2+} ion in a bidentate fashion, according to the η^2-type.[46-50] The quantum-mechanical study[51] has shown that formation of molecules

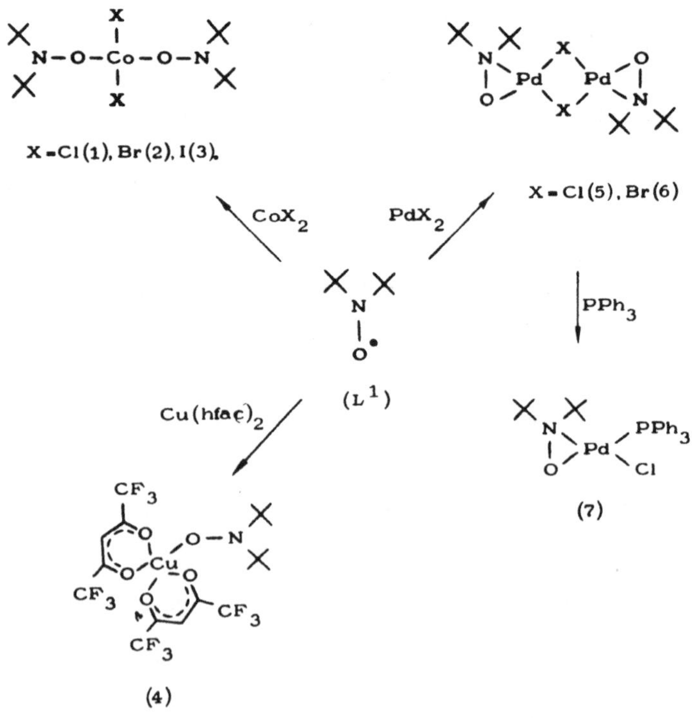

SCHEME 1.

from L^1 or L^2 and [Pd(PH$_3$)Cl] should proceed with an essential charge transfer from the metal to the NG π^*-orbital, leading to formation of a strong covalent bond. Another case of η^2-type coordination has been reported for compound Cu^{2+}-**10** (Scheme 2).[52] The strong antiferromagnetic interaction between the unpaired electrons of Cu^{2+} and the NG has been attributed to the favorable mutual position of magnetic orbitals in the coordination sphere of the complex. The quantum-mechanical calculation[51] for Cu(L^1)Br$_2$ with a coordination node isoelectronic to complex **10** has shown the bonding electrons of five atoms of the node: NG, copper, and two bromines to be essentially delocalized throughout the whole node, the singlet-triplet splitting value being 5000 cm^{-1}.

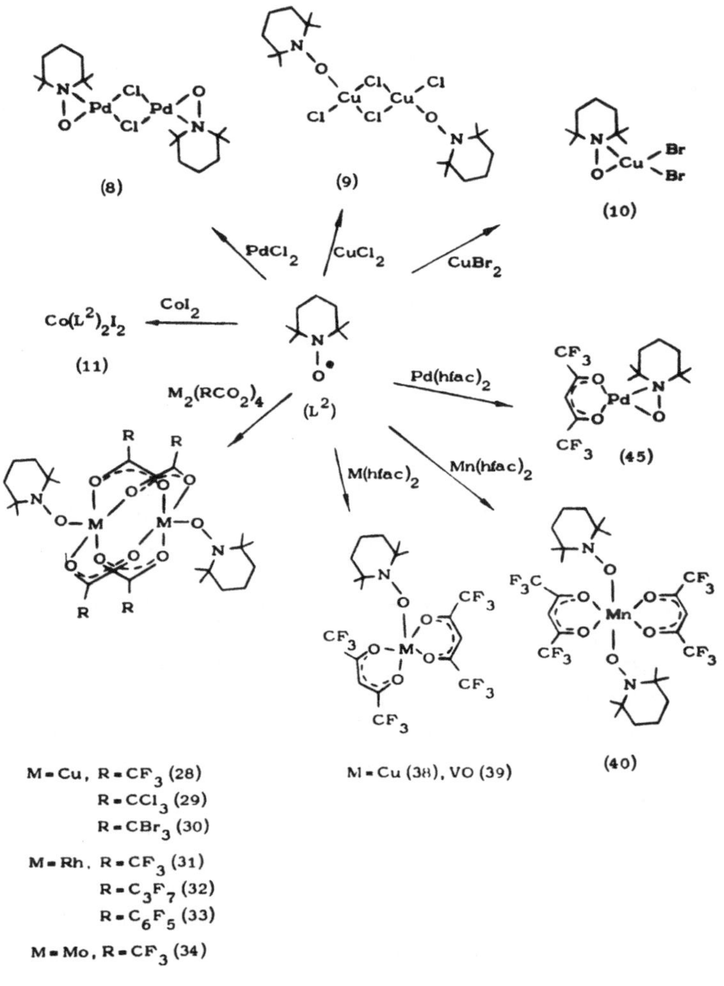

SCHEME 2.

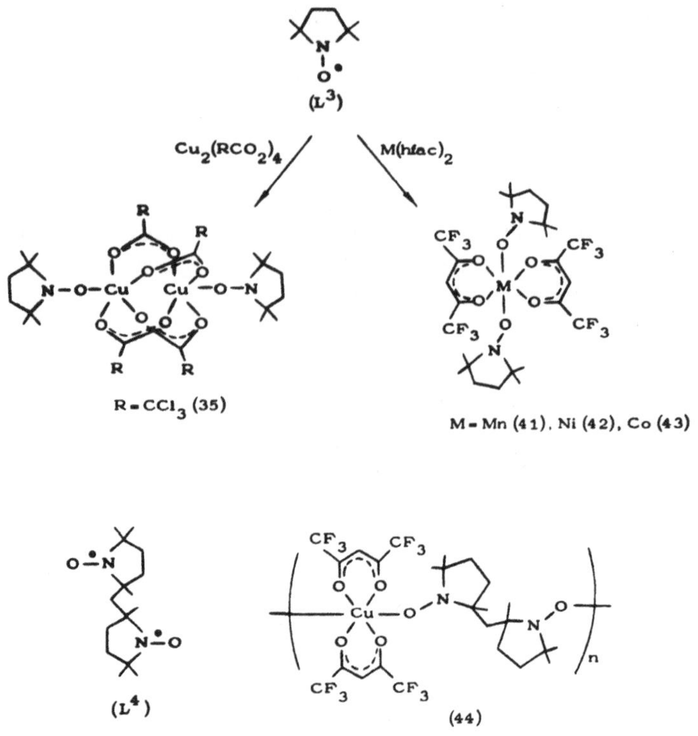

SCHEME 3.

In the reactions of nitroxides with Cu^{2+} compounds, especially in alkaline media, as will be shown later, redox processes may occur. Thus,[56] the reaction of $Cu(ClO_4)_2$ with L^2 proceeds as a multistep redox process eventually leading to metallic copper, a mixture of copper oxides, 2,2,6,6-tetramethylpiperidinium perchlorate, and the products of reduction of L^2. The reactions of L^2 with perchlorates of other metals $M(ClO_4)_n \cdot mH_2O$ in triethylorthoformiate gave bi- and polynuclear complexes $M(ClO_4)_2(L^2)$, M = Co(12), Ni(13); $M(ClO_4)_2(L^2)_2$, M = Fe(14), Zn(15),[53] and $M(ClO_4)_3(L^2)_2(H_2O)_3$; M = Y, La, Ce, Pr, Nd, Sm, Eu, Gd, Dy, Ho, Er, and Yb (16 to 27),[54,55] and no redox processes occurred in these reactions. For complexes **12-27**,, NG coordination by metal ion has been suggested, but their structures have not been examined.

Among compounds with monofunctional nitroxides most structural data have been obtained for metal carboxylates with L^2 and L^3. The dinuclear metal carboxylates are rather strong acceptors that easily form the mixed-ligand coordination compounds $M_2(RCO_2)_4L_2$. Substitution of the hydrocarbon group R (Scheme 2) in the carboxylate anion by its halogenated analogs enhances accepting ability of the metal.[59-61] A nontrivial molecular structure has been reported for complexes **29** and **35** (Schemes 2 and 3), which represents a new structural variant of Cu^{2+} carboxylate dimers. Every copper atom in the dimer molecule has a trigonally bipyramidal surrounding formed by oxygen atoms of bridge carboxylic groups and oxygen of the $>N{\dot-}O$ fragment. The coordination polyhedron of one copper atom in this structure is practically orthogonal to that of another copper atom inside the dimer. The Cu–Cu lengths equal to 3.526 Å in complex **29** and 3.197 Å in complex **35** exceed by far the distance of 2.6 Å typical of classical "lanterns"; the Cu^{2+} carboxylate dimers.[63,64] The CuON angle value is close to 120°, which is favorable for efficient overlapping of magnetic orbitals of Cu^{2+} and the nitroxyl fragment. This accounts for the diamagnetism of compounds **28-30** and **35**. In the synthesis of metal carboxylate complexes **28-37**, even insignificant amounts of moisture lead to the formation of hydrated products. Thus, in the synthesis of complex **36** (Scheme 4), the presence of small amounts of water in the starting L^3 has led to water transfer into the end product.[65] Because NG is a weak donor, the case of joint coordination of NG and the H_2O molecule by metal ion found in molecule **36** is rather unusual. It is unusual because NG, which is a weak donor, should be ousted from the coordination sphere by water molecules.

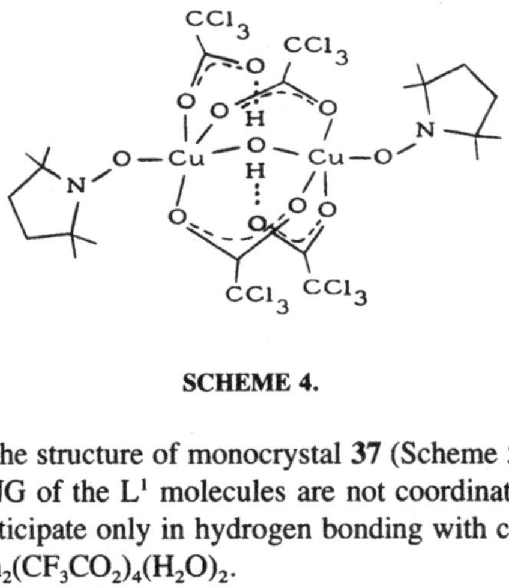

SCHEME 4.

Investigation of the structure of monocrystal **37** (Scheme 5) grown in moist air has shown that the NG of the L^1 molecules are not coordinated by the Rh atoms. The NG oxygens participate only in hydrogen bonding with coordinate water molecules of lanterns $Rh_2(CF_3CO_2)_4(H_2O)_2$.

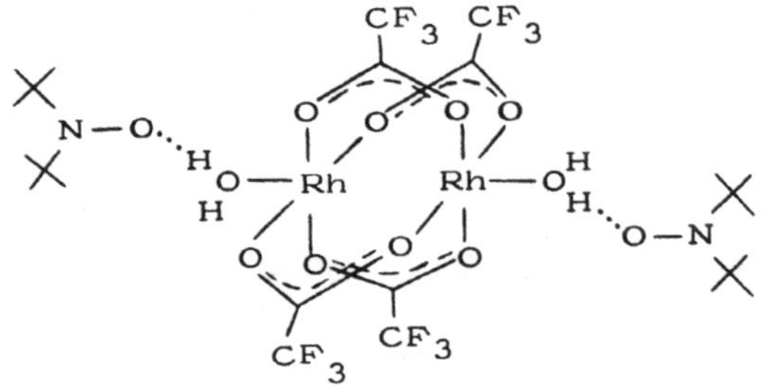

SCHEME 5.

Studies of mixed-ligand complexes on the basis of metal hexafluoroacetyl-acetonates with L^1 to L^3 have shown that in the case of Cu^{2+}, complexes of stoichiometry $Cu(hfac)_2$: $L = 1:1$ are formed (compounds **4**, **38**), whereas with other metals (Mn^{2+}, Ni^{2+}, Co^{2+}), those of stoichiometry $M(hfac)_2$: $L = 1:2$ (compounds **40–43**) are formed. This is quite a general tendency for mixed-ligand complexes of metal acetylacetonates associated with the electronic d^9-configuration of Cu^{2+}. Two NGs may be coordinated by one Cu^{2+} ion, as a rule, only with bidentate-bridge coordination of the NG, as exemplified by coordination of the nitroxyl biradical L^4 in complex **44** (Scheme 3). The solid phase of complex **44** is formed by the polymer chains $[Cu(hfac)_2(L^4)]_n$ where the NG L^4 is coordinated by various copper atoms.[74]

An attempt[71] was made to synthesize a mixed-ligand complex $Pd(hfac)_2$ with L^2. In the molecules of product **45** (Scheme 2), the square environment of Pd^{2+} is formed by two oxygen atoms of the hfac-anion and by oxygen and nitrogen of the η^2-coordinated hydroxylamine anion (Scheme 2). It is yet unclear how L^2 is reduced to hydroxylamine in the presence of Pd^{2+} or in the synthesis of complexes **5–8**. Based on structural data for complex **45** and the literature data, the authors of Reference 71 have suggested a structural criterion to determine the oxidation level of the coordinated fragment N–O in 2,2,6,6-tetramethylpiperidine derivatives. If the N–O distance is about 1.27 Å and deviation of nitrogen from the plane of oxygens and carbons at the 2- and 6-positions of the heterocycle does not exceed 0.15 Å, then it is the NG that is coordinated. However, if the length is of an order of 1.35 Å, while nitrogen deviates more from the plane, for example, by 0.35 Å in the molecules of **45,** then it is the anion of the corresponding hydroxylamine that is coordinated. It should be noted that for the oxammonium salts **46** and **47** (Scheme 6), the N–O distance is essentially smaller, being 1.22[72] and 1.19 Å,[73] respectively.

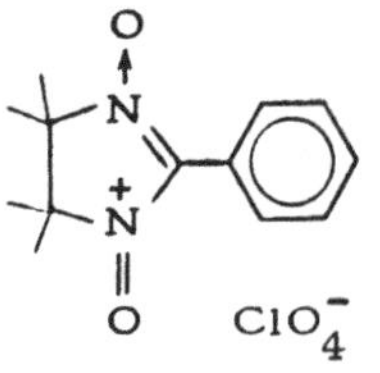

(46) 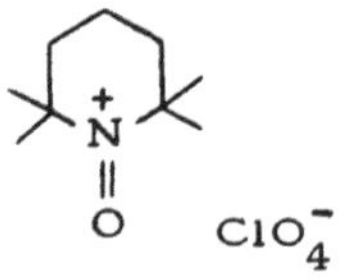　　　　　(47)

SCHEME 6.

III. MONOFUNCTIONAL NITRONYL NITROXIDE

The structure of nitronyl nitroxides is favorable for bridge coordination of paramagnetic ligands, being a symmetric structure with actually two equivalent donor N–O groups. Nevertheless, a number of mixed-ligand complexes **48–61** have been reported (Scheme 7) in which the solid phase involves the monodentate co-ordination of nitronyl nitroxide molecules.[80-92] The results of these studies may be summarized as follows

1. In the syntheses of mixed-ligand complexes, mainly hexafluoroacetyl-acetonates[80-87] and fluorinated metal carboxylates[88-90] were used as metal-containing matrices. Of other metal compounds, only the cupric chloride complex has been reported.[77]
2. In most cases, removal of water from the reaction mixture in the synthesis of nitronyl nitroxide complexes is not required, because they possess greater basicity[77-79] as compared to nitroxides of type L[1] to L[3]. Scheme 7 shows that compounds **48, 52–60** were synthesized using hydrogen-containing reagents.
3. In the solid phase, complexes **48–61** have molecular structure.

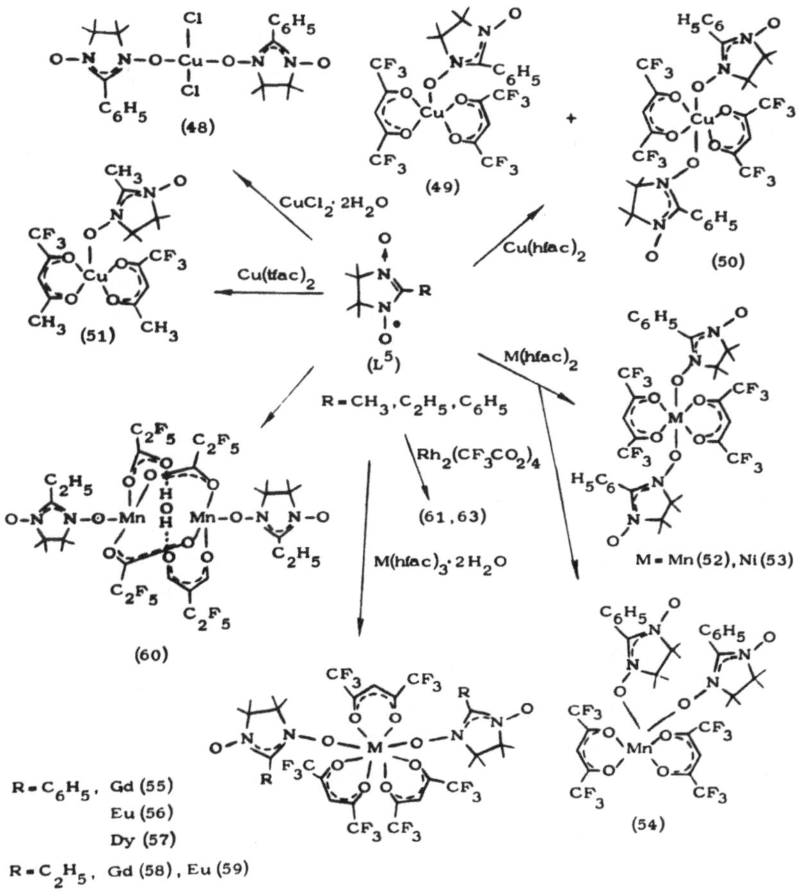

SCHEME 7.

We should note the following specific features found in studies on metal complexes of monofunctional nitronyl nitroxides. It is well known that of all paramagnetic metal ions of the first transition series, only Cu^{2+} is apt to form complexes with square coordination of the metal. Along with complexes **9** and **10** discussed in Section II, these include the monofunctional nitronyl nitroxide complex **48** (Scheme 7). Among $M(hfac)_2$, Cu^{2+} is less inclined to increase its coordination number up to 6 as compared to other transition metals. Therefore, in the reactions with monofunctional nitroxides or nitronyl nitroxides, $Cu(hfac)_2$ usually forms monomer complexes with a coordination number equal to 5, for example, complexes **4** (Scheme 1), **38** (Scheme 2), **49**, and **51** (Scheme 7). The coordination Cu^{2+} polyhedron may be similar both to a square pyramid and to a trigonal bipyramid. The same compound can exist in both conformations. Thus, as a result of the reaction of $Cu(hfac)_2$ with L^5 (R = C_6H_5), compounds **49** and **50** crystallized simultaneously from the reaction mixture and were separated mechanically.[81] A structural peculiarity of **49** was found to be the presence of two molecules of different conformations in one elementary unit of this compound. In one of them the coordination CuO_5 polyhedron is closer to the square pyramid, while in another, to the trigonal bipyramid. Such situations strongly complicate the analysis of magnetic properties of a compound, but in the molecules of **49** the L^5 nitroxyl oxygens in both conformations lie in the equatorial plane of the coordination polyhedron, which always leads to a strong antiferromagnetic interaction between the lone electrons of the Cu^{2+} ion and the NG. Therefore the molecules of both conformations are diamagnetic. The coordination polyhedron of complex **50** represents a tetragonally distorted octahedron in which the donor $>N\overset{\cdot}{-}O$ oxygens occupy apical positions. Analysis of the temperature curve of magnetic sensitivity of the polycrystalline complex **50** has shown, in agreement with theoretic concepts,[76] the presence of ferromagnetic exchange interactions between the lone electrons of Cu^{2+} and the NG. In the solid phase of complex **51**, the $>N\overset{\cdot}{-}O$ oxygen occupies an apical position and complements the square coordination of Cu^{2+} in the $Cu(tfac)_2$ fragment to the square-pyramidal one.[83] Such a geometry favors a ferromagnetic exchange interaction, which is also in accordance with theoretical calculation results.[76] It should be pointed out that compound **51** is an interesting case of a mixed-ligand complex on the basis of an asymmetric metal-containing matrix of cupric *cis*-bis(trifluoroacetylacetonate). Due to this matrix, the highest known energy value of the ferromagnetic exchange interaction between the lone electrons of Cu^{2+} and the NG equal to 65 cm^{-1} has been achieved. On NG coordination by the paramagnetic ions of other d-metals of the first transition series: Mn^{2+}, Ni^{2+}, and Co^{2+} (**52–54**) the exchange interaction is of an antiferromagnetic character, which is general for all complexes of Mn^{2+}, Ni^{2+}, and Co^{2+} (**1–3, 11, 40–43**, and **52–54**).

An interesting example of the effect of substituent R in L^5 on the mixed-ligand complex geometry is represented by complexes **52** and **54**. In the molecules of **52** (R = C_6H_5), L^5 occupies the *trans*-position, while in the molecules of **54** (R = CH_3), the *cis*-position. Taking into account the data of References 113 and 115, the R substituent effect on the *cis*- or *trans*-coordination of L^5 has also been found in the polymer complexes $Mn(hfac)_2$ with L^5 (Section IV.D).

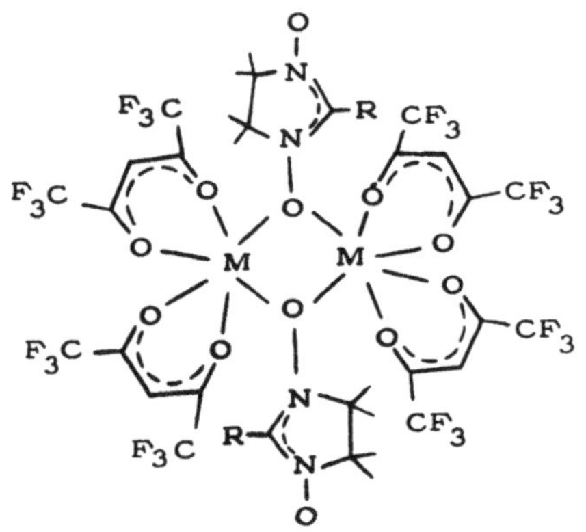

$R = C_2H_5$, $M = Ni (65)$, $Co (66)$

$R = CH_3$, $M = Mn (81)$

SCHEME 8.

A rare example of complexes with the 1,1-bridge function of L^5 is represented by the binuclear complexes **65, 66,** and **81** (Scheme 8) reported in Reference 91 and 92.

Conditions of synthesis and magnetic properties of rare-earth metal complexes **55–59** are similar.[85-87,122] The most interesting results of investigations of complexes **55–59** proved to be the fact that the exchange interaction between lone electrons of a rare-earth metal ion and the coordinated nitronyl nitroxide is, first, ferromagnetic in character and, second, it is smaller in absolute value than the energy of the antiferromagnetic exchange interaction between lone electrons of two coordinated L^5. In the synthesis of compounds on the basis of fluorinated metal carboxylates, the reagent ratio is critical to the $M_2(RCO_2)_4:L$ ratio in the end product and, consequently, to the compound structure. Thus, the reaction of $Rh_2(CF_3CO_2)_4$ with nitronyl nitroxides or the imino nitroxide at the reagent ratio 1:2 gives **61** and **62** (Scheme 9) a molecular structure, while at the reagent ratio 1:1 the same reaction leads, owing to the necessity to fill all coordination sites, to formation of solid-phase polymers **63** and **64.**[89,90]

Studies on the magnetism of complex **61** have established the lone electrons of coordinated nitronyl nitroxides to be in a strong antiferromagnetic exchange interaction ($J = -83.6$ cm^{-1}). An efficient spin density "transfer" leading to a strong exchange interaction is favored by the Rh–Rh system, but if the paramagnetic centers are wide apart, as, for example, in **62** where there is no direct NG coordination by metal ion, then the exchange interaction energy value drops ($J = -5.2$ cm^{-1}). It is interesting that the same exchange interaction energy order is preserved in the polymer complexes **63** ($J = -98.8$ cm^{-1}) and **64** ($J = -2.3$ cm^{-1}). The synthesis and properties of polymer mixed-ligand metal complexes with nitroxides are discussed at length in Section IV.D.

Summing up the discussion in Sections II and III, the data obtained unambiguously confirm the weak donating ability of NGs. For the synthesis of complexes with metal-NG coordination with monofunctional nitroxides, strong acceptors should

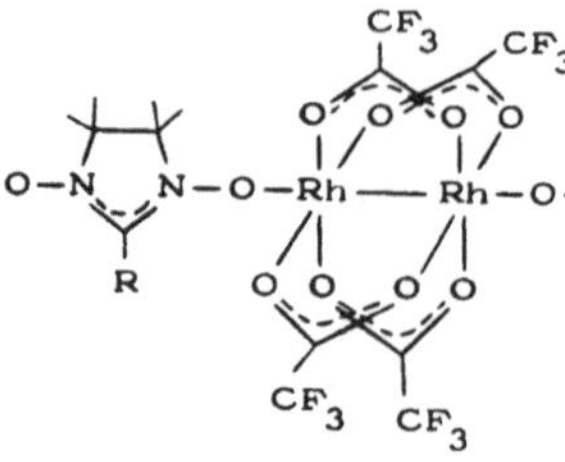

$R = C_6H_5$ (64)

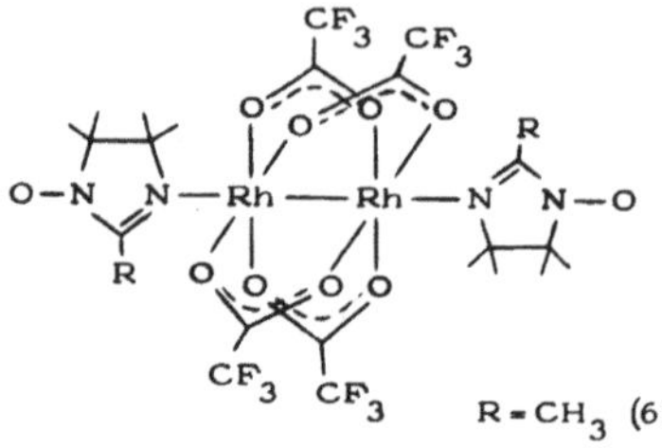

$R = CH_3$ (62)

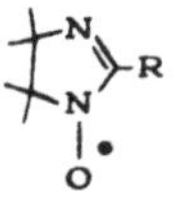

(L^{11})

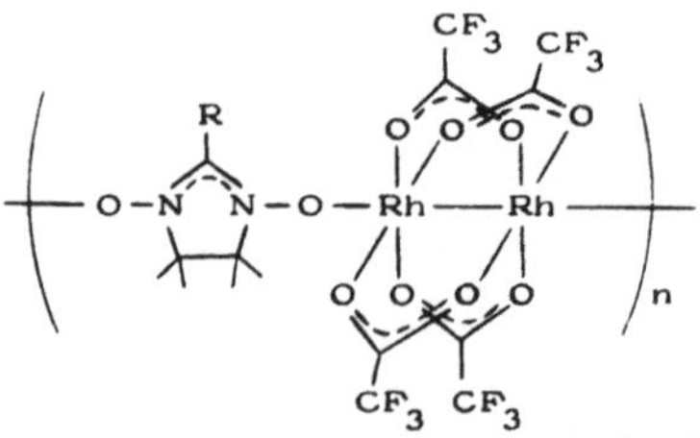

$R = CH_3$ (63)

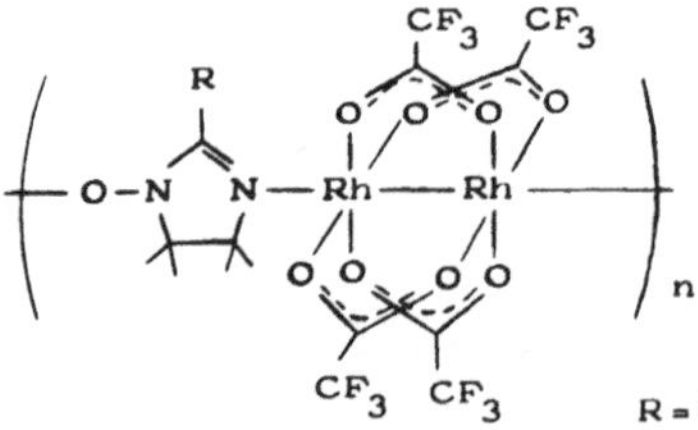

$R = CH_3$ (64)

SCHEME 9.

be used, such as, for example, fluorinated acetylacetonates or metal carboxylates. In most complexes with monofunctional nitroxides, the NG is coordinated monodentately at the oxygen atom, and more rarely bidentately according to the η^2-type. Both of these coordination methods lead to formation of solid-phase molecular compounds, i.e., they hinder formation of dimensional structures necessary for the design of molecular ferromagnetic synthesis. Synthesis of solid-phase n-dimensional structures could be promoted by the 1,1-bridge coordination of the NG (Scheme 8). Up until now only a few attempts to realize this type of coordination were successful.[91,92] The use of bi- and trifunctional nitroxides for the synthesis of n-dimensional complexes proved to be much more efficient.

IV. BIFUNCTIONAL NITROXIDE WITH A BRIDGE FUNCTION

A. SPECIFICS OF THE APPROACH

Intensive research on bifunctional nitroxide complexes with NG coordination owes much of its origin to Reference 15. It reported synthesis, structure, and magnetic properties of complex **67** obtained by the reaction of $Cu(hfac)_2$ with L^6 in CH_2Cl_2. As some data of Reference 15 which were obtained for the first time are of major importance, we will discuss them in more detail. The solid phase of the complex is formed by the $[Cu(hfac)_2(L^6)]_n$ polymer chains (Scheme 10).

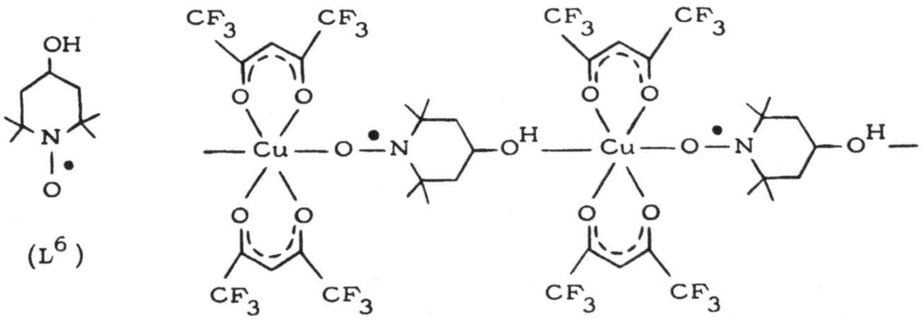

SCHEME 10.

The copper atoms of chelate $Cu(hfac)_2$ matrices inside the chains are bonded with nitroxyl and hydroxy oxygens of different L^6 molecules possessing the bridge function, with coordinated L^6 being in several conformations. A remarkable preparative feature of synthesis of complex **67** was found to be the possibility of using water-containing reagents.

Another nontrivial result of studies reported in Reference 15 was the absence of spin pairing of the Cu^{2+} ion and the coordinated NG. The energy of ferromagnetic exchange interaction in the exchange cluster $Cu^{2+}-O\dot{-}N<$ was found to be 13 cm^{-1} (the intercluster exchange interaction $J'/k = -78$ mK^{93}).

B. LIMITATIONS OF THE APPROACH

The methodological approach to synthesis of compounds with a metal-NG bond by synthesis of complexes with bridge function of the paramagnetic ligand is attractive due to a possibility of varying the components of the mixed-ligand complex formation within wide limits. At the same time, attempts to realize the $M-O\dot{-}N<$ coordination can encounter complications.

First, along with donor atoms of the functional group, other donor molecules that are present in the reaction mixture can participate in metal ion coordination.

Thus, the use of water-containing solvents and $Cu(hfac)_2$ or the anhydrous solvents and $Cu(hfac)_2 \cdot H_2O$ in the reaction with L^7 has led to formation of complex **68** where Cu^{2+} is surrounded by the tetragonally distorted octahedron of hfac-anion oxygens, nitrile nitrogen, and water oxygen[94] (Scheme 11). The water molecule is attached by a hydrogen bond to oxygen of the NG L^7, which leads to formation of polymer chains $[(H_2O)Cu(hfac)_2(L^7)]_n$, but there is no direct coordination of $Cu^{2+}–O–N<$.

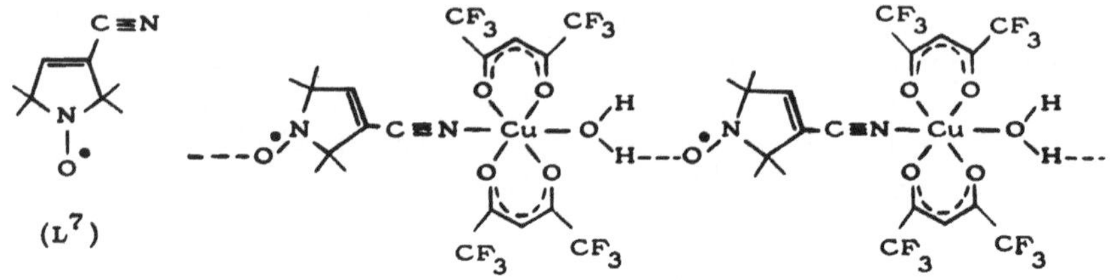

SCHEME 11.

Separation of paramagnetic centers in the chain by water molecules was the reason for the fact that measurement of the $\chi(T)$ dependence in the temperature range of 81 to 294 K, just as would be expected, did not result in detection of any coupling between lone electrons in complex **68.**

Second, NG coordination may be hindered by donor atoms of functional groups occupying coordination sites. For example, in the dimer molecules **69**, both axial positions are occupied by oxygen atoms of the hydroxy group of L^6 (Scheme 12).[59]

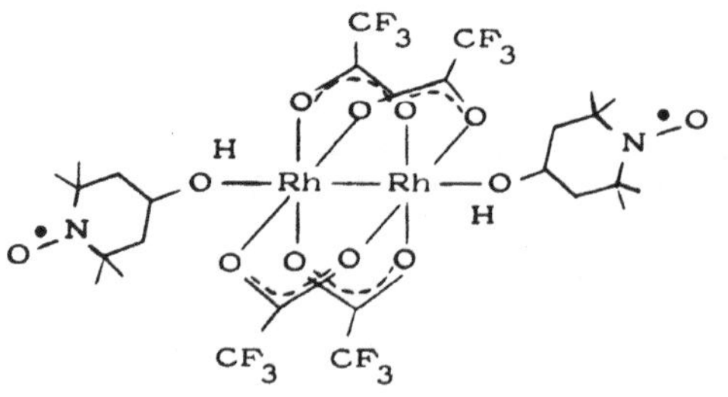

(69)

SCHEME 12.

Occupation of coordination sites by donor atoms of functional groups having generally the higher donating ability than the NG can also take place when the potential coordination capability of the metal remains unrealized. Thus, the crystalline structures of complexes **70, 71** (Scheme 13) proved to be molecular because the coordination number of Cu^{2+} is equal to 5 and the NG does not participate in metal ion coordination.

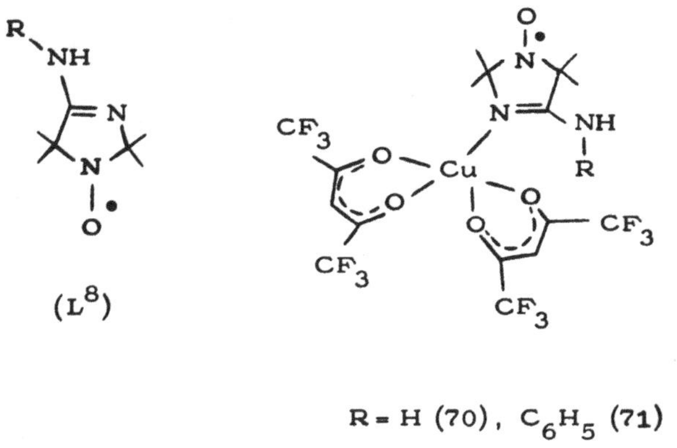

R = H (70), C_6H_5 (71)

SCHEME 13.

Third, NG coordination may be hindered by formation of compounds having the clatrate-like structure. An example of such a compound is Pd(hfac)$_2$(L^7)$_2$ (72) described in Reference 99. Its structure is unusual among metal complexes with nitroxides. It consists of the alternating layers of Pd(hfac)$_2$ and L^7 molecules. The shortest contact (3.047 Å) between atoms of different layers in structure 72 is the distance between NG oxygens in L^7 and the γ-carbon of the hfac anion of the chelate. Thus, there is no chemical bond between the Pd(hfac)$_2$ molecules and L^7 in crystals 72. In spite of this, the compound sublimates in vacuum without decomposition. As expected, magnetic susceptibility measurements did not reveal any appreciable interactions between the unpaired electrons on the L^7 molecules.

C. DERIVATIVES OF THE 3-IMIDAZOLINE NITROXIDE

The synthetic route to complexes with an M–O$\dotplus$N< coordination by affecting a bridging coordination of a paramagnetic organic ligand proposed in Reference 15 has been found to be fruitful in the synthesis of complexes with derivatives of the imidazoline nitroxide.[10] In Scheme 14 are examples of such complexes that are polymeric in the solid state 73,74. Among the studied mixed-ligand complexes of metals with derivatives of the imidazoline radical,[96,97] complex 74 is the only example of a compound where the imine nitrogen atom of the imidazoline heterocycle remains uncoordinated. In all other complexes with derivatives of the 3-imidazoline nitroxide, the imine nitrogen atom was always observed to be coordinated.

For many derivatives of L^9 one notes a remarkable property to form mixed-ligand complexes with Cu(hfac)$_2$ having a Cu(hfac)$_2$ to L^9 ratio of 3:2 of which complexes 75–77 are an example. Structural studies on complexes 75–78 indicated that their crystals are formed by trinuclear molecules (Schemes 14 and 15) with L^9 acting as a bidentate bridge. The imine nitrogen atoms of the 3-imidazoline heterocycle in these trinuclear molecules are coordinated to the "terminal" Cu atoms whereas the NG oxygen atoms of the two L^9 or L^{10} molecules are coordinated to

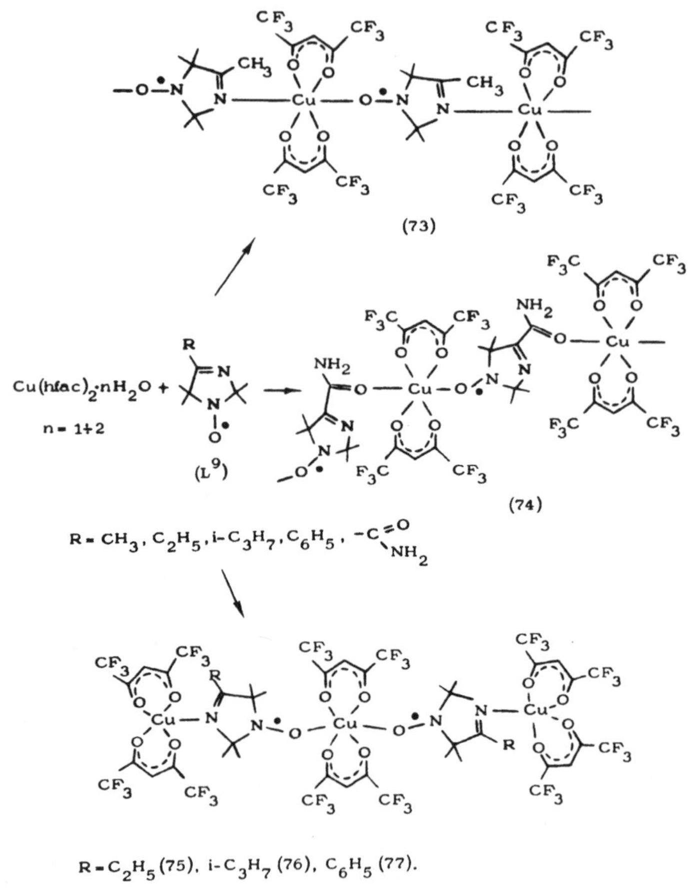

(73)

Cu(hfac)$_2 \cdot n$H$_2$O +   (L^9) $\longrightarrow$

$n = 1 \div 2$

(74)

R = CH$_3$, C$_2$H$_5$, i-C$_3$H$_7$, C$_6$H$_5$, $-$C$\stackrel{=O}{\underset{NH_2}{}}$

R = C$_2$H$_5$ (75), i-C$_3$H$_7$ (76), C$_6$H$_5$ (77).

SCHEME 14.

the "central" Cu atom. As a result, the coordination number of the central atom is 6 and that of the terminal Cu atoms is 5.[100-104] It has been found recently that a trinuclear complex **80** (Scheme 15) analogous to **75–78** can be formed with derivatives of 2-imidazoline.[105] A mixed-ligand complex **79** with a nitronyl nitroxide (Scheme 15) of the stoichiometry 3:2 was also reported.[106] One notes that in the molecules of **79** the exchange interaction between the unpaired electrons on the paramagnetic centers is of an antiferromagnetic character while in **75–78, 80** it is ferromagnetic.

The evidence in References 100 to 106 suggests that compounds with a 3:2 stoichiometry are rather common in the series of mixed-ligand complexes of Cu(hfac)$_2$ with bifunctional radicals performing a bridging function. The fact that 3:2 compounds have been thus far reported only for mixed-ligand complexes of Cu(hfac)$_2$ can be related to the ability of Cu^{2+} bischelates to form complexes with coordination

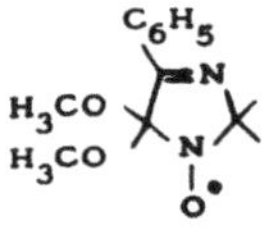

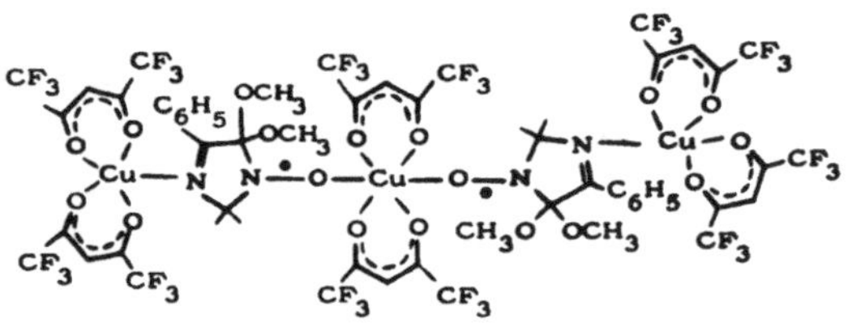

(78)

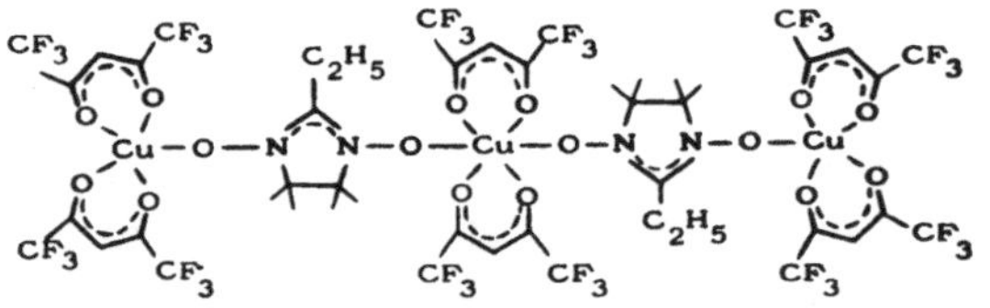

(79)

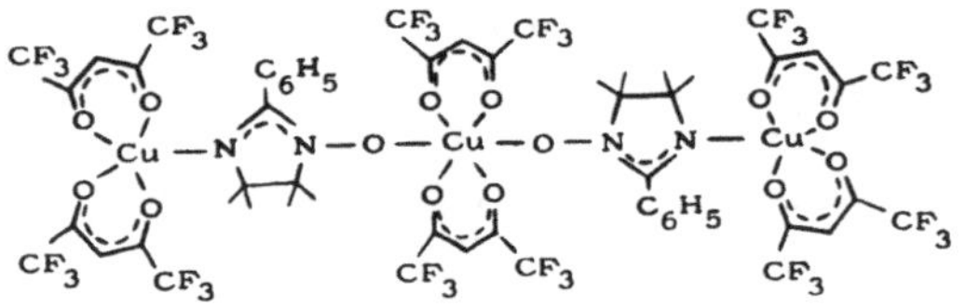

(80)

SCHEME 15.

numbers of both 5 and 6. It is very difficult to find an explanation of the reason for such an easy coordination of NG of the [Cu(hfac)$_2$L] fragments by the central copper atom in the presence of water in the reaction mixture. This may be due to the trinuclear molecules being formed from mononuclear Cu(hfac)$_2$L molecules and it is the steric factors and the specific solvatation of the species formed in the solution that have a decisive influence of whether structures molecular in the solid state, the polymeric complexes [Cu(hfac)$_2$L]$_n$, or type 3:2 complexes will be formed from Cu(hfac)$_2$L present in the solution. Thus for L^9 (Scheme 14) only in the case of R = CH$_3$ was the complex (73) found to be a polymer in the solid state, whereas R = C$_2$H$_5$, i-C$_3$H$_7$, C$_6$H$_5$ give rise to 3:2 compounds and the solid phases of complexes **70, 71** are formed by mononuclear molecules.

For N-oxide derivatives of the 3-imidazoline nitroxide L^{12} (Scheme 16) no 3:2 mixed-ligand complexes have been observed to form in the interaction with Cu(hfac)$_2$ · nH$_2$O. With the derivatives of L^{12}, complexes of the composition Cu(hfac)$_2$(L^{12}); R = CH$_3$ **(86)**, C$_6$H$_5$ **(87)** have been isolated.[108] Considering this fact, as well as the stoichiometry data for the mixed-ligand complexes with nitronyl nitroxides

$$(L^{12})$$

$$(L^9H)$$

$R = CH_3, C_6H_5.$

$R = CH_3, C_2H_5, i\text{-}C_3H_7, C_6H_5.$

SCHEME 16.

(Sections III and IV), it can be suggested that it is the asymmetric structure of the nitroxide molecule that favors formation of 3:2 complexes.

Derivatives of the 3-imidazoline nitroxide differ from those of piperidine, pyrroline, pyrrolidine, or nitronyl nitroxides by an easy accessibility for items of the corresponding hydroxylamines, L^9H (Scheme 16).

This has greatly contributed to the fact that it was in a study of the interaction products of L^9H and $Cu(hfac)_2 \cdot nH_2O$ that a specific (2:3:2) type of mixed-ligand complexes of metal hexafluoroacetylacetonates with nitroxides having the direct $Cu^{2+}\text{-}O\dot{-}N<$ bond could be revealed, which are impossible to obtain by the direct reaction of $Cu(hfac)_2 \cdot nH_2O$ with nitroxides (Scheme 17).

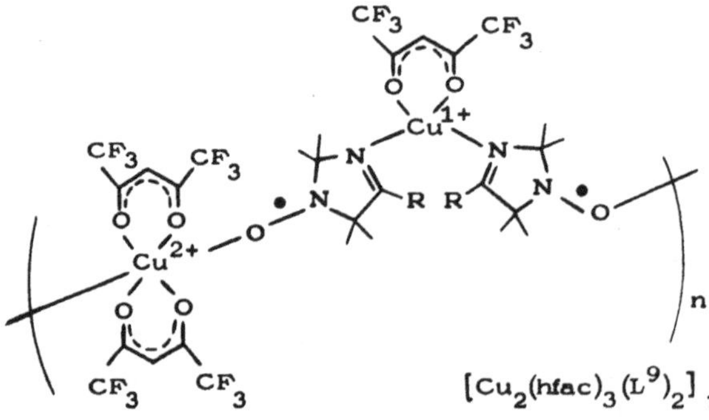

$$[Cu_2(hfac)_3(L^9)_2]_n$$

$R = CH_3 (82), C_2H_5 (83), i\text{-}C_3H_7 (84), C_6H_5 (85).$

SCHEME 17.

In the reaction of $Cu(hfac)_2 \cdot nH_2O$ with L^9H in an inert atmosphere upon heating a redox process takes place

$$2\ Cu(hfac)_2 + 2\ L^9H \rightarrow Cu_2(hfac)_3(L^9)_2 + Hfac + H$$

resulting in the separation of a mixed-valence complex of Cu^{2+} and Cu^{1+} as a solid phase that is formed in the solid state by the polymeric chains $[Cu_2(hfac)_3(L^9)_2]_n$. Examples of polymeric complexes with alternating Cu^{2+} and Cu^{1+} ions in the solid state have been reported for complexes with diamagnetic organic ligands,[110] but among mixed-ligand complexes of metals with nitroxides such polymers were first described in References 13, 17, and 107. It is important to emphasize the necessity of heating the reaction mixture in the synthesis of complexes **82–85** because in the absence of heating complexes with hydroxylamines $Cu(hfac)_2(L^9H)$ (Scheme 18) are usually formed. The use of an inert atmosphere[107-109] is necessary because of a sharp acceleration of the oxidation of L^9H to L^9 by the air oxygen in the presence of $Cu(hfac)_2$, resulting in separation of the same solid products as in the interaction of $Cu(hfac)_2$ with L^9. Scheme 18 summarizes the composition of the products formed in the interaction of $Cu(hfac)_2 \cdot nH_2O$ with the derivatives of 3-imidazoline nitroxide and its analogous hydroxylamine containing hydrocarbon substituent in the 4th position of the heterocycle.

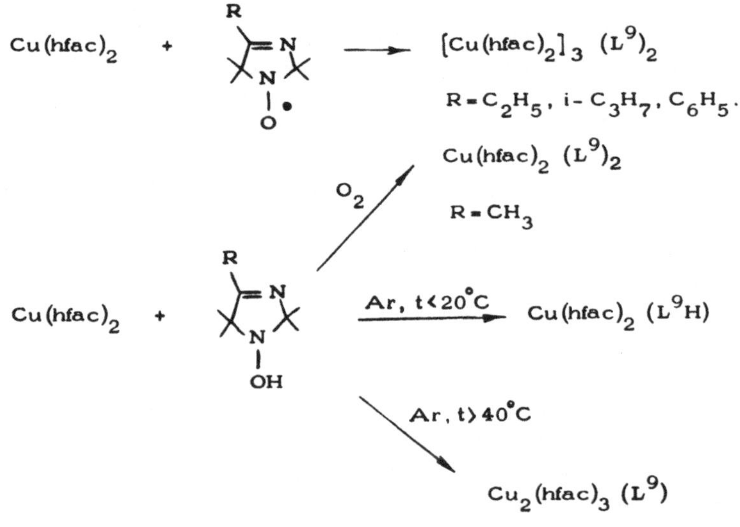

SCHEME 18.

In conclusion, note that in most of the $Cu^{2+}-O\dot{-}N<$ coordinated complexes considered in this section (**67, 73–78, 80, 82–85**) the exchange interaction in the exchange clusters $Cu^{2+}-O\dot{-}N<$ (or $>N\dot{-}O-Cu^{2+}-O\dot{-}N<$) is of a ferromagnetic character. The ferromagnetic character of exchange interactions in the exchange cluster $Cu^{2+}-O\dot{-}N<$ first documented in Reference 15 for complex **67**, seemed initially unusual and for a long time there has been no sufficiently clear interpretation of the molecular magnetism in the compounds under discussion. All explanations were, as a rule, of a qualitative character and were based on the concept of direct exchange interaction, i.e., an interaction occurring by direct overlap of the orbitals of unpaired electrons on the paramagnetic centers of Cu^{2+} (the $3d_{xy}$-AO orbital) and NG (the π^*-antibonding MO). The molecular orbitals in the studied complexes being normally nonorthogonal, it was difficult to understand the ferromagnetic character of the exchange interactions in the exchange cluster $Cu^{2+}-O\dot{-}N<$. Only recently was it possible to propose a satisfactory explanation.

A detailed quantum-chemical study has been performed[16,76,111] of the different factors determining the magnitude and sense of the spin-exchange interactions J in axially coordinated complexes of Cu^{2+} with nitroxide. An analysis of the direct exchange interactions in compounds of such a type based on nonempirical calculations of J parameters indicated that this interaction mechanism is not the main one in the formation of the molecular magnetism. The mechanism that dominates is a ferromagnetic interaction occurring due to delocalization of an unpaired electron from π^*-MO of the nitroxide to an AO of Cu^{2+} with subsequent intra-atomic interaction with an unpaired electron on the Cu^{2+} $3d_{xy}$-AO.

D. DERIVATIVES OF THE 2-IMIDAZOLINE NITRONYL NITROXIDE

It was previously noted that the structure of nitronyl nitroxide L^5 is a favorable one for the bridging coordination as L^5 contains two equivalent N–O donor groups in their symmetrical structure. At the present time, a rather large group of polymeric mixed-ligand complexes on the basis of hexafluoroacetylacetonates of metals and L^5 are known in which the nitronyl nitroxide performs a bridging function (Scheme 19). Formation of complexes with such coordination of L^5 is favored by the use in the synthesis of an $M(hfac)_m$ to L^5 ratio of 1:1. With an excess of L^5 formation of $M(hfac)_m (L^5)_2$ with a monodentate coordination of the nitronyl nitroxide is possible (see Scheme 7).

Complexes on the basis of $Mn(hfac)_2$ have been studied in the most detail in References 112 to 117. It was shown that by a fine adjustment of the synthesis conditions it is possible to obtain compounds with different stoichiometries and structures.[82,112-115] Synthesis of Mn^{2+} complexes should be performed under mild conditions because prolonged boiling of $Mn(hfac)_2$ solutions with L^5 results in reduction of L^5 to the corresponding imino nitroxides and formation of mixed-ligand complexes of $Mn(hfac)_2$ with L^{11} or with L^5 and L^{11} simultaneously, as was convincingly demonstrated in the case of the reaction of $Mn(hfac)_2 \cdot 2H_2O$ with L^5 ($R = C_6H_5$).[116,117] The complex of $Mn(hfac)_2$ with L^5 ($R = C_6H_5$) has been obtained as two polymorphs (**88** and **89**). The solid phase of modification **88** is formed by the hexanuclear molecules $[Mn(hfac)_2(L^5)]_6$ while that of modification **89** by the polymeric chains $[Mn(hfac)_2(L^5)]_n$ (Scheme 19). The crystals of **90–93**, as of **89**, consist of polymeric chains. The nature of R has a substantial influence on the structure of the polymeric chains. For $R = i-C_3H_7$, complex **90** develops in the solid state a *trans*-coordination[113] of L^5 whereas for $R = C_2H_5$, $n-C_3H_7$ in complexes **92,93** there is a *cis*-coordination[115] of L^5. It will be recalled that for the complexes $Mn(hfac)_2(L^5)_2$ with a monodentate coordination of L^5, also, formation of both *cis*-(**54**) and *trans*-derivatives (**52**) has been observed.[82] The structure of **89, 91,** and

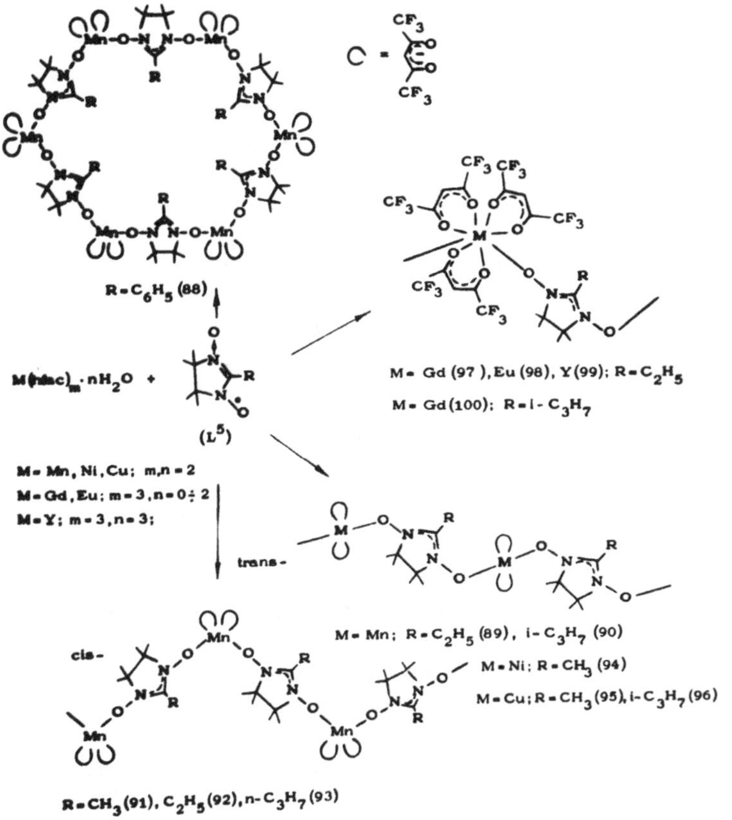

SCHEME 19.

94 (Scheme 19) has not been determined and the complexes have been assumed to have a chain structure analogous to that of **90**, **92**, and **93** on the basis of similarity of their magnetic properties. In the exchange clusters $M^{2+}-O\dot{-}N<$ of complexes **88–94** there is a strong exchange interaction of an antiferromagnetic character.

In the polymeric chains of complexes **95**, **96** the bidentate L^5 bridges are in a *trans*-coordination to each other (Scheme 19). In contrast to complexes **88–94**, in the exchange clusters $Cu^{2+}-O\dot{-}N<$, as in the previously considered complexes of Cu^{2+} with an apical coordination of the nitroxyl group **67**, **73–78**, **82–85**, the exchange interaction is ferromagnetic.[118,119] In the trinuclear complex **79** (Scheme 15) with a bidentate bridging coordination of L^5, NGs are coordinated to the terminal Cu^{2+} atoms in an equatorial position and, as expected, this results in a strong antiferromagnetic exchange interaction between the spins of Cu^{2+} and L^5. The remnant magnetic moment of the complex that corresponds to one unpaired electron is due to the central Cu^{2+} ion of the molecules.[106]

For the polymeric in the solid-state complexes of rare-earth elements **97–100** the exchange interaction between the unpaired electrons on the metal ion and on L^5 has been found to be of a ferromagnetic character, though of a small magnitude, whereas the exchange interaction between the unpaired electrons on the coordinated L^5 is antiferromagnetic and of a larger magnitude.[120-122]

Such an exchange interaction was also observed for complexes **55–59** (Scheme 7) of rare-earth elements[85-87] with a monodentate coordination of L^5 and has been confirmed by comparison of magnetic properties of complexes **97, 98,** and **99**.[120,121] Complex **99** is isostructural with complexes **97** and **98** but the central atom in **99** is diamagnetic, so that the exchange interaction in complex **99,** which is only due to the interaction of an unpaired electron on L^5, actually closely approximates that of compounds **97, 98,** and **100**. It is of an antiferromagnetic character and is close in magnitude to that found for compounds **97, 98,** and **100**.

Note the simultaneous formation of complex **100** and a complex of the composition $Gd(hfac)_3(L^5)(H_2O)$ in the reaction of $Gd(hfac)_3 \cdot 2H_2O$ with L^5 (R = i-Pr). In the latter complex L^5 is coordinated in a monodentate fashion, with the 8th coordination site being occupied by the oxygen atom of a water molecule, and it is the coordination of water that prevents formation of a polymeric structure. The $Gd(hfac)_3(L^5)(H_2O)$ complex as well as the isostructural complex $Eu(hfac)_3(L^5)(H_2O)$ represent a rare case of simultaneous coordination of a nitroxide and a water molecule.[148]

In summarizing the material presented in Section IV., it should be said that the approach to the synthesis of compounds with a metal-NG coordination using a bridging bifunctional nitroxide as a synthon is much superior to the approach that uses a ligand with only the NG as a donor group as to the possibility they give for the variation of the starting components in the reactions of mixed-ligand complex formation. Preparation of complexes with bifunctional nitroxides usually does not require such precautions as removal of moisture from the reaction mixture, performing the synthesis in an inert atmosphere. Solid phases of complexes with bifunctional nitroxides are considerably more stable in storage and in handling in a study than complexes with monofunctional nitroxides. It should be noted that the use of bifunctional nitroxides in the synthesis of M–O∸N< coordination compounds, although being a favorable factor, is in itself not sufficient for a bridging coordination of the nitroxide to occur. As shown in Section III, in solid mixed-ligand complexes even those that are symmetrical with respect to their donor groups, ligands like nitronyl nitroxides may be coordinated in a monodentate fashion. Nevertheless, most of the known complexes with a metal-NG coordination contain a bridging bifunctional radical as a synthon. One should also not forget that bridging coordination of ligands favors formation of dimensional structures in the solid state of the complexes which, in turn, promotes the occurrence of magnetic-phase transitions. Among complexes with a bidentate bridging coordination of their radicals, chain polymers have been revealed in which solid phases exhibit a one-dimensional ordering at low temperatures and in several cases there was a three-dimensional ordering occurring by a dipole-dipole interaction of the moments. The results of studies in this direction undoubtedly make a contribution to the progress in an actual problem of today — the synthesis of molecular ferromagnetics — and are discussed in more detail in Section VI.

V. TRIFUNCTIONAL NITROXIDE

A. Cu²⁺ CHELATES CONTAINING A DONOR GROUP IN AN α-POSITION TO THE NITROXYL GROUP

It appears, at first sight, that complexes containing a metal-NG bond can be easily synthesized by using a paramagnetic ligand containing in its structure a donor functional group in an α-position to the $>N\text{-}O$ group. Such a situation is depicted in Scheme 20 for the general case **(A)** and as examples of compounds **(101, 118)**.

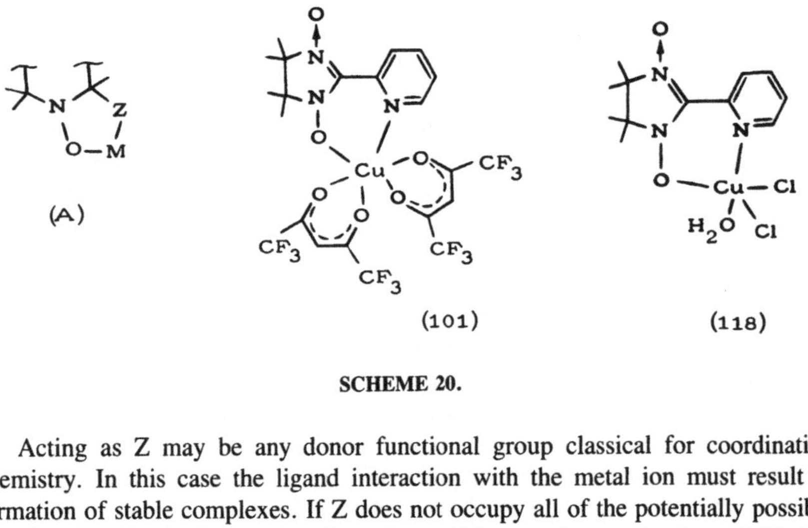

SCHEME 20.

Acting as Z may be any donor functional group classical for coordination chemistry. In this case the ligand interaction with the metal ion must result in formation of stable complexes. If Z does not occupy all of the potentially possible coordination sites in the coordination sphere of the metal, coordination of NG may take place along with the coordination of the Z donor atoms. Coordination of the NG occurs as if forcefully, due to Z being in an α-position to the $>N\text{-}O$ group. Thus, in **101** the pyridine fragment nitrogen, having strong donor properties, easily coordinates to Cu²⁺ thereby creating favorable conditions for the coordination of the oxygen atom of the nitronyl-nitroxyl fragment. As a result the chelate coordination of the paramagnetic organic ligand takes place.[123] If such a ligand as L¹³ does not occupy all potential coordination sites at its chelate coordination, other donor molecules present in the reaction mixture may enter the coordination sphere along with the anions (see complex **118** in Scheme 20[149]). In the case when the nitrogen atom is in a *meta-* or *para*-position (L¹⁴ and L¹⁵, Scheme 21) coordination only occurs at the nitrogen atom of the pyridine fragment and NG is not coordinated.[123,132,133]

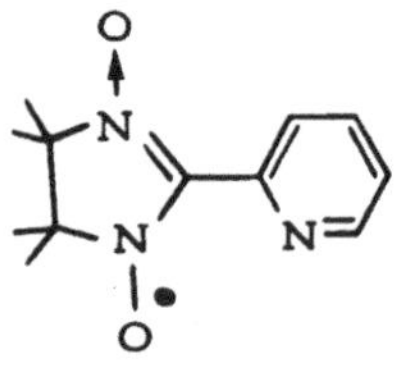

(L^{13}) 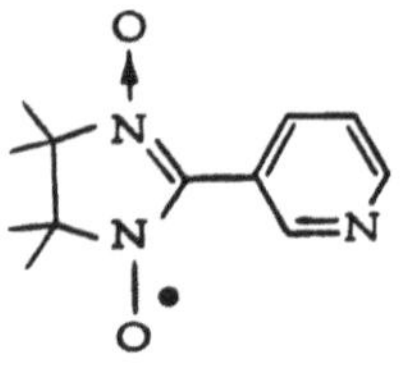   (L^{14}) 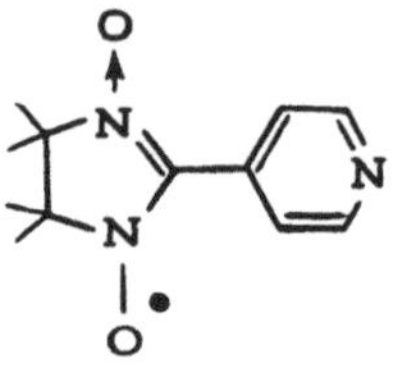   (L^{15})

SCHEME 21.

Taking into account the known regularities of the chemistry of coordination compounds and the large possibilities of the synthetic organic chemistry, it would be possible to choose, depending on the metal, such functional groups Z, which would create favorable conditions for formation of stable complexes with the chosen metal. In spite of the large potential of this approach, complexes of paramagnetic metal ions with such ligands are not numerous. One reason for preventing development of studies in this direction is that most of the synthesized nitroxides lying in the α-position to NG are hydrocarbon groups and all known complexes of this type (**101–107**) are diamagnetic, i.e., the exchange interactions in these complexes between the unpaired electrons on the paramagnetic centers are strongly antiferromagnetic.

Scheme 22 shows examples of complexes with a Cu–NG bond containing a donor group in an α- (**107**) or β-position (**102–106**) to the NG.[124–131,134] It has been found that complexes **102–107** are highly reactive in solution, for example, they readily react with acetone, CHCl$_3$, and C$_2$H$_5$OH. A decisive influence on the processes taking place in the reactions is the pH of the medium.[134] Thus, in the complexation reaction of Cu^{2+} with thiosemicarbazone, L^{18}H, in CH$_3$OH, CHCl$_3$, CH$_2$Cl$_2$, and acetone the nitroxyl fragment of L^{18}H is reduced to a hydroxylamine, which results in formation of **108**. In the reaction with the analogous semicarbazone, L^{19}H, such a process was not revealed, i.e., in this case the presence of a semicarbazone group instead of a thiosemicarbazone one does not promote reduction of the >N–O fragment to >N–OH.[135] Reduction of nitroxides to the corresponding hydroxylamines has been mentioned in Reference 1 (compounds **5–8**, and **45**) for L^1 and L^2 in their interaction with Pd^{2+} salts. It has been shown[134,135] that such a process is also possible in the presence of Cu^{2+} compounds. In contrast to L^1 and L^2, which did not have any functional groups other than NG, in the case of L^{18}H and L^{19}H the possibility of occurrence of such a process is also associated, along with the pH of the medium, with the nature of the functional group (Z) being coordinated. On the contrary, in alkali media in the presence of Cu^{2+} salts the hydroxylamine fragment of L^{16}H quickly oxidizes to a nitroxide one and complex **107** is produced (Scheme 22).

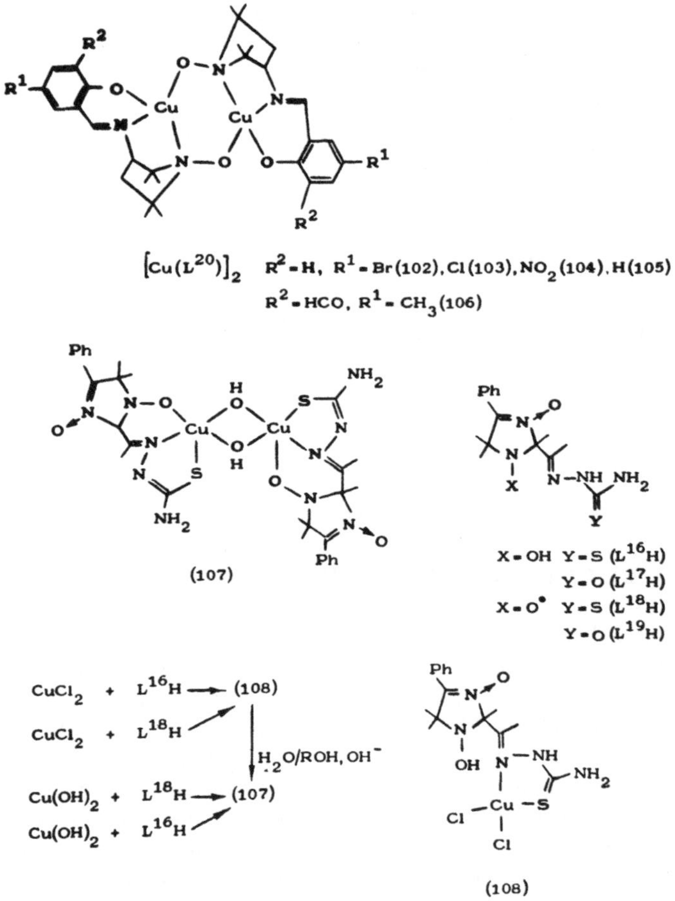

$$[Cu(L^{20})]_2 \quad R^2 = H, \; R^1 = Br\,(102), Cl\,(103), NO_2\,(104), H\,(105)$$
$$R^2 = HCO, \; R^1 = CH_3\,(106)$$

(107)

X = OH Y = S $(L^{16}H)$
Y = O $(L^{17}H)$
X = O· Y = S $(L^{18}H)$
Y = O $(L^{19}H)$

$$CuCl_2 \; + \; L^{16}H \longrightarrow (108)$$
$$CuCl_2 \; + \; L^{18}H$$
$$\Big| H_2O/ROH, OH^-$$
$$Cu(OH)_2 \; + \; L^{18}H \longrightarrow (107)$$
$$Cu(OH)_2 \; + \; L^{16}H$$

(108)

SCHEME 22.

Thus the possibility of the occurrence of redox processes in the systems under consideration is greatly influenced by pH of the medium, the metal ion, the structure of the nitroxide, and the nature of its donor functional groups. The diamagnetism of complexes **101–107** is consistently explained by a strong antiferromagnetic exchange interaction between the unpaired electrons on Cu^{2+} and the coordinated NG. It was shown that at a small separation between the donor functional group (Z) and NG in the nitroxide their simultaneous coordination to the Cu^{2+} ion may give rise to complexes having a high reactivity.

Among the compounds containing an NG in close vicinity to the metal ion are also spin-labeled derivatives of ferrocene **137–139**, (Scheme 23) of which the synthesis and properties have been described in References 162 to 165.

The exchange interactions between the unpaired electrons on the paramagnetic metal ion of the sandwich fragment and the NG would undoubtedly be of interest, but Fe(II) in compounds **137–139** are in a low-spin d^6 state, i.e., diamagnetic. Attempts to oxidize spin-labeled ferrocenes **137–139** to obtain the corresponding nitroxyl-containing Fe(III) derivatives were unsuccessful because the formation of carbocations of type **140** stabilized by the π-system of the sandwich takes place.[162]

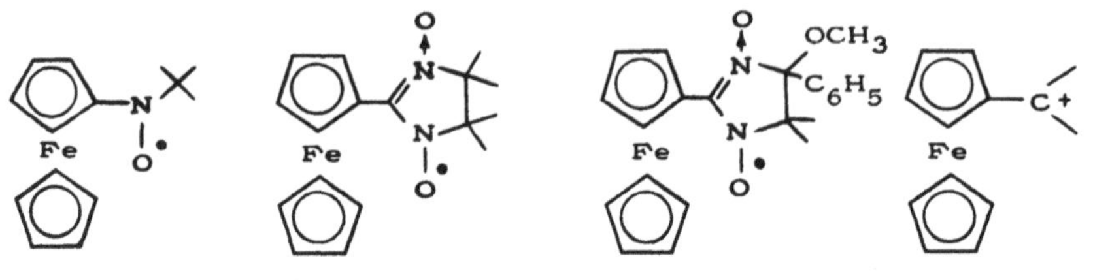

(137)   (138)   (139)   (140)

SCHEME 23.

B. METAL BISCHELATES WITH A BRIDGING COORDINATION OF THE NITROXIDE

Metal bischelates are the main group of complexes of transition metals with trifunctional nitroxides exhibiting a metal-NG bond in the solid phase. In the chelates of the previous section it was the structure of the nitroxide that made the occurrence of such a coordination possible. In contrast to this, in the case of ML_2 bischelates it is practically impossible to predict in advance whether the compounds will develop such a coordination in the crystallization of chelates from solution. At the present time accumulation of experimental evidence on the types of bischelates realizing such a coordination is taking place. The known examples are the complexes with nitroxides of two types: complexes with deprotonated β-diketonates and complexes with deprotonated enamineketones.

1. Copper β-Diketonates

Representatives of this group of complexes (**109–111**) are shown in Scheme 24. The coordination of the oxygen atoms of NG of neighboring molecules completing the environment around Cu^{2+} to a tetragonally distorted octahedron is shown by the dashed line. The apical distances $Cu-O_{N-O}$ in bischelates **109, 110** equal to 2.58 to 3.16 Å are considerably greater than the equatorial distances $Cu-O$ in β-diketonates which are normally equal to 1.9 Å. The solid phases of complexes **109** and **110** can be considered as coordination polymers in which two different zigzag chains are intersecting on each copper atom.[136,137,140,141]

The experimental $\chi(T)$ data for these complexes most closely fits the result of a calculation made for the case of ferromagnetic exchange clusters.[138,139,141] It is shown that a correct analysis of magnetic properties of complex **110** is complicated by the presence of two conformations of the $Cu(L^{22})_2$ molecule in the unit cell of the compound.[141]

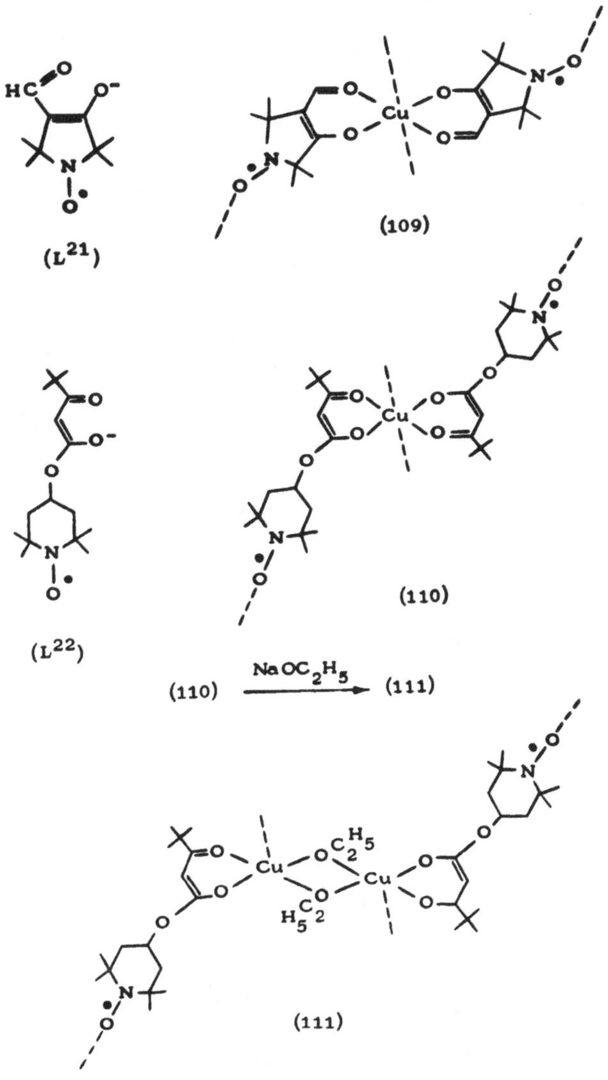

SCHEME 24.

Heating of Cu(L²²)₂ with sodium ethanolate in an inert atmosphere or boiling of a Cu(L²²)₂ solution in a large amount of ethanol is accompanied by a process that is rare for the metal β-diketonates resulting in formation of compound **111** (Scheme 24). In the binuclear molecules [Cu(L²²)(OC₂H₅)]₂ the square environment around each copper atom defined by two oxygen atoms of L²² and two oxygen atoms of the bridging ethoxy groups is completed to a square pyramidal one at the expense of oxygen atoms of the NG of neighboring binuclear molecules.[142] In the exchange clusters $Cu^{2+}-O\dot{-}N<$ in the solid phase of **111**, as in the solid phases of **109, 110**, the exchange interaction has a ferromagnetic character.

2. Metal Enamineketonates

References 13 and 143 to 147 describe the synthesis, structure, and magnetic properties of bischelates $M(L^{23})_2$ and $M(L^{24})_2$ (Scheme 25). These complexes are easily formed in the reaction of metal acetates with enamineketone derivatives of the 3-imidazoline nitroxide in alkali media. An X-ray structure study of single crystals of **112, 113,** and **115** indicated the complexes to have a layered polymeric structure which occurs due to the metal ion coordination, along with the oxygen and nitrogen atoms of the deprotonated enamineketone group, and oxygen atoms of NGs from two adjacent molecules.[13,97] The structures of complexes **114** and **116**, as determined from X-ray diffraction data, are analogous to those of **113** and **115**, respectively. The coordination of the NG by the metal ion is largely favored by a rather high acceptor ability of R $=$ CF_3, $C(O)OC_2H_5$.

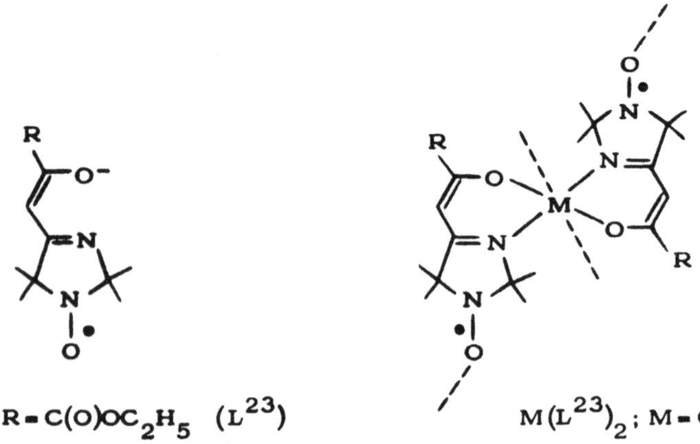

$R = C(O)OC_2H_5$ (L^{23})

$R = CF_3$ (L^{24})

$M(L^{23})_2$; $M =$ Cu (112), Co (113), Ni (114).

$M(L^{24})_2$; $M =$ Co (115), Ni (116).

SCHEME 25.

In the solid phases analogous to $M(L^{23})_2$ and $M(L^{24})_2$ containing R CH_{3-}, C_2H_{5-}, and *tert*-C_4H_9 groups the coordination of NG of the neighboring molecules is absent.[97] In complexes of Ni^{2+} NG coordination is absent not only with such bulky substituents as R $=$ *tert*-C_4H_9, C_6H_5, that create considerable sterical hindrances when determining the tetrahedral coordination of the metal ion, but it is also absent in the case of R $=$ CH_3, C_2H_5 in which complexes in the solid state have a square *trans*-coordination. In spite of all this, the acceptor ability of R and its structure are not the only factors that favor or prevent the coordination of the NG of the neighboring molecules. The nature of the metal is also of importance.

Comparing the series of bischelates $M(L^{23})_2$ and $M(L^{24})_2$ in Scheme 25 one can see that the series of complexes $M(L^{24})_2$ that have a layered polymeric structure in the solid state do not include the complexes of $Cu^{2+}-Cu(L^{24})_2$ **(117)**. This is because the crystal structure of **117** is a molecular one with no NG coordination from the neighboring molecules.[144] Besides, complexes, for example **115**, may exist as several polymorphs, one with this added coordination and the other without.

In the exchange clusters $>N\dot{-}O-M^{2+}-O\dot{-}N<$ of complexes of Co^{2+} and Ni^{2+} **(113–116)** the exchange interactions are of a strong antiferromagnetic character whereas in Cu^{2+} complex **112**, as in the previously considered Cu^{2+} bischelates **109–111**, they are ferromagnetic, a fact that is undoubtedly of interest for designing molecular ferromagnetics.

Thus the use of tri- (and higher) functional nitroxides for the synthesis of complexes having a metal-NG coordination deserves the closest attention. First, because this produces very stable complexes. Secondly, it is possible to design such nitroxides that will fill a necessary number of the coordination sites, balance out the positive charge of the metal, and provide for sufficiently high acceptor properties of the metal in the resultant coordination matrices. In developing methods for the synthesis of molecular ferromagnetics, the introduction into the synthesis of nitroxides of the type L^{21} to L^{24} will help get rid of ''ballast'' diamagnetic anions, such as hexafluoroacetylacetonate, haloidated carboxylate, or halogenide. By using polyfunctional nitroxides containing a special set of donor groups it will be possible to obtain heterospin coordination polymers with an electronic structure that would be favorable for an effective conduction of spin density which is, in turn, of no small importance when designing molecular ferromagnetics.

VI. MOLECULAR FERROMAGNETICS ON THE BASIS OF COMPLEXES WITH NITROXIDES

A. ZERO-DIMENSIONAL SYSTEMS

Classified as zero-dimensional systems should be the coordination compounds of metals with nitroxides in which both intramolecular (intracluster) and intermolecular (intercluster) exchange interactions between the unpaired electrons on the paramagnetic centers are of a ferromagnetic character, i.e., the energies J and $J'z$ are positive (the latter implies writing down the spin-Hamiltonian as $H = -J \cdot S_1 \cdot S_2$). Of the complexes discussed in the previous sections meeting these requirements are the complexes listed in Table 1.

It could be expected that with decreasing temperature the solid phases of complexes **77, 117** will sooner or later realize a magnetic-phase transition to the ferromagnetic state. However, the values of $J'z$ are so small that such a transition can be detected only at temperatures of the order of 0.1 K, which was not attained in References 101 and 144.

Still, even if registered, such a transition would hardly be of any substantial interest because its T_c would have been very low.

TABLE 1
Examples of Complexes With Positive Values of J and J'z

Compound	No.	J,cm^{-1}	J'z,cm^{-1}	Ref.
[Cu(hfac)$_2$]$_3$(L^9)$_2$,R=C$_6$H$_5$	77	12.0	0.05	101
Cu(L^{24})$_2$	177	13.9	0.06	104

The data in Table 1 is of interest in the following respects. First, they show that the occurrence of a magnetic-phase transition to the ferromagnetic state in complexes with nitroxides is possible in principle. Second, they confirm the general rule: for the magnetic-phase transition to occur at higher temperatures, compounds of a higher dimensionality are necessary in which both J and J'z would be of a rather large magnitude of the same order. It is obvious that compounds having a molecular crystal structure, i.e., zero-dimensional systems, do not meet these requirements because always for them J > J'z and |J'z| ≤ 1 cm^{-1}.

The next stage toward creation of molecular ferromagnetics on the basis of complexes with nitroxides would be, therefore, development of the synthesis of one-dimensional complex compounds containing bifunctional nitroxides as a para-magnetic synthon with a bridging function. For the spin interaction in such an exchange cluster to be large (J) it is always desirable that a metal-NG coordination be realized.

B. ONE-DIMENSIONAL SYSTEMS

Among the complexes of paramagnetic metals with nitroxides possessing a metal-NG bond there is a large number of compounds consisting in the solid state of polymeric chains. Of the complexes discussed in the previous sections there are complexes of Cu^{2+} (**44, 67, 73, 74, 82–85, 95, 96**), Mn^{2+} (**89–93**), Ni^{2+} (**94**), Rh^{2+} (**63, 64**), and Gd^{3+}, Eu^{3+}, and Y^{3+} (**97–100**). The most progress in the designed synthesis of ferromagnetics using chain polymers has been achieved by Caneschi, Gatteschi, Sessoli, and Rey.[14,98,150]

They synthesized a series of mixed-ligand complexes of Mn(hfac)$_2$ with L^5 and studied their structure and magnetic properties.

In complexes of Mn^{2+} (**89–93**) and Ni^{2+} (**94**) formed by the polymeric chains [Mn(hfac)$_2$ (L^5)]$_n$, the bidentate bridging coordination of L^5 results in strong exchange interactions of an antiferromagnetic character[84,113-115] between the unpaired electrons of the metal ion and L^5. For complexes of Mn^{2+} **89–93** J = − 210 to − 330 cm^{-1} and for Ni^{2+} complex **94** J = − 424 cm^{-1}. The strong antiferromagnetic exchange interaction in the heterospin chains [M(hfac)$_2$(L^5)]$_n$ determines that complexes **89–93** behave as typical one-dimensional ferromagnetics (Scheme 26) which was confirmed by the studies of the anisotropy of magnetic properties of the complexes of single crystals at low temperatures.[14,113-115,150]

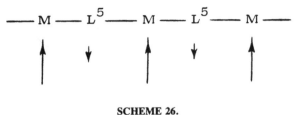

SCHEME 26.

At 7.6 to 8.6 K for complexes of Mn^{2+} and at 5.3 K for the complex of Ni^{2+} magnetic-phase transitions to the ferromagnetic state have been revealed. A three-dimensional magnetic ordering in complexes **89–94** has been related by the authors of References 113 to 115 to a dipole-dipole interaction of the moments of the chains. This allowed a rational explanation of the reason for the lower value of T_c in complex of Ni^{2+} **94** as compared with complexes of Mn^{2+} **89–93**: as a result of the strong antiferromagnetic interaction between the spins, the moment in the Ni^{2+} complex chains is substantially lower than that in the Mn^{2+} complex chains because the spin of Ni^{2+} (S = 1) is smaller than that of Mn^{2+} (S = 5/2) and it is this circumstance that results in a decrease in T_c. The structure of complexes of Cu^{2+} **95,96** is also analogous to that of complexes **89–94**. Moreover, exchange interactions in the exchange clusters $Cu^{2+}–O\dot{-}N<$ are of a ferromagnetic character. However, they are small in magnitude (J $\sim$ 25 to 30 cm^{-1} within the chain and $|J'z|\sim$0.002 to 0.046 cm^{-1} between the chains) and no magnetic ordering effects could be found for complexes **95,96** at temperatures above 1 K.[119] For the same reason, i.e., because of the small values of the exchange interaction energies no magnetic phase transitions could be revealed in complexes of Cu^{2+} (**44, 67, 73, 74, 82–85**), Rh^{2+} (**64**), and rare-earth elements (**97–100**). In addition to this, the interchain interaction of the moments in the solid phases of these complexes is of an antiferromagnetic character and a decrease in temperature results in a decrease in the magnetic moment or its complete disappearance.

It was observed in References 14, 98, and 150 that magnetics **89–94** in the solid state are essentially one-dimensional. The bulky hfac ligands prevent effective interaction between the chains. A three-dimensional ordering appears in this case only due to the fact that as a result of a strong antiferromagnetic interaction between nitronyl nitroxides and Mn^{2+} and Ni^{2+} there arises a strong uncompensated magnetic moment in the polymeric chain. In this connection it is believed[14,98,150] that an increase in the Curie temperature for such compounds can be achieved by increasing the intrachain spin interaction and/or by decreasing the interchain distance, i.e., by increasing the interchain interaction of the magnetic moments. Because further substantial increase in interchain exchange interaction for complexes of Mn^{2+} and Ni^{2+} does not seem possible as it is already very large — 300 to 400 cm^{-1} — in complexes **89–94** (attempts to achieve an increase in the intrachain exchange interaction energy of a ferromagnetic character would be reasonable in the polymeric complexes of Cu^{2+}), it is better to replace the hfac ligands by that which would promote a more effective interchain dipole-dipole interaction. Therefore, mixed-ligand complexes of Mn^{2+} perfluorocarboxylates with nitronyl nitroxides have been synthesized and their magnetic properties studied.[14,151,152]

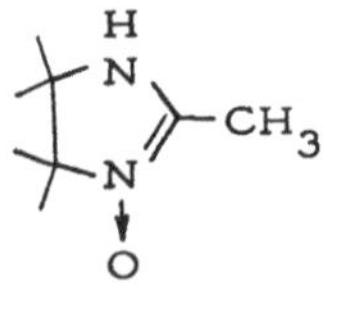

$$Mn(C_2F_5CO_2)_2 \ (L^5), \ R = CH_3 \ (119)$$

$$[Mn(C_2F_5CO_2)_2]_2 \ (L^5)(L^{25}), \ R = CH_3 \ (120)$$

$$Mn(C_6F_5CO_2)_2 \ (L^5), \ R = CH_3 \ (121)$$

$$[Mn(C_6F_5CO_2)_2]_2(L^5), \ R = C_2H_5 \ (122)$$

(L^{25})

SCHEME 27.

By the reaction of $Mn(R'CO_2)_2 \cdot 2H_2O$ with L^5 ($R = CH_3$, C_2H_5), mixed-ligand complexes **119–122** (Scheme 27) have been obtained. For complexes **121** and **122** in the solid-state, magnetic-phase transitions to the ferromagnetic state at 24 and 20.5 K, respectively, have been found, which no doubt, is a great success. For compound **121** a hysteresis loop has been revealed at 4.2 K with a coercive field of 240 G and a spontaneous magnetization of 4200 $cm^3 \cdot G \cdot mol^{-1}$. Between the unpaired electrons on the paramagnetic centers of complex **121** there are strong antiferromagnetic interactions. The magnetization equal at 500 G to about 50 $cm^3 \cdot G \cdot mol^{-1}$ increases to 10,000 $cm^3 \cdot G \cdot mol^{-1}$ with the increase in field to 10,000 G.[14,151] Unfortunately, the structures of complexes **121** and **122** have not been determined. Based on the results of a structural study for complex **119** that indicated the possibility of two-dimensional systems in this complex, the higher T_c values for complexes **121** and **122** as compared with complexes **89–94** have been suggested[152] to be due to a two-dimensional structure of their solid phases.

C. TWO-DIMENSIONAL SYSTEMS

Belonging to the two-dimensional metal complexes with a metal-NG coordination are metal bischelates with a bridging coordination of the trifunctional nitroxide **109–116.** The solid phases of these complexes should be considered as coordination layered polymers. The first example of complexes with such a structure was a complex of Cu^{2+} **110** (Scheme 24).[141] It is recalled that the exchange clusters Cu^{2+}–O–$N<$ **109–112** are of a ferromagnetic character and this is, in principle, favorable for the occurrence of a magnetic-phase transition to a ferromagnetic state.

The exchange interaction of a ferromagnetic character between the spins of the paramagnetic centers within a layer is a favorable but not sufficient factor for a magnetic-phase transition to take place. It is also necessary that the exchange interaction energy in the exchange channels of the layer are not small. A ferromagnetic ordering of spins in strong magnetic fields at 4.2 K has been only detected for complexes of Cu^{2+} with imidazoline nitroxide derivatives **112**[146,147] and **123** (Scheme 28).[158]

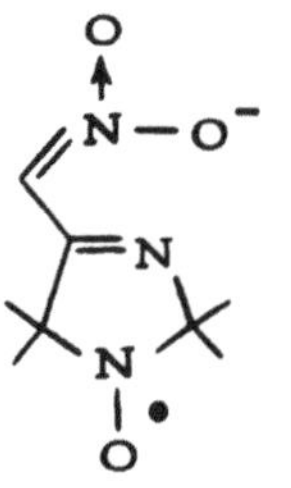

(L²⁶)   (123)

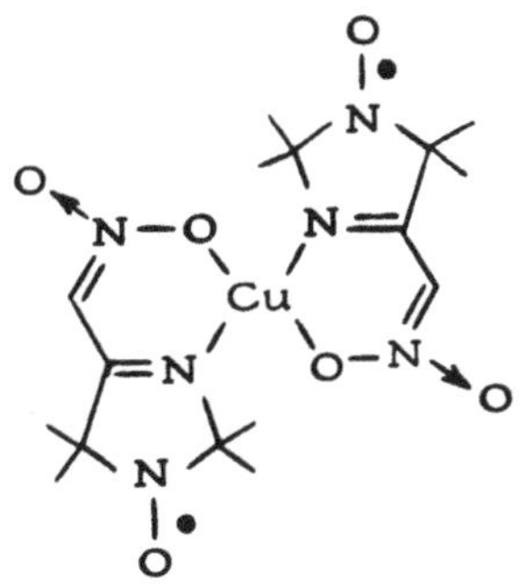

SCHEME 28.

Observation of the ferromagnetic ordering for complexes with the derivatives of the imidazoline nitroxide was possible due to a high efficiency of the imidazoline heterocycle in "conducting the spin density" both in solution[154,155] and in the solid state.[10,13] The structure of complex **123** could not be determined and the layered polymeric structure of complex **112** has been proved by X-ray structure analysis.[97] A projection of the polymeric layer in the structure of complex **112** is given in Figure 1.

One of the ways toward an increased exchange interaction energy between the spins of paramagnetic centers in structures analogous to that in Figure 1 would be replacing Cu^{2+} by a metal ion having a higher electron spin then Cu^{2+}. This was achieved by the synthesis of complexes Ni^{2+} and Co^{2+} **113–116**. The structure of these complexes in the solid state is practically the same as in complex **112**. The exchange interaction energy in the exchange clusters $>N\dot{-}O–M^{2+}–O\dot{-}N<$ of complexes **113–116** has a magnitude of about $-(100 \div 200)$ cm^{-1}, i.e., it is considerably higher than in the exchange clusters $>N\dot{-}O–Cu^{2+}–O\dot{-}N<$ of complexes **109–112** (10 to 30 cm^{-1}). Here there arises another problem: the antiferromagnetic character of the exchange interactions in complexes **113–116** results in a decrease in the magnetic moments of the complexes with decreasing temperature. As an example, Figures 2 and 3 show temperature dependence of the molar paramagnetic susceptibility χ_M and the effective magnetic moment for complexes **116** and **115**. The broad maxima in the $\chi(T)$ curves for both compounds indicate considerable antiferromagnetic interaction between the paramagnetic centers.

Deviating from the subject let us give a closer inspection to Figure 3. It illustrates the fact that complex **115** exists in two polymorphs α-Co(L²⁴)₂ and β-Co(L²⁴)₂. The synthesis conditions of these modifications have been described in Reference 145. Magnetic properties discussed above refer to the β-Co(L²⁴)₂ modification having in the solid state a layered polymeric structure with a $>N\dot{-}O–Co^{2+}–O\dot{-}N<$ coordination. The α-Co(L²⁴)₂ modification has a molecular crystal structure with no $Co^{2+}–O\dot{-}N<$ coordination. It can be seen from Figure 3 how considerable the

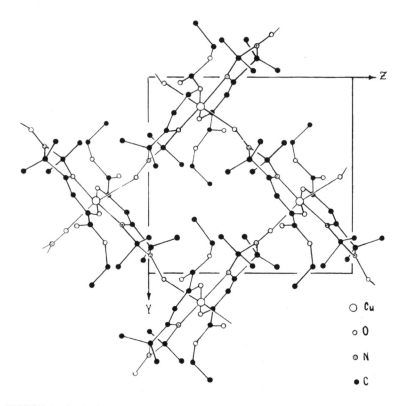

FIGURE 1.   Projection of a layer in the structure of complex Cu(L²³)₂ onto the **(100)** plane.

differences may be in magnetic properties between the polymorphs of one and the same compound. The behavior of β-Co(L²⁴)₂ is similar to that of Ni(L²⁴)₂. The dependence $\chi_M(T)$ for this compound has a maximum at 140 K indicating an essentially ferromagnetic behavior of β-Co(L²⁴)₂. On the contrary, for α-Co(L²⁴)₂ with decreasing temperature there is an increase in μ_{eff} indicating the presence of ferromagnetic exchange interactions in α-Co(L²⁴)₂ (Figure 3; this figure also shows the phase transition from β-Co(L²⁴)₂ to α-Co(L²⁴)₂ occurring upon heating to 144°C). It should be emphasized once more that it is only from X-ray data that a conclusion can be made at the present time about the existence of the M–O≐N< coordination in the solid state. One should be extremely careful when using data from indirect methods for this purpose, including the magnetic ones. Thus, as in the case of complex **115**, we have succeeded in obtaining two polymorphs for complex **124**, α-Cu(L²⁷)₂ and β-Cu(L²⁷)₂ (Scheme 29).

Magnetic properties of these modifications are (Figure 4) similar to those of α- and β-Co(L²⁴)₂. With decreasing temperature μ_{eff} for α-Cu(L²⁷)₂ increases (J = 14.3 cm⁻¹, J′z = 0.12 cm⁻¹) and for β-Cu(L²⁷)₂ decreases (J = −5.7 cm⁻¹, J = 1.2 cm⁻¹). An X-ray study of α- and β-modifications of complex **124** indicated, however, that they both have a molecular crystal structure and in the solid phases

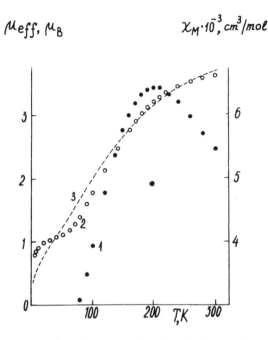

FIGURE 2. The dependences χ_M (T) − 1, μ_{eff} (T) − 2, and a theoretical μ_{eff} (T) curve − 3 for the complex Ni(L^{24})$_2$.

of both α- and β-modification there is no Cu^{2+}–O$\div$N< coordination. The different magnetic properties of the modifications of complex **24** are due to the differences in the type of the central atom coordination in α- and β-Cu(L^{27})$_2$. In the molecules of the α-modification the coordination polyhedron is a distorted tetrahedron, whereas in the β-modification the environment of the central atom is a square.[157,158] In principle, the existence of different polymorphs of complexes provides a valuable basis for a deeper insight into the magneto-structural correlations in complexes of metals with nitroxides, but at the moment we are only interested in those that have a metal-NG coordination.

Coming back to the discussion of magnetic properties of complexes **113–116** it should be said that no magnetic-phase transitions have been revealed for these complexes in the temperature region 4.2 to 300 K. This may be explained by the fact that because of the strong antiferromagnetic exchange interactions in the exchange clusters >N$\div$O–Ni^{2+}–O$\div$N< and >N$\div$O–Co^{2+}–O$\div$N< the remnant spin is very small. In other words, magnetic properties of complexes **113–116** are an example of such a situation where antiferromagnetic exchange interactions turned out to be too large for the transition to the ferromagnetic state to occur because the remnant magnetic moments in the exchange clusters have become very small. A decrease in the antiferromagnetic exchange interaction and, hence, an increase in the remnant spin in complexes **113–116** could be achieved by "moving apart" the paramagnetic centers. This was achieved for complexes **115, 116** by inserting

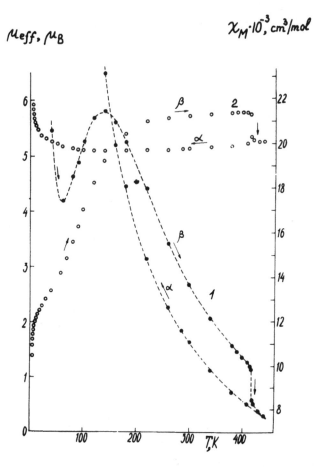

FIGURE 3. The dependences χ_M (T) − 1, μ_{eff} (T) − 2 for α- and β-modifications of the complex $Co(L^{24})_2$.

(L^{27})

(124)

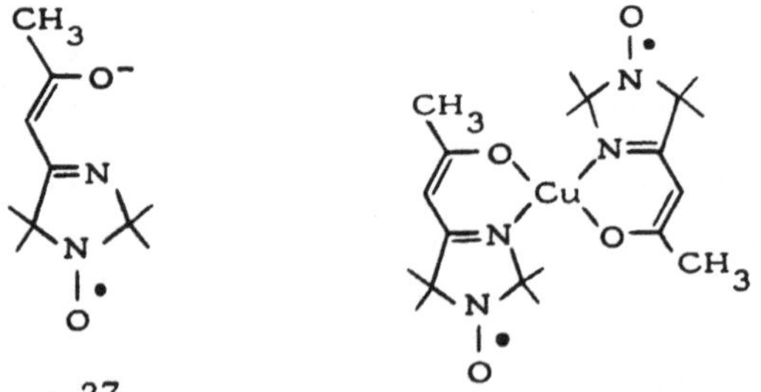

SCHEME 29.

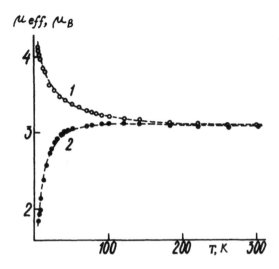

FIGURE 4. The μ_{eff} (T) dependence for α-(1) and β-(2) modifications of $Cu(L^{27})_2$.

alcohol molecules between their paramagnetic centers, i.e., by going from the exchange clusters $>N\dot{-}O\text{--}M^{2+}\text{--}O\dot{-}N<$ to the exchange clusters

$$>N\dot{-}O\text{---}HO\text{--}M^{2+}\underset{\underset{R}{\diagdown}}{\overset{\overset{R}{\diagdown}}{\text{------}}}OH\text{---}O\dot{-}N<$$

Such a transition was possible because among all studied bischelates with enamineketone derivatives of the imidazoline nitroxide complexes $Ni(L^{24})_2$ and $Co(L^{24})_2$ have the greatest acceptor ability and recrystallizing from alcohols, for example from methanol or ethanol, they separate into the solid phase with two alcohol molecules. When stored in air $M(L^{24})_2$ $(R^1OH)_2$ (Scheme 30) gradually, within several days, lose their alcohol but when kept under a layer of a saturated solution in alcohol or under an atmosphere of the corresponding alcohol or under a glue film they remain stable for months.[13,159-161]

According to X-ray data complexes $Ni(L^{24})_2(CH_3OH)_2$ **(125)**, $Ni(L^{24})_2(C_2H_5OH)_2$ **(126)**, and $Co(L^{24})_2(CH_3OH)_2$ **(127)** are isostructural. Their crystals are formed by centrosymmetrical molecules $M(L^{24})_2(R^1OH)_2$. The metal atom has an octahedral environment formed by two oxygen and two nitrogen atoms of bidentately coordinated L^{24} in the equatorial plane and by two oxygen atoms of the alcohol molecules in axial positions. In the crystal lattice the molecules are arranged in layers. Within the layers the molecules are connected to each other by hydrogen bonds between the oxygen atoms of NG and the oxygen atoms of the OH groups of the coordinated

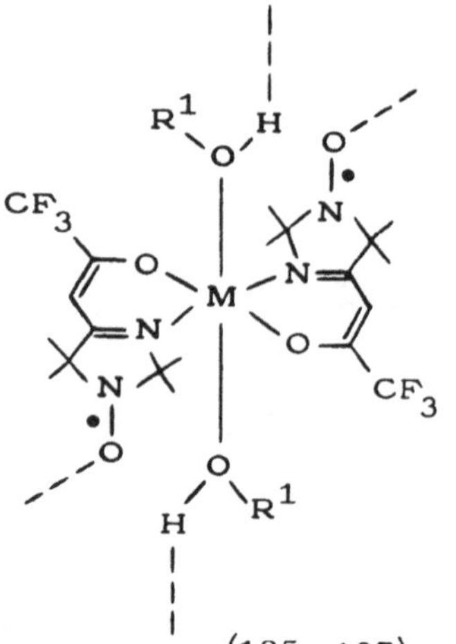

(125 - 127)

SCHEME 30.

alcohol molecules in the adjacent molecules. This is shown in Figure 5 by the example of the crystal structure of complex **125**. The arrangement of the paramagnetic centers in the layer is shown in Figure 6.

In the temperature region of 5 to 8 K polycrystalline samples of complexes **125–127** as well as of complexes $Co(L^{24})_2(C_2H_5OH)_2$ (**128**) and $Ni(L^{24})_2(R^1OH)_2$ where R^1 = C_3H_5 (**129**), n-C_3H_7 (**130**), n-C_4H_9 (**131**), and n-C_5H_{11} (**132**) undergo magnetic-phase transitions characteristic of layered antiferromagnetics in the case of complex **126** and of ferromagnetics in the case of complexes **125, 127–132**.[159-161] In magnetic fields higher than 2 kOe complex **126** also becomes ferromagnetic, i.e., it is a typical metamagnetic. The magnitudes of spontaneous magnetization and the values of Curie temperatures for these complexes are given in Table 2.

It can be seen from Table 2 that the spontaneous magnetizations are below the values expected for these compounds. As pointed out in References 159 to 161 this is due to the fact that the magnetic phase transition in these complexes consists of a ferromagnetic ordering of the "remnant" resultant magnetic moments of individual exchange clusters

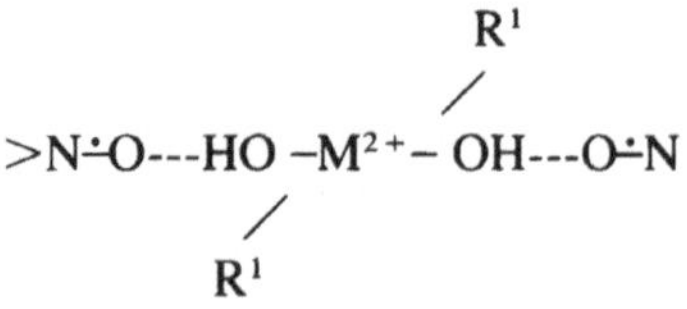

$$>N \dot{-} O \text{---} HO - M^{2+} - OH \text{---} O \dot{-} N \overset{\displaystyle R^1}{\underset{\displaystyle R^1}{\diagup\diagdown}}$$

within which paramagnetic centers interact in an antiferromagnetic fashion. It was the consideration of this circumstance that has determined transition from the synthesis of compounds of Ni^{2+} that were the first to show ability for magnetization[160]

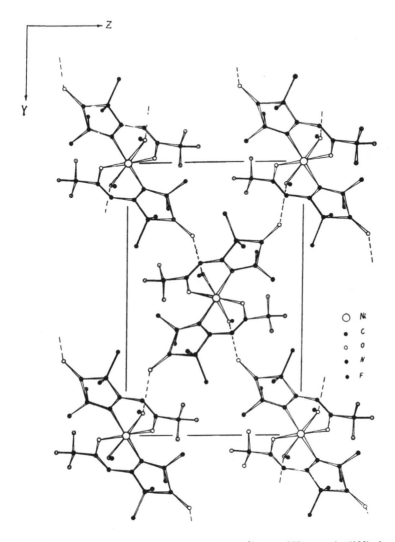

FIGURE 5. Projection of the crystal structure of $Ni(L^{24})_2$ $(CH_3OH)_2$ onto the **(100)** plane.

to the synthesis of Co^{2+} compounds.[161] Cobalt has a larger spin than nickel and it was natural to expect that for Co^{2+} compounds having the same structure as compounds of Ni^{2+} the magnitudes of magnetization would be higher. Indeed, data in Table 2 indicate that transition to Co^{2+} compounds to obtain increased σ_0 values was justified.

Discussing magnetic properties of mixed-ligand metal complexes with the derivatives of the imidazoline nitroxide let us mention one more important result obtained in the studies of these complexes. In Reference 153 the synthesis, structure, and magnetic properties of complexes $M(L^{26})_2(CH_3OH)_2$ where M = Ni **(134)** and

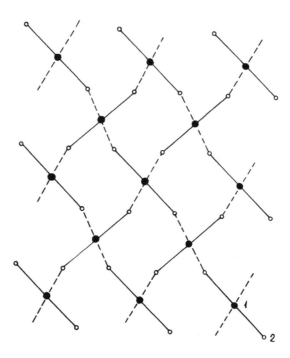

FIGURE 6. The arrangement of the paramagnetic centers in the layer: Ni atoms − 1; O atoms of the NG − 2; and intramolecular contacts Ni ... ‒O are shown by solid lines. Intermolecular contacts are shown by dotted lines.

TABLE 2
Magnetic Phase Transition Temperatures and
Spontaneous Magnetizations for Complexes
125–133

Compound	No.	T_c,K	σ_o (4.2 K), $\dfrac{G \cdot cm^3}{mol}$
$Ni(L^{24})_2(CH_3OH)_2$	125	5.8	230.295[b]
$Ni(L^{24})_2(C_2H_5OH)_2$	126	6.2	610.690[b]
$Ni(L^{24})_2(C_3H_5OH)_2$	129	6.2	365
$Ni(L^{24})_2(n\text{-}C_3H_7OH)_2$	130	6.0	277
$Ni(L^{24})_2(n\text{-}C_4H_9OH)_2$	131	5.9	535
$Ni(L^{24})_2(n\text{-}C_5H_{11}OH)_2$	132	6.3	436
$Ni(L^{24})_2(HO(CH_2)_4OH)$	133	6.4	452.495[b]
$Co(L^{24})_2(CH_3OH)_2$[a]	127	7.1	2500
$Co(L^{24})_2(C_2H_5OH)_2$[a]	128	8.1	2100.2150[b]

[a] Magnetizations at 4.3 K.
[b] For References 125, 126, 128, and 133, the values of σ_o are given for samples of complexes from different syntheses.

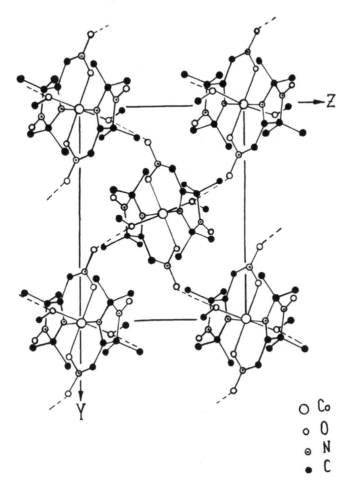

FIGURE 7. Projection of the structure of the $Co(L^{26})_2(CH_3OH)_2$ complex onto the **(100)** plane.

Co (**135**) were reported (see Scheme 28). The crystal structure of complexes **134**, **135** is extremely close to that for complexes **125–127**. The most significant feature of the structure layers in complexes **125–127** is that polymerization of the $ML_2(CH_3OH)_2$ molecules occurs via hydrogen bonds between the alcohol molecules coordinated by the metal ion and the oxygen atoms of deprotonated L^{26} nitrogroups of adjacent molecules, whereas the oxygen atoms of the NG do not participate in formation of intermolecular bonds (Figure 7). No magnetic-phase transitions have been found for complexes **134**, **135**. Comparison of the structure and magnetic properties of complexes **134**, **135** with those of complexes **125**, **127** unambiguously indicates that exclusion of the NG from the system of the polymeric bonds, in the present case destruction of the bond, makes the realization of a magnetic-phase

$$M\text{--O H---O}\overset{\cdot}{=}N<$$
$$/$$
$$R^1$$

transition impossible, at least at temperatures above that of the liquid gelium.

D. THREE-DIMENSIONAL SYSTEMS

Evidence for three-dimensional complexes with nitroxides is extremely limited. The only known complexes are $M(L^{24})_2(HO(CH_2)_4OH)$ where $M = Ni$ **(133)**, Co **(136)**.[166] Their synthesis has been set up on the basis of an analysis of the structure of compounds $M(L^{24})_2(R^1OH)_2$. The above-discussed complexes **125–132** are unstable under normal conditions. They gradually lose their alcohol changing to $M(L^{24})_2$ **(115, 116)**, becoming incapable of undergoing magnetic-phase transition to a ferromagnetic state. In order to obtain compounds that are more stable under normal conditions than $M(L^{24})_2(R^1OH)_2$, it was reasonable to use 1,4-butanediol in the synthesis of mixed-ligand complexes. Why 1,4-butanediol? Analysis of the structure of the complex $Ni(L^{24})_2(C_2H_5OH)_2$ indicated that the nearest distance between the carbon atoms of the CH_3 groups of the ethanol molecules in the adjacent layers are of the order of 4 Å. Hence, in order to retain the layered structure exhibited by complex **126** and at the same time to sew together the layers with chemical bonds, it was reasonable to use just 1,4-butanediol. Indeed, in the interaction of complexes **115, 116** with 1,4-butanediol in acetone it was possible to obtain mixed-ligand complexes $Ni(L^{24})_2[HO(CH_2)_4OH]$ and $Co(L^{24})_2[HO(CH_2)_4OH]$ having a framework structure that was stable under normal conditions for several years. Figure 8 shows the projection of the structure of complex **136** onto the **(100)** plane showing the sewing of the layers by the 1,4-butanediol molecules. Complex Ni^{2+} **(133)** is isostructural to Co^{2+} complex **(136)**. For complex **133** there is a magnetic-phase transition (Table 2) and for complex **136** the nonlinear character of the $\sigma(H)$ dependence below 6 K is not only described by the Brillouin function, but represents the sum of contributions from spontaneous magnetization and the interaction of magnetic moments with the external field.[159]

Thus, the search for materials exhibiting magnetic-phase transition to a ferromagnetic state that is based on the synthesis of transition metal complexes with nitroxides involving coordination of the NG already gave promising results. Complexes **(89–94, 118, 121, 122, 125–133)** have been obtained exhibiting magnetic-phase transition at temperatures above the liquid gelium one (4.2 K). It has also been shown that such transitions may occur not only in the presence of the $M\text{--O}\overset{\cdot}{=}N<$ bond but also in the presence of the bond. T_c values of the order of

$$M\text{--O H---O}\overset{\cdot}{=}N<$$
$$/$$
$$R^1$$

25 K have been attained.

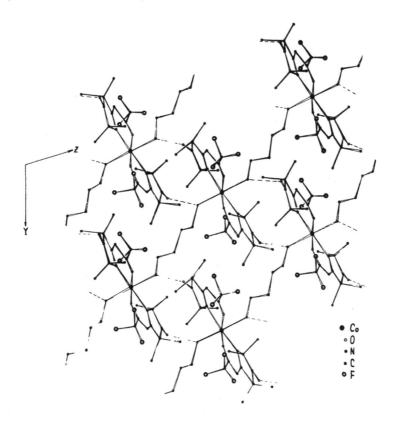

FIGURE 8. Projection of the structure of the Co(L²⁴)₂[HO(CH₂)₄OH] complex onto the **(100)** plane.

In this connection, further studies in this direction appear promising. Development of the crystal design of two- and three-dimensional systems on the basis of transition metal complexes with nitroxides with an NG involved in the polymeric bond system gives hope for the development of a designed synthesis of complexes with rather high T_c values. At the present time the problem is obtaining complexes with a T_c exceeding the liquid nitrogen temperature. For all complexes with nitroxides which have exhibited magnetic-phase transition, strong exchange interactions of a ferromagnetic type were observed. It may be noted that no magnetic-phase transition to a ferromagnetic state could thus far be observed for complexes in which interactions in the solid state are only ferromagnetic because the energies in them, at least of intercluster exchange interactions, are small. As a rule, these are complexes of Cu^{2+}. Of course, it would be of interest to realize magnetic phase transition on Cu^{2+} complexes with a nitroxide because copper is a nonmagnetic metal.

REFERENCES

1. **Eaton, S. S. and Eaton, G. R.**, Interaction of spin labels with transition metals, *Coord. Chem. Rev.*, 26, 207, 1978.
2. **Abakumov, G. A.**, Complexes of stable free radicals and the problem of spin labels in coordination chemistry, *Zh. Vses. Khim. Ova.*, 24(2), 156, 1979; *Chem. Abstr.*, 91, 4708, 1979.
3. **Drago, R. S.**, Free radicals reactions of transition metal systems, *Coord. Chem. Rev.*, 32, 7, 1980.
4. **Volodarsky, L. B., Grigor'ev, I. A., and Sagdeev, R. Z.**, Stable imidazoline nitroxides, in *Biological Magnetic Resonance*, Vol. 2, Berliner, L. J. and Reuben, J., Eds., Plenum Press, New York, 1980, 169.
5. **Parmon, V. N., Kokorin, A. I., and Zhidomirov, G. M.**, *Stable (Organic) Biradicals*, Nauka, Moscow, 1980, 36; *Chem. Abstr.*, 94, 191572m, 1981.
6. **Larionov, S. V.**, Coordination compounds of metal with stable nitroxyl radicals, *Zh. Strukt. Khim.*, 23(4), 125, 1982; *Chem. Abstr.*, 97, 192172e, 1982.
7. **Zolotov, Yu. A., Petrukhin, O. M., Nagy, V. Yu., and Volodarsky, L. B.**, Stable free-radical complexing reagents in applications of electron spin resonance to the determination of metals, *Anal. Chim. Acta*, 115, 1, 1980.
8. **Larionov, S. V.**, Imidazoline nitroxides in coordination chemistry, in *Imidazoline Nitroxides, Vol. 2, Applications*, Volodarsky, L. B., Ed., CRC Press, Boca Raton, FL, 1988, 81.
9. **Eaton, S. S. and Eaton, G. R.**, Interaction of spin labels with transition metals. Part 2, *Coord. Chem. Rev.*, 83, 29, 1988.
10. **Ovcharenko, V. I., Gel'man, A. B., and Ikorskii, V. N.**, Complexes with metal-nitroxyl bonds, *Zh. Strukt. Khim.*, 30(5), 142, 1989; *Chem. Abstr.*, 112, 150464z, 1990.
11. **Larionov, S. V.**, Coordination compounds of platinum metals with nitroxyl radicals of 3-imidazoline, 3-imidazolidine and 1,2-hydroxylamino oxime, *Blagorod. Met. Khim. Tekhnol.*, Novosibirsk, 1989, 164; *Chem. Abstr.*, 112, 209656n, 1990.
12. **Larionov, S. V.**, Some new tendencies in chemistry of metal coordination compounds with nitroxides, *Izv. Sib. Otd. Akad. Nauk S.S.S.R., Ser. Khim. Nauk*, 3, 34, 1990.
13. **Ovcharenko, V. I., Ikorskii, V. N., Vostrikova, K. E., Burdukov, A. B., Romanenko, G. V., Pervukhina, N. V., and Podberezskaya, N. V.**, Low temperature ferromagnetics based on imidazoline nitroxide complexes, *Izv. Sib. Otd. Akad. Nauk S.S.S.R., Ser. Khim. Nauk*, 5, 100, 1990.
14. **Caneschi, A., Gatteschi, D., Sessoli, R., and Rey, P.**, Toward molecular magnets: the metal-radical approach, *Acc. Chem. Res.*, 22, 392, 1989.
15. **Anderson, O. P. and Kuechler, T. C.**, Crystal and molecular structure of a nitroxyl-radical complex of copper(II): bis(hexafluoroacetylacetonato) (4-hydroxy-2,2,6,6-tetramethylpiperidinyl-*N*-oxy)copper(II), *Inorg. Chem.*, 19, 1417, 1980.
16. **Musin, R. N., Schastnev, P. V., and Malinovskaya, S. A.**, Delocalization mechanism of ferromagnetic exchange interactions in the complexes of Cu(II) with nitroxyl radicals, in *Trends in Applied Theoretical Chemistry*, Kluwer Academic Publishers, 1992, 167.
17. **Burdukov, A. B., Ovcharenko, V. I., Ikorskii, V. N., Pervukhina, N. V., Podberezskaya, N. V., Grigor'ev, I. A., Larionov, S. V., and Volodarsky, L. B.**, A new type of mixed-ligand complexes with nitroxyl radicals, *Inorg. Chem.*, 30, 972, 1991.
18. **Miller, J. S., Epstein, A. J., and Reiff, W. M.**, Molecular/organic ferromagnets, *Science*, 240, 40, 1988.
19. **Buchachenko, A. L.**, Organic and molecular ferromagnets: achievements and problems, *Usp. Khim.*, 59, 529, 1990; *Chem. Abstr.*, 112, 244529b, 1990.
20. **Kaim, W.**, The transition metal coordination chemistry of anion radicals, *Coord. Chem. Rev.*, 76, 187, 1987.
21. **Kabachnik, M. I., Bubnov, N. N., Solodovnikov, S. P., and Prokof'ev, A. I.**, Structure and stereochemistry of paramagnetic complexes of metal with *o*-quinones, *Usp. Khim.*, 53, 487, 1984; *Chem. Abstr.*, 100, 166561s, 1984.

22. **Schott, A., Schott, H., Wilke, G., Brant, J., Hoberg, H., and Hoffman, E. G.**, Reactionen von Nickel(O) mit Radicalen, *Libigs Ann. Chem.*, 508, 1973.
23. **Milaev, A. G. and Okhlobystin, O. Yu.**, Free organometallic radicals (nontransition metals), *Usp. Khim.*, 49, 1892, 1980; *Chem. Abstr.*, 94, 3325m, 1981.
24. **Pierpont, C. G. and Buchanan, R. M.**, Transition metal complexes of *o*-benzoquinone, *o*-semiquinone and catecholate ligands, *Coord. Chem. Rev.*, 38, 45, 1981.
25. **Milaeva, E. R., Rubezhov, A. Z., Prokof'ev, A. J., and Okhlobystin, O. Yu.**, Unpaired electron in transition metal complexes, *Usp. Khim.*, 51, 1638, 1982; *Chem. Abstr.*, 98, 4583p, 1983.
26. **More, K. M., Eaton, G. R., and Eaton, S. S.**, Metal-nitroxyl interaction. 25. Exchange interaction via saturated linkages in spin-labeled porphirines, *Can. J. Chem.*, 60, 1392, 1982.
27. **Gans, P., Buisson, G., Duee, E., Marchon, J. C., Erler, B. S., Scholz, W. F., and Reed, C. A.**, High-valent iron porphyrins: synthesis, X-ray structures, π-kation radical formulation and notable magnetic properties of chloro(meso-tetraphenylporphinato)iron(III) hexachloroantimonate and bis(perchlorato)(meso-tetraphenylporphinato)iron(III), *J. Am. Chem. Soc.*, 108, 1223, 1986.
28. **Abakumov, G. A.**, *Organometallic Compounds and Radicals*, Nauka, Moscow, 1985.
29. **Song, H., Reed, C. A., and Scheidt, W. R.**, Metallolporphyrin π-cation radicals: intrinsically ruffled or planar core conformation. Molecular structure of (mesitylporphinato) copper(II) hexachloroantimonate, *J. Am. Chem. Soc.*, 111, 6865, 1989.
30. **Rakitin, Yu. V., Nemtzev, N. V., Lobanov, A. V., Abakumov, G. A., Novotortsev, V. M., and Kalinnikov, V. T.**, Ferromagnetic complex of copper(II) with paramagnetic ligands, *Koor. Khim.*, 1189, 1982; *Chem. Abstr.*, 97, 229071z, 1982.
31. **Kahn, O., Prins, R., Reedijk, J., and Thompson, J. S.**, Orbital symmetries and magnetic interaction between copper(II) ions and the *O*-semiquinone radical. Magnetic studies of (di-2-pyridylamin)(3,5-di-*tert*-butyl-*O*-semiquinonato) copper(II) perchlorate and bis(bis(3,5-di-*tert*-butyl-*O*-semiquinonato)-copper(II), *Inorg. Chem.*, 26, 3557, 1987.
32. **Volodarsky, L. B., Grigor'ev, I. A., Dikanov, S. A., Reznikov, V. A., and Stchukin, G. I.**, *Imidazoline Nitroxides*, Vol. 1, Volodarsky, L. B. Ed., CRC Press, Boca Raton, FL, 1988.
33. **Rozantsev, E. G. and Sholle, V. D.**, *Organic Chemistry of Free Radicals*, Khimija, Moscow, 1979.
34. **Buchachenko, A. L. and Vasserman, A. M.**, *Stable Radicals*, Khimija, Moscow, 1973.
35. **Volodarsky, L. B. and Vainer, L. M.**, Synthesis and properties of stable nitroxyl radicals of imidazolines, *Khim. Farm. Zh.*, 17, 524, 1983; *Chem. Abstr.*, 99, 70590k, 1983.
36. **Golubev, V. A., Sen, V. D., and Rozantsev, E. G.**, Paramagnetic 1-oxyl-3-imidazolinium salts, *Izv. Akad. Nauk S.S.S.R., Ser. Khim.*, 2773, 1974; *Chem. Abstr.*, 82, 125319k, 1975.
37. **Lim, Y. Y. and Drago, R. S.**, Lewis basicity of a free-radical base. II. *Inorg. Chem.*, 11, 1334, 1972.
38. **Boymel, P. M., Eaton, G. R., and Eaton, S. S.**, Metal-nitroxyl interactions. 14. Bis(hexafluoroacetylacetonato)copper(II) adducts of spin-labeled pyridines, *Inorg. Chem.*, 19, 727, 1980.
39. **Bilgren, C., Drago, R. S., Stahlbush, J. R., and Kuechler, T. C.**, Use of a spin-label adduct of Rh$_2$ (pfb)$_4$ to probe the inductive effect of coordination through the metal-metal bond, *Inorg. Chem.*, 24, 4268, 1985.
40. **Boymel, P. M., Braden, G. A., Eaton, G. R., and Eaton, S. S.**, Metal-nitroxyl interactions. 17. Spin-labeled adducts of bis(hexafluoroacetylacetonato)copper(II), *Inorg. Chem.*, 19, 735, 1980.
41. **Drago, R. S., Kuechler, T. C., and Kroeger, M.**, Lewis acidity of bis(1,1,1,5,5,5-hexafluoropentane-2,4-dionato)oxovanadium(IY). Spin pairing of electrons in the 2,2,6,6-tetramethyl-piperidinyl-*N*-oxy adduct, *Inorg. Chem.*, 18, 2337, 1979.
42. **Beck, W., Schmidtner, K., and Keller, H. J.**, Kobalt(II)-halogenide Komplexe mit dem Di-*tert*-butylstickstoffoxid-Radikal, *Chem. Ber.*, 100, 503, 1967.

43. **Brown, D. C., Maier, T., and Drago, R. S.,** Complexes of the free-radical ligand di-*tert*-butyl nitroxide, *Inorg. Chem.,* 10, 2804, 1971.

44. **Jahr, D., Rebhan, K. H., Schwarzhans, K. E., and Wiedemann, J.,** Metallkomplexe mit paramagnetischen Derivaten des Piperidin-1-oxyls, *Z. Naturforsch.,* 28b, 55, 1973.

45. **Beck, W. and Schmidtner, K.,** Palladiumhalogenid-Komplexe mit dem Di-*tert*-butylstickstof-foxid-Radical [XPdON(C(CH₃)₃)₂]₂ (X=Cl, Br), *Chem. Ber.,* 100, 3363, 1967.

46. **Okunaka, M., Matsubayashi, G., and Tanaka, T.,** Preparation and reaction of some di-*tert*-butylnitroxido-palladium(II) complexes, PdX(Bu₂NO)L [L=PPh₃, AsPh₃, P (OPh)₃; X=Cl, Br, I], *Bull. Chem. Soc. Jpn,* 48, 1826, 1975.

47. **Okunaka, M., Matsubayashi, G., and Tanaka, T.,** Palladium(II) complexes of 2,2,6,6-tetramethylpiperidine-*N*-oxyl radical, *Bull. Chem. Soc. Jpn.,* 50, 907, 1977.

48. **Okunaka, M., Matsubayashi, G., and Tanaka, T.,** Preparation and properties of some cationic palladium(II) complexes of the di-*tert*-butylnitroxide radical, *Bull. Chem. Soc. Jpn.,* 50, 1070, 1977.

49. **Okanaka, M., Matsubayashi, G., and Tanaka, T.,** Chloro(di-*tert*-butylnitrox-ido)(phenacylide)palladium(II). Evidence for the three-membered ring formation in the palladium-nitroxide moiety, *Inorg. Nucl. Chem. Lett.,* 12, 813, 1976.

50. **Dickman, M. H. and Doedens, R. J.,** Structure of chloro(2,2,4,4-tetramethylpiperidinyl-1-oxo-*O*, *N*)(triphenylphosphine)palladium(II), a metal complex of a reduced nitroxyl radical, *Inorg. Chem.,* 21, 682, 1982.

51. **Rohmer, M. M., Grand, A., and Bernard, M.,** The metal-nitroxyl interaction in MNO metallocycles (M = Cu, Pd). An *ab initio* SCF/CI study, *J. Am. Chem. Soc.,* 112, 2875, 1990.

52. **Caneschi, A., Grand, A., Laugier, J., Rey, P., and Subra, R.,** Three-center binding of a nitroxyl free radical to copper(II) bromide, *J. Am. Chem. Soc.,* 110, 2307, 1988.

53. **Karayannis, N. M., Paleos, C. M., and Mikulski, C. M.,** Some divalent 3D metal perchlorate complexes with the 2,2,6,6-tetramethylpiperidine nitroxide free radical, *Inorg. Chim. Acta,* 7, 74, 1973.

54. **Mikulski, C. M., Skryantz, J. S., and Karayannis, N. M.,** Bond-order of the NO group in 2,2,6,6-tetramethylpiperidine nitroxide radical complexes with metal perchlorates, *Inorg. Nucl. Chem. Lett.,* 11, 259, 1975.

55. **Mikulski, C. M., Gelfand, L. S., Pytlewski, L. L., Skryantz, J. S., and Karayannis, N. M.,** Yttrium(III) and lanthanide(III) perchlorate complex with the 2,2,6,6-tetramethylpiper-idine nitroxide free radical, *Transition Met. Chem.,* 3, 276, 1978.

56. **Paleos, C. M., Karayannis, N. M., and Labes, M. M.,** Redaction of 2,2,6,6-tetramethyl-piperidine nitroxide radical via complex formation with copper(II) perchlorate, *J. Chem. Soc. Chem. Commun.,* 195, 1970.

57. **Zelonka, R. A. and Baird, M. C.,** Electron spin exchange between di-*tert*-butyl nitroxide and copper(II) β-diketone complexes, *J. Am. Chem. Soc.,* 93, 6066, 1971.

58. **Zelonka, R. A. and Baird, M. C.,** Spin exchange between paramagnetic metal complex and aliphatic nitroxides: ¹⁹F nuclear magnetic resonance spectroscopy of copper(II) complexes, *J. Chem. Soc. Chem. Commun.,* 1448, 1970.

59. **Cotton, F. A. and Felthouse, T. R.,** Structures of two dirhodium(II) compounds containing hydrogen-bonded nitroxyl groups: tetrakis(trifluoroacetato)bis(2,2,6,6-tetramethyl-4-hydroxy-piperidinyl-1-oxy)dirhodium(II) and tetrakis(triflurocetato)-diaquadirhodium(II) di-*tert*-butyl ni-troxide solvate, *Inorg. Chem.,* 21, 2667, 1982.

60. **Richman, R. M., Kuechler, T. C., Tanner, S. R., and Drago, R. S.,** Electron paramagnetic resonance study of the adduct of a nitroxide and rhodium trifluoroacetate dimer, *J. Am. Chem. Soc.,* 99, 1055, 1977.

61. **Porter, L. C., Dickman, M. H., and Doedens, R. J.,** A novel variation on a classical dimeric structure type. Preparation and structure of the metal-nitroxyl complex [Cu(O₂CCCl₃)₂(Tempo)]₂, *Inorg. Chem.,* 22, 1962, 1983.

62. **Porter, L. C., Dickman, M. H., and Doedens, R. J.,** Nitroxyl adducts of copper(II) complexes with a novel dimeric structure, *Inorg. Chem.,* 25, 678, 1986.

63. **Doedens, R. J.**, Structure and metal-metal interactions in copper(II) carboxylate complexes, *Prog. Inorg. Chem.*, 21, 209, 1976.

64. **Porai-Koshits, M. A.**, Crystal chemistry and stereochemistry of carboxylates, *Itogi Nauki Tekh., Ser. Crystallochemistry*, Moscow, VINITI, 15, 3, 1981; *Chem. Abstr.*, 96, 96492t, 1982.

65. **Porter, L. C. and Doedens, R. J.**, Preparation and crystal structure of a diamagnetic copper(II) trichloroacetate complex containing a nitroxyl radical ligand, *Inorg. Chem.*, 24, 1006, 1985.

66. **Dong, T.-Yu., Hendrickson, D. N., Felthouse, T. R., and Shieh, H.-S.**, Bis(nitroxyl)radical adducts of rhodium(II) and molybdenum(II) carboxylate dimers: magnetic exchange interactions propagated by metal-metal bonds, *J. Am. Chem. Soc.*, 106, 5373, 1984.

67. **Felthouse, T. R., Dong, T.-Yu., Hendrickson, D. N., Shieh, H.-S., and Thompson, M. R.**, Magnetic exchange interactions in nitroxyl biradicals bridged by metal-metal bonded complexes: structural, magnetochemical and electronic spectral characterization of tetrakis(perfluorocarboxylato)dimetal(II) complexes of rhodium and molybdenum containing axially coordinated nitroxyl radicals, *J. Am. Chem. Soc.*, 108, 8201, 1986.

68. **Dickman, M. H. and Doedens, R. J.**, Structure of bis(hexafluoroacetylacetonato) (2,2,6,6-tetramethylpiperidinyl-1-oxy)copper(II), a Cu(II)-nitroxyl radical complex with substantial magnetic coupling, *Inorg. Chem.*, 20, 2677, 1981.

69. **Dickman, M. H., Porter, L. C., and Doedens, R. J.**, Bis(nitroxyl) adducts of bis(hexafluoroacetylacetonato)manganese(II). Preparation, structure and magnetic properties, *Inorg. Chem.*, 25, 2595, 1986.

70. **Porter, L. C., Dickman, M. H., and Doedens, R. J.**, Bis(nitroxyl) adducts of cobalt and nickel hexafluoroacetylacetonates. Preparation, structures and magnetic properties of $M(F_6acac)_2(proxyl)_2$ ($M = Co^{2+}$, Ni^{2+}), *Inorg. Chem.*, 27, 1548, 1988.

71. **Porter, L. C. and Doedens, R. J.**, (Hexafluoroacetylacetonato-*O,O'*) (1-hydroxy-2,2,6,6-tetramethylpiperidinato-*O,N*)palladium(II), [Pd($C_5HF_6O_2$) ($C_9H_{18}NO$)], a metal complex containing a reduced nitroxyl radical, *Acta Crystallogr.*, C41, 839, 1985.

72. **Caneschi, A., Laugier, J., and Rey, P.**, Oxidation of a nitronyl introxide by copper(II) ions: synthesis and X-ray diffraction characterization of 4,4,5,5-tetramethyl-1-oxo-2-phenyl-4,5-dihydroimidazolium perchlorate *N*(3)-oxide, *J. Chem. Soc. Perkin Trans. 1.*, 1077, 1987.

73. **Atovmyan, L. O., Golubev, V. A., Golovina, N. I., and Klitskaya, G. A.**, Synthesis and structure of 2,2,6,6-tetramethyl-1-oxopiperidinium perchlorate, *Zh. Strukt. Khim.*, 16, 92, 1975; *Chem. Abstr.*, 82, 163631p, 1975.

74. **Matsunaga, P. T., McCall, D. T., Carducci, M. D., and Doedens, R. J.**, An adduct with one-dimensional chain structure from bis(hexafluoroacetylacetonato)copper(II) and an aliphatic nitroxyl diradical, *Inorg. Chem.*, 29, 1655, 1990.

75. **Gel'man, A. B. and Ikorskii, V. N.**, Magnetic susceptibility of nonuniform quasi-one-dimensional paramagnets, *Zh. Strukt. Khim.*, 30(4), 81, 1989; *Chem. Abstr.*, 111, 223878d, 1989.

76. **Musin, R. N., Schastnev, P. V., and Malinovskaya, S. A.**, Delocalization mechanism of ferromagnetic exchange interaction in the copper(II) complexes with nitroxyl radicals, *Int. Conf. Nitroxyl Radicals*, September 18–23, 1989, Novosibirsk, Russia, p. 89P.

77. **Osiecki, J. H. and Ullman, E. F.**, Studies of free radicals. 1. α-Nitronyl nitroxides, a new class of stable radicals, *J. Am. Chem. Soc.*, 90, 1078, 1968.

78. **Ullman, E. F., Osiecki, J. H., Boocock, D. G. B., and Darcy, R.**, Studies of stable free radicals. X. Nitronyl nitroxide monoradicals and biradicals as possible small molecule spin labels, *J. Am. Chem. Soc.*, 94, 7049, 1972.

79. **Helbert, J. N., Kopf, P. W., Poindexter, E. H., and Wagner, B. E.**, Complexing and protonaton of free-radical imidazoline-1-oxyl and imidazoline-1-oxyl-3-oxide ligands: a magnetic-resonance investigation, *J. Chem. Soc., Dalton Trans.*, 998, 1975.

80. **Laugier, J., Rey, P., Benelli, C., Gatteschi, D., and Zanchini, C.**, Unusual magnetic properties of the adduct of copper chloride with 2-phenyl-4,4,5,5-tetramethylimidazoline-1-oxyl-3-oxide, *J. Am. Chem. Soc.*, 108, 6931, 1986.

81. **Gatteschi, D., Laugier, J., Rey, P., and Zanchini, C.**, Crystal and molecular structures and magnetic properties of the adducts of copper(II) hexafluoroacetylacetonate with the nitroxide ligand 2-phenyl-4,4,5,5-tetramethylimidazoline-1-oxyl-3-oxide, *Inorg. Chem.*, 26, 938, 1987.

82. **Caneschi, A., Gatteschi, D., Laugier, J., Pardi, L., Rey, P., and Zanchini, C.,** Structure and magnetic properties of two bis(nitronylnitroxide)adductsofbis(hexafluoroacetylacetonato)manganese(II). Molecular orbital interpretation of the coupling in manganese-nitroxide complexes, *Inorg. Chem.,* 27, 2027, 1988.

83. **Caneschi, A., Gatteschi, D., Grand, A., Laugier, J., Pardi, L., and Rey, P.,** Moderate ferromagnetic exchange between copper(II) and a nitronyl nitroxide in a square-piramidal adduct. MO interpretation of the mechanism of exchange in copper(II)-nitroxide complex, *Inorg. Chem.,* 27, 1031, 1988.

84. **Caneschi, A., Gatteschi, D., Renard, J. P., Rey, P., and Sessoli, R.,** Magnetic coupling in zero- and one-dimensional magnetic systems formed by nickel(II) and nitronyl nitroxides. Magnetic phase transition of a ferromagnetic chain, *Inorg. Chem.,* 28, 2940, 1989.

85. **Benelli, C., Caneschi, A., Gatteschi, D., Laugier, J., and Rey, P.,** Struktur und magnetische Eigenschaften eines Addukts aus Gadolinium-hexafluoroacetylacetonate und dem Radikal 4,4,5,5-Tetramethyl-2-phenyl-4,5-dihydro-1*H*-imidazol-3-oxide-1-oxyl, *Angew. Chem.,* 99, 958, 1987.

86. **Benelli, C., Caneschi, A., Gatteschi, D., Guillon, O., Pardi, L., and Rey, P.,** Magnetic properties of a dysprosium(III) complex with a nitronyl nitroxide, *Inorg. Chim. Acta,* 160, 1, 1989.

87. **Benelli, C., Caneschi, A., Gatteschi, D., Pardi, L., Rey, P., Shum, D. P., and Carlin, R. L.,** Magnetic properties of lanthanide complexes with nitronyl nitroxides, *Inorg. Chem.,* 28, 272, 1989.

88. **Caneschi, A., Ferraro, F., Gatteschi, D., Melandri, M. C., Rey, P., and Sessoli, R.,** Synthesis, structure and magnetic properties of a dinuclear manganese(II) complex with one μ-aqua and two μ-carboxylato bridges, *Angew. Chem. Int. Ed. Engl.,* 28, 1365, 1989.

89. **Cogne, A., Grand, A., Rey, P., and Subra, R.,** σ-Assisted exchange interaction in linear adducts of nitroxides with dirhodium tetrakis(trifluoroacetate), *J. Am. Chem. Soc.,* 109, 7927, 1987.

90. **Cogne, A., Grand, A., Rey, P., and Subra, R.,** Coordination chemistry of the nitronyl and imino nitroxides. Linear chain adducts with rhodium(II) trifluoroacetate dimer, *J. Am. Chem. Soc.,* 111, 3230, 1989.

91. **Benelli, C., Caneschi, A., Gatteschi, D., Melandri, M. C., and Rey, P.,** Ground S=4 state in a manganese(II)-nitronyl nitroxide ferromagnetic ring, *Inorg. Chim. Acta,* 172, 137, 1990.

92. **Caneschi, A., Gatteschi, D., Laugier, J., Rey, P., and Sessoli, R.,** Structure and magnetic properties of chains of diamonds of four spin formed by metal(II) hexafluoroacetylacetonates (metal = cobalt, nickel) and the nitronyl nitroxide radical 4,4,5,5-tetramethyl-2-ethyl-4,5-dihydro-1*H*-imidazolyl-1-oxyl, *Inorg. Chem.,* 27, 1553, 1988.

93. **Benelli, C., Gatteschi, D., Carnegie, D. W. Jr., and Carlin, R. L.,** A linear chain with alternating ferromagnetic and antiferromagnetic exchange: Cu(hfac)$_2$ · TEMPOL, *J. Am. Chem. Soc.,* 107, 2560, 1985.

94. **Dickman, M. H. and Doedens, R. J.,** Preparation and crystal structure of aqua(3-cyano-2,2,5,5-tetramethylpyrrolinyl-1-oxy)bis(hexafluoroacetylacetonato)copper(II), a metal complex of a nitroxide radical having a hydrogen-bonded, *Inorg. Chem.,* 22, 1591, 1983.

95. **Pervukhina, N. V., Podberezskaya, N. V., and Ovcharenko, V. I.,** Molecular and crystal structure of (4-amine-2,2,5,5-tetramethyl-3-imidazoline-1-oxyl)bis(hexafluoroacetylacetonato)copper(II), *Zh. Strukt. Khim.,* 32(3), 92, 1991.

96. **Pervukhina, N. V.,** Crystallo-Structural Investigation of Copper Coordination Compounds with Stable 3-Imidazoline Nitroxyl Radical Derivatives, Dissertation, Institution of Inorganic Chemistry, Novosibirsk, 1990.

97. **Romanenko, G. V.,** X-ray Investigation of Co(II), Ni(II), Cu(II) and Zn(II) Complexes with Enaminoketone Derivatives of the 3-Imidazoline Nitroxyl Radical, Dissertation, Institution of Inorganic Chemistry, Novosibirsk, 1990.

98. **Caneschi, A., Gatteschi, D., and Rey, P.,** The chemistry and magnetic properties of metal nitronyl nitroxide complexes, in *Progress in Inorganic Chemistry,* Vol. 39, Lippard, S. J., Ed., John Wiley & Sons, 1991, 331.

99. **Porter, L. C. and Doedens, R. J.**, Co-crystallized bis(hexafluoroacetylacetonato) palladium (II) and 3-cyano-2,2,5,5-tetramethyl-2,5-dihydropyrrolyl-1-oxyl, $C_{10}H_2F_{12}O_4Pd \cdot 2C_9H_{13}N_2O$, *Acta Crystallogr.*, C40, 1814, 1984.

100. **Ovcharenko, V. I., Ikorskii, V. N., Podberezskaya, N. V., Pervukhina, N. V., and Larionov, S. V.**, Preparation and magnetic properties of trinuclear complex of bis(hexafluoroacetylacetonato) copper (II) with 4-phenyl-2,2,5,5-tetramethyl-3-imidazoline-1-oxyl, *Zh. Strukt, Khim.*, 32(6), 1403, 1987; *Chem. Abstr.*, 107, 189497j, 1987.

101. **Pervukhina, N. V., Podberezskaya, N. V., Ovcharenko, V. I., and Larionov, S. V.**, Crystal and molecular structure of the mixed-ligand complex of bis(hexafluoroacetylacetonato) copper (II) with nitroxyl radical 4-phenyl-2,2,5,5-tetramethyl-3-imidazoline-1-oxyl, *Zh. Strukt. Khim.*, 32(1), 123, 1991.

102. **Ovcharenko, V. I., Ikorskii, V. N., Gel'man, A. B., Burdukov, A. B., and Larionov, S. V.**, Preparation and magnetic properties of the adducts of bis(hexafluoroacetylacetonato) copper (II) and cobalt (II) with the stable nitroxyl imidazoline radicals, *Proc. Conf. Coord. Chem.*, 11th ed., 1987, 299; *Chem. Abstr.*, 108, 178934u, 1988.

103. **Burdukov, A. B., Ovcharenko, V. I., Pervukhina, N. V., Podberezskaya, N. V., and Ikorskii, V. N.**, Tri-nuclear Cu(F₆acac)₂ complexes with 3-imidazoline nitroxyl radical derivatives, *Proc. Conf. Coord. Chem.*, 12th ed., 1989, 69; *Chem. Abstr.*, 112, 209734t, 1990.

104. **Burdukov, A. B., Ovcharenko, V. I., Pervukhina, N. V., Ikorskii, V. N., and Podberezskaya, N. V.**, New examples of tri-nuclear bis(hexafluoroacetylacetonato) copper (II) complexes with imidazoline nitroxides, *Zh. Neorg. Khim.*, 36, 2058, 1991.

105. **Luneau, D., Rey, P., Laugier, J., Fries, P., Caneschi, A., Gatteschi, D., and Sessoli, R.**, N-bonded copper (II)-imino-nitroxide complexes exhibiting large ferromagnetic interactions, *J. Am. Chem. Soc.*, 113, 1245, 1991.

106. **Caneschi, A., Gatteschi, D., Sessoli, R., Hoffman, S. K., Laugier, J., Pardi, L., and Rey, P.**, Crystal and molecular structure, magnetic properties and EPR spectra of a trinuclear copper (II) complex with bridging nitronyl nitroxides, *Inorg. Chem.*, 27, 2390, 1988.

107. **Burdukov, A. B., Ovcharenko, V. I., Ikorskii, V. N., Larionov, S. V., and Volodarsky, L. B.**, Product of copper(II) hexafluoro-acetylacetonate reaction with 1-hydroxy-3-imidazolines, the nearest precursors of nitroxyl radicals, *Izv. Akad. Nauk S.S.S.R., Ser. Khim.*, 1693, 1988; *Chem. Abstr.*, 110, 154218d, 1989.

108. **Ovcharenko, V. I., Ikorskii, V. N., Gel'man, A. B., and Larionov, S. V.**, Preparation and magnetic properties of polycrystalline complexes of copper (II) and cobalt (II) hexafluoroacetylacetonates with nitroxyl radicals of imidazoline, *Zh. Neorg. Khim.*, 33, 1735 1988; *Chem. Abstr.*, 109, 203721c, 1988.

109. **Burdukov, A. B., Ovcharenko, V. I., Ikorskii, V. N., Semyannikov, P. P., Larionov, S. V., and Volodarsky, L. B.**, Copper (II) and cobalt (II) hexafluoroacetylacetonate mixed-ligand complexes with 2,2,4,5,5-pentamethyl-3-imidazoline-1-oxyl and the corresponding 1-hydroxy derivative, *Zh. Neorg. Khim.*, 34, 366, 1989; *Chem. Abstr.*, 110, 224394j, 1989.

110. **Dunaj-Jurko, M., Ondrejovic, C., Melnik, M., and Garaj, J.**, Mixed-valence copper (II) compounds: analysis and classification of crystallographic data, *Coord. Chem. Rev.*, 83, 1, 1988.

111. **Musin, R. Z., Schactnev, P. V., and Malinovskaya, S. A.**, Delocalization mechanism of ferromagnetic exchange interactions in the complexes of copper (II) with nitroxyl radicals, *Inorg. Chem.*, 31, 4118, 1992.

112. **Caneschi, A., Gatteschi, D., Laugier, J., Rey, P., Sessoli, R., and Zanchini, C.**, Preparation, crystal structure and magnetic properties of an oligonuclear complex with 12 coupled spin and an S=12 ground state, *J. Am. Chem. Soc.*, 110, 2795, 1988.

113. **Caneschi, A., Gatteschi, D., Rey, P., and Sessoli, R.**, Structure and magnetic properties of ferromagnetic chain formed by manganese (II) and nitronyl nitroxides, *Inorg. Chem.*, 27, 1756, 1988.

114. **Caneschi, A., Gatteschi, D., Renard, J. P., Rey, P., and Sessoli, R.**, Magnetic phase transition and low-temperature EPR spectra of a one-dimensional ferromagnet formed by manganese (II) and a nitronyl nitroxide, *Inorg. Chem.*, 28, 1976, 1989.

115. **Caneschi, A., Gatteschi, D., Renard, J. P., Rey, P., and Sessoli, R.,** Ferromagnetic phase transitions of two one-dimensional ferromagnets formed by manganese (II) and nitronyl nitroxides *cis* octahedrally coordinated, *Inorg. Chem.,* 28, 3314, 1989.

116. **Carducci, M. D. and Doedens, R. J.,** Dimeric complex of a reduced nitroxyl radical with bis(hexafluoroacetylacetonato) manganese (II), *Inorg. Chem.,* 28, 2492, 1989.

117. **Caneschi, A., Gatteschi, D., Laugier, J., Rey, P., and Zanchini, C.,** Synthesis, X-ray crystal structure and magnetic properties of two dinuclear manganese (II) compounds containing nitronyl nitroxides and their reduced derivatives, *Inorg. Chem.,* 28, 1969, 1989.

118. **Caneschi, A., Gatteschi, A., Laugier, J., and Rey, P.,** Ferromagnetic alternating spin chains, *J. Am. Chem. Soc.,* 109, 2191, 1987.

119. **Cabello, C. I., Caneschi, D., Carlin, R. L., Gatteschi, D., Rey, P., and Sessoli, R.,** Structure and magnetic properties of ferromagnetic alternating spin chains, *Inorg. Chem.,* 29, 2582, 1990.

120. **Benelli, C., Caneschi, A., Gatteschi, D., Pardi, L., and Rey, P.,** One-dimensional magnetism of a linear chain compound containing yttrium (III) and a nitronyl nitroxide radical, *Inorg. Chem.,* 28, 3230, 1989.

121. **Benelli, C., Caneschi, A., Gatteschi, D., Pardi, L., and Rey, P.,** Structure and magnetic properties of linear-chain complex of rare-earth ions (gadolinium, europium) with nitronyl nitroxides, *Inorg. Chem.,* 28, 275, 1989.

122. **Benelli, C., Caneschi, A., Gatteschi, D., Pardi, L., and Rey, P.,** Linear-chain gadolinium (III) nitronyl nitroxide complexes with dominant next-nearest-neighbour magnetic interactions, *Inorg. Chem.,* 29, 4223, 1990.

123. **Richardson, P. F. and Kreilick, R. W.,** Copper complex with free-radical ligand, *J. Am. Chem. Soc.,* 99, 8183, 1977.

124. **Mamedova, Yu. G., Medzhidov, A. A., and Kolomina, L. N.,** Copper (II) chelates with paramagnetic Schiff bases, *Zh. Neorg. Khim.,* 17, 2946, 1972; *Chem. Abstr.,* 78, 51902p, 1973.

125. **Aliev, V. S., Medzhidov, A. A., Mamedova, P. S., and Mamedova, Yu. G.,** Activation of transition metal-free radical systems during electron spins coupling, *Dokl. Akad. Nauk Az. S.S.R.,* 29(5), 24, 1973; *Chem. Abstr.,* 80, 70631g, 1974.

126. **Medzhidov, A. A., Kutovaya, T. M., Rodin, N. P., Guseinova, M. K., Timakov, I. A., and Mamedov, Kh. S.,** Binuclear complexes of copper (II) with Schiff bases, derivatives of *O*-hydroxyaromatic aldehydes and 2,2,5,5-tetramethylpyrrolidine-1-oxyl, *Koord. Khim.,* 5, 1433, 1979; *Chem. Abstr.,* 91, 221664a, 1979.

127. **Timakov, I. A.,** Preparation and Physico-Chemical Investigation of Mono- and Binuclear Copper (II) Complexes with Nitroxyl Radical Derivatives, Institution of Physical Chemistry, Dissertation, Baku, 1982.

128. **Medzhidov, A. A. and Timakov, I. A.,** Magnetic properties of complexes of copper(II) with paramagnetic Schiff bases, *Koord. Khim.,* 8, 1043, 1982; Chem. Abstr., 97, 228716b, 1982.

129. **Atovmyan, L. O., Golovina, N. I., Klitskaya, G. A., Medzhidov, A. A., Zvarykina, A. V., Strukov, V. B., and Fedutin, D. N.,** Structure and magnetic properties of a complex of copper(II) with a Schiff base derived from 2,2,6,6-tetramethyl-4-amino-piperidine-1-iminoxyl, *Zh. Strukt. Khim.,* 16(4), 624, 1975; *Chem. Abstr.,* 83, 186701w, 1975.

130. **Rodin, N. P., Medzhidov, A. A., Kutovaya, T. M., and Guseinova, M. K.,** Structure of the complexes of transition metals with paramagnetic ligands, *Zh. Strukt. Khim.,* 22(1), 85, 1981; *Chem. Abstr.,* 94, 201152r, 1981.

131. **Shnulin, A. N., Struchkov, Yu. T., Mamedov, Kh. S., Medzhidov, A. A., and Kutovaya, T. M.,** Stereochemistry of a copper(II) atom in salicylaldiminate complexes, *Zh. Strukt. Khim.,* 18(6), 1006, 1977; *Chem. Abstr.,* 88, 113706v, 1978.

132. **Caneschi, A., Ferraro, F., Gatteschi, D., Rey, P., and Sessoli, R.,** Ferro- and antiferromagnetic coupling between metal ions pyridine-substituted nitronyl nitroxides, *Inorg. Chem.,* 29, 4217, 1990.

133. **Caneschi, A., Ferraro, F., Gatteschi, D., Rey, P., and Sessoli, R.,** Crystal structure and magnetic properties of a copper(II) chloride nitronyl nitroxide complex containing six exchange-coupled $S = 1/2$ spins, *Inorg. Chem.,* 29, 1756, 1990.

134. **Ovcharenko, V. I., Sorokina, T. N., Ikorskii, V. N., Larionov, S. V., and Volodarsky, L. B.,** Coordination compounds of Cu(II) with the nitroxyl radical 2-acetyl-2,5,5-trimethyl-4-phenyl-3-oxide-1-oxyl-thiosemicarbazone and the corresponding 1-hydroxy derivative, *Izv. Akad. Nauk S.S.S.R., Ser. Khim.,* 162, 1986; *Chem. Abstr.,* 104, 179060k, 1986.

135. **Owtcharenko, W. I., Schnell, E., and Schwarzhans, K. E.,** Kupfer(II)- Komplexe mit para- und diamagnetischen Semicarbazonliganden, *Monash. Chem.,* 118, 773, 1987.

136. **Espie, J. C., Ramasseul, R., Rassat, A., and Rey, P.,** Complexes metaux-nitroxydes, *Bull. Soc. Chim. Fr.,* Part II, 33, 1983.

137. **Laugier, J., Ramasseul, R., Rey, P., Espie, J. C., and Rassat, A.,** Complexes metaux-nitroxydes, structures de deux complexes du cuivre II avec des β-cetyoaldehydes a faible interaction magnetique, *Nouv. J. Chim.,* 7(1), 11, 1983.

138. **Benelli, C., Gatteschi, D., Zanchini, C., Latour, J. M., and Rey, P.,** Weak exchange interaction between nitroxides and copper(II) ions monitored by EPR spectroscopy, *Inorg. Chem.,* 25, 4242, 1986.

139. **Espie, J. C., Laugier, J., Ramasseul, R., Rassat, A., and Rey, P.,** Nitroxydes: etude par rpe d'un chelate de cuivre (^{63}Cu) nitroxyde. Systeme a trois centres radicalaires pratiquement alignes, *Nouv. J. Chim.,* 4(4), 205, 1980.

140. **Briere, R., Giroud, A. M., Rassat, A., and Rey, P.,** Nitroxydes. 93. Complexes cuivriques de β-ceto-esters, *Bull. Soc. Chim. Fr.,* (3–4), Part II, 147, 1980.

141. **Grand, A., Rey, P., and Subra, R.,** X-ray diffraction characterization and electronic structure of the complex of copper(II) with a nitroxyl-β-keto ester: bis[(1-oxy-2,2,6,6-tetramethylpiperidine-4-yl)pivaloylacetato]copper(II), *Inorg. Chem.,* 22, 391, 1983.

142. **Drillon, M., Grand, A., and Rey, P.,** Structural and magnetic properties of a bis(nitroxyl) binuclear copper(II) complex, *Inorg. Chem.,* 29, 771, 1990.

143. **Ovcharenko, V. I., Larionov, S. V., Mohosoeva, V. K., and Volodarsky, L. B.,** Complexes of metals with fluorine-containing paramagnetic enamine ketone, *Zh. Neorg. Khim.,* 28, 151, 1983; *Chem. Abstr.,* 98, 118506w, 1983.

144. **Pervukhina, N. V., Ikorskii, V. N., Podberezskaya, N. V., Nikitin, P. S., Gel'man, A. B., Ovcharenko, V. I., Larionov, S. V., and Bakakin, V. V.,** Crystal and molecular structure and magnetic properties of bis[4-(2'-oxo-3',3',3'-trifluoropropylidene)-2,2,5,5-tetramethyl-3-imidazoline-1-oxyl]copper(II), *Zh. Strukt. Khim.,* 27(4), 61, 1986; *Chem. Abstr.,* 106, 11395r, 1987.

145. **Ikorskii, V. N. and Ovcharenko, V. I.,** Magnetic properties of nickel(II) and cobalt(II) bis-chelates with paramagnetic ligand, *Zh. Neorg. Khim.,* 35, 2093, 1990.

146. **Ovcharenko, V. I., Vostrikova, K. E., Ikorskii, V. N., Romanenko, G. V., Podberezskaya, N. V., Reznikov, V. A., and Volodarsky, L. B.,** The first example of the coordination of N⁻O group in the solid bischelates with imidazoline nitroxide, *XXVIII Int. Conf. Coord. Chem.,* Gera, GDR, Vol. 1, 1990, 6.

147. **Ovcharenko, V. I., Vostrikova, K. E., Romanenko, G. V., Podberezskaya, N. V., Reznikov, V. A., and Volodarsky, L. B.,** The first example of the coordination of N⁻O group in the solid innercomplex bischelates with imidazoline nitroxide derivatives, *XYII All-Union Conf. Coord. Chem.,* Minsk, U.S.S.R., Vol. 3, 1990, C-31.

148. **Benelli, C., Caneschi, A., Fabretti, A. C., Gatteschi, D., and Pardi, L.,** Ferromagnetic coupling of gadolinium(III) ions and nitronyl nitroxide radicals in an essentially isotropic way, *Inorg. Chem.,* 29, 4153, 1990.

149. **Rey, P., Laugier, J., Caneschi, A., and Gatteschi, D.,** Synthesis of high-spin molecular species using nitroxide organic radicals and transition metal ions, *Mol. Cryst. Liq. Cryst.,* 176, 337, 1989.

150. **Caneschi A., Gatteschi, D., Sessoli, R., and Rey, P.,** Magnetic materials formed by metal ions and nitroxides, *Mol. Cryst. Liq. Cryst.,* 176, 329, 1989.

151. **Caneschi, A., Gatteschi, D., Renard, J. P., Rey, P., and Sessoli, R.,** Magnetic phase transition in manganese(II) pentafluorobenzoate adducts with nitronyl nitroxides, *J. Am. Chem. Soc.,* 111, 785, 1989.

152. **Caneschi, A., Gatteschi, D., Melandri, M. C., Rey, P., and Sessoli, R.**, Structure and magnetic properties of manganese(II) carboxylate chains with nitronyl nitroxides and their reduced amidino-oxide derivatives. From random-exchange one-dimensional to two-dimensional magnetic materials, *Inorg. Chem.*, 29, 4228, 1990.

153. **Ovcharenko, V. I., Ikorskii, V. N., Romanenko, G. V., Reznikov, V. A., and Volodarsky, L. B.**, Synthesis and magnetic properties of Cu(II), Ni(II) and Co(II) complexes with nitroenamine — a stable nitroxide imidazoline radical. Crystal structure of CoL_2 $(CH_3OH)_2$, *Inorg. Chim. Acta*, 187(1), 67, 1991.

154. **Volodarsky, L. B., Kutikova, G. A., Kobrin, V. S., Sagdeev, R. Z., and Molin, Yu. N.**, Synthesis and spectral studies of stable iminoxyl radicals. Derivatives of 2,2,5,5-tetramethyl-3-imidazoline-1-oxyl-4-carboxylic acid and amide, salt and copper(II) complex, *Izv. Sib. Otd. Akad. Nauk S.S.S.R., Ser. Khim. Nauk*, N7(3), 101, 1971.

155. **Sagdeev, R. Z., Molin, Yu. N., Sadikov, R. A., Volodarsky, L. B., and Kutikova, G. A.**, The EPR and NMR investigation of free radical complexes with transition metals, *J. Magn. Res.*, 9, 13, 1973.

156. **Gel'man, A. B. and Ikorskii, V. N.**, Magnetic properties of nickel(II) chelates with a paramagnetic imidazolinecarboxylic acid, *Koord. Khim.*, 14, 491, 1988; *Chem. Abstr.*, 108, 230670w, 1988.

157. **Ovcharenko, V. I., Patrina, L. A., Boguslavski, E. G., Ikorskii, V. N., Durasov, V. B., and Larionov, S. V.**, New group of volatile metal chelates with spin-labeled enaminoketones, *Zh. Neorg. Khim.*, 32(5), 1129, 1987; *Chem. Abstr.*, 107, 108077x, 1987.

158. **Ikorskii, V. N., Ovcharenko, V. I., Vostrikova, K. E., Pervukhina, N. V., and Podberezskaya, N. V.**, Magnetic properties of copper(II) bischelates with the enaminoketone derivatives of the imidazoline nitroxyl radicals, *Zh. Neorg. Khim.*, 37, 1177, 1992.

159. **Ovcharenko, V. I., Vostrikova, K. E., Ikorskii, V. N., Romanenko, G. V., Podberezskaya, N. V., Larionov, S. V., and Sagdeev, R. Z.**, Synthesis, structure and magnetic properties of low-temperature ferromagnets based on nickel(II) and cobalt(II) complexes with imidazoline nitroxyl radicals, Preprint 89-2, Novosibirsk, Russia, 1989.

160. **Ovcharenko, V. I., Vostrikova, K. E., Romanenko, G. V., Ikorskii, V. N., Podberezskaya, N. V., and Larionov, S. V.**, Synthesis, crystal structure and magnetic properties of di(methanol) and di(ethanol)-bis[2,2,5,5-tetramethyl-1-oxyl-3-imidazoline-4-(3',3',3'-trifluoromethyl-1'-propenyl-2'-oxyato)]nickel(II). A new type of low-temperature ferromagnetics, *Dokl. Akad. Nauk S.S.S.R.*, 306, 115, 1989; *Chem. Abstr.*, 111, 208064s, 1989.

161. **Ovcharenko, V. I., Vostrikova, K. E., Ikorskii, V. N., Larionov, S. V., and Sagdeev, R. Z.**, Low-temperature ferromagnetics bis[2,2,5,5-tetramethyl-1-oxyl-3-imidazoline-4-(3',3',3'-trifluoromethyl-1'-propenyl-2'oxyato)]cobalt(II) dimethanol solvate, $CoL_2(CH_3OH)_2$, *Dokl. Akad. Nauk S.S.S.R.*, 306, 660, 1989; *Chem. Abstr.*, 111, 165872j, 1989.

162. **Forrester, A. R., Hepburn, S. P., Dunlop, R. S., and Mills, H. H.**, *t*-Butylferrocenylnitroxide, a stable ferrocenyl radical, *J. Chem. Soc., Chem. Commun.* (D), 698, 1969.

163. **Owtcharenko, W. I., Huber, W., and Schwarzhans, K. E.**, Synthese von 4,4,5,5-tetramethyl-3-oxide-2-ferrocenyl-imidazoline-1-oxyl, *Z. Naturforsch.*, 41B, 1587, 1986.

164. **Owtcharenko, W. I., Huber, W., and Schwarzhans, K. E.**, Synthese eines Ferrocen-Spinlabels, *Monash. Chem.*, 118, 955, 1987.

165. **Ovcharenko, V. I. and Podoplelov, A. V.**, Products of ferrocenaldehyde with hydroxylamine derivatives, *Metalloorg. Khim.*, 2, 1136, 1989; *Chem. Abstr.*, 112, 235541x, 1990.

Chapter 5

NEW APPROACHES TO THE GENERATION OF HETEROCYCLIC NITROXIDES USING ORGANOMETAL REAGENTS

Chapters 2 and 3 discussed the addition reactions of reactive short-lived radicals and organometal reagents to the nitrone group, including that in heterocycle, as a method for generating a nitroxyl radical center. The use of this approach essentially expands the possibilities of synthetic chemistry of nitroxides and permits synthesis of not only new heterocyclic nitroxides but also acyclic nitroxides with a nontraditional environment of the radical center. This section overviews recent data obtained by the authors on methods for generating some heterocyclic mono- and biradicals and acyclic radicals using the addition reaction of organometal compounds at the nitrone group in heterocycle.

Thus, extremely sterically hindered 3-imidazoline nitroxides **1** have been obtained by the reaction of 2H-imidazole-1-oxide **2** with methyl- and phenyllithium and subsequent oxidation of intermediate hydroxylamines **3**.[1]

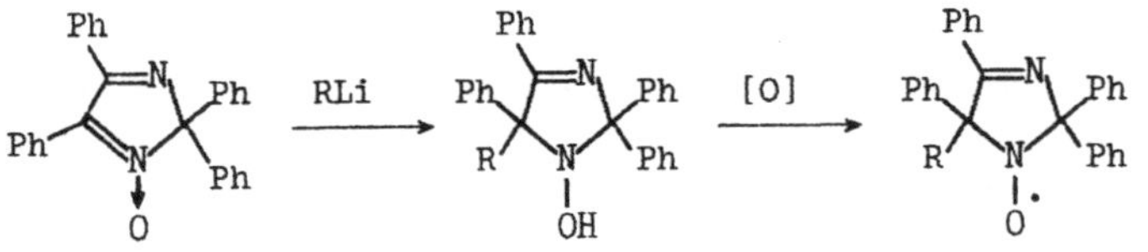

SCHEME 1.

Upon the reaction of 4H-imidazole-3-oxides **4** with phenyllithium excess, only one mole of the reagent adds predominately at the phenylnitrone group. Subsequent oxidation with MnO_2 leads to 3-imidazoline nitroxides **5**. This method makes it possible to introduce into the 2-position of heterocycle substituents which may not be introduced at the stage of 3-imidazoline heterocycle construction. This is exemplified by compounds **5** ($R^2 = ortho$-HO-C_6H_4), which are of interest as paramagnetic chelating reagents (compare to Reference 2). Thus, the reaction of 4H-imidazole-3-oxides with organolithium compounds is a convenient new method for the synthesis of 3-imidazoline nitroxides, which is similar to the method used to prepare pyrrolidine nitroxides (3.III.C.2).[1]

The interaction of 4H-imidazole-3-oxides **4** with phenyllithium generally leads to imino nitroxides **6** being the products of addition at the phenylimino group ($R^1 = C_6H_5$), in small yields. This pathway is favored, possibly, because the imino group is conjugated with the nitrone group (compare to Reference 3). The yields of imino nitroxides formed in this reaction are essentially lower than of 3-imidazoline nitroxides **5**.[1] In contrast, the reaction of 4H-imidazole-3-oxides **4** with $R^1 = $ 2-

pyrrolyl, with methyllithium proceeds exclusively as the addition at the imino group, forming imino nitroxides **6**, which are of interest as paramagnetic ligands.[20]

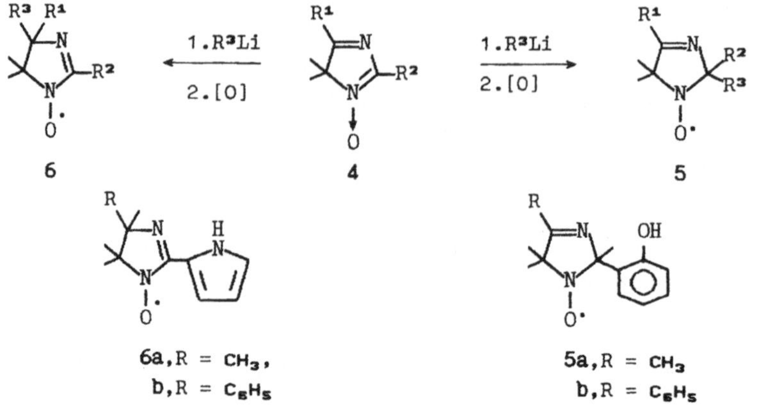

6a, R = CH₃,
b, R = C₆H₅

5a, R = CH₃
b, R = C₆H₅

SCHEME 2.

The reaction of 4*H*-imidazole dioxide **7** (R = CH₃) with phenyllithium excess leads to the addition of only one mole of reagent, forming after oxidation the nitronyl nitroxide **8**. This reaction pathway seems to be due to metallation of the methylnitrone group which hinders the addition of phenyllithium at it (compare to Reference 6). It was unexpectedly found that 3-imidazoline-3-oxide **9**, the product of addition of PhLi at the methylnitrone group, may be isolated, and its oxidation by MnO₂ leads to nitroxide **10**.[1]

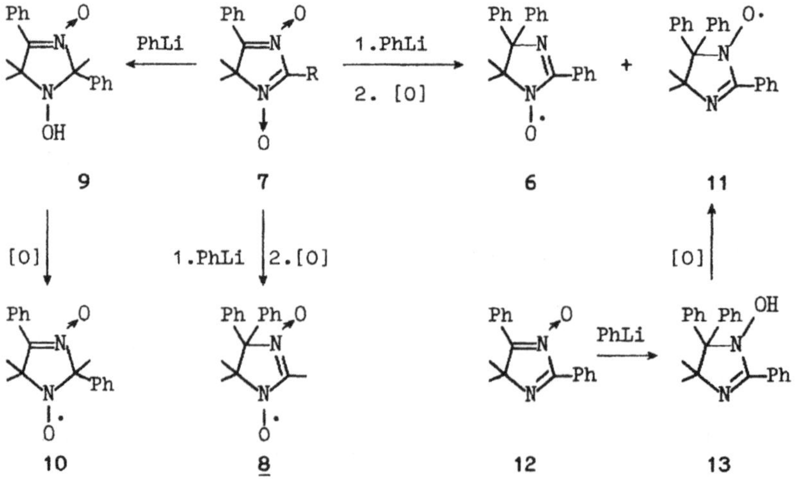

SCHEME 3.

In the reaction of 4*H*-imidazole dioxide **7** (R = C_6H_5) containing two phenylnitrone groups with an equimolar amount of phenyllithium, two isomeric imino nitroxides **6** and **11** are formed. Oxygen position in heterocycle was determined by the alternative synthesis of nitroxide **11** by the reaction of 4*H*-imidazole-1-oxide **12** (R = C_6H_5) with phenyllithium excess (compare to Reference 5) and subsequent oxidation of the hydroxylamino derivative **13**, where compound **11** is formed in a high yield. It should be noted that phenyllithium fails to add at the C=N bond of the phenylimino group (compare to Reference 6).[1]

In the reaction of 4*H*-imidazole dioxide (R = C_6H_5) with phenyllythium excess, the addition occurs at both nitrone groups, but the dihydroxy derivative of imidazolidine **14** formed is unstable and gradually decomposes both in a solid state and when heated in organic solvents. When compound **14** oxidized with MnO_2, the corresponding biradical **15** could not be isolated.[12] It should be noted, that practically the only biradical with two nitroxyl groups in one heterocycle is known.[7]

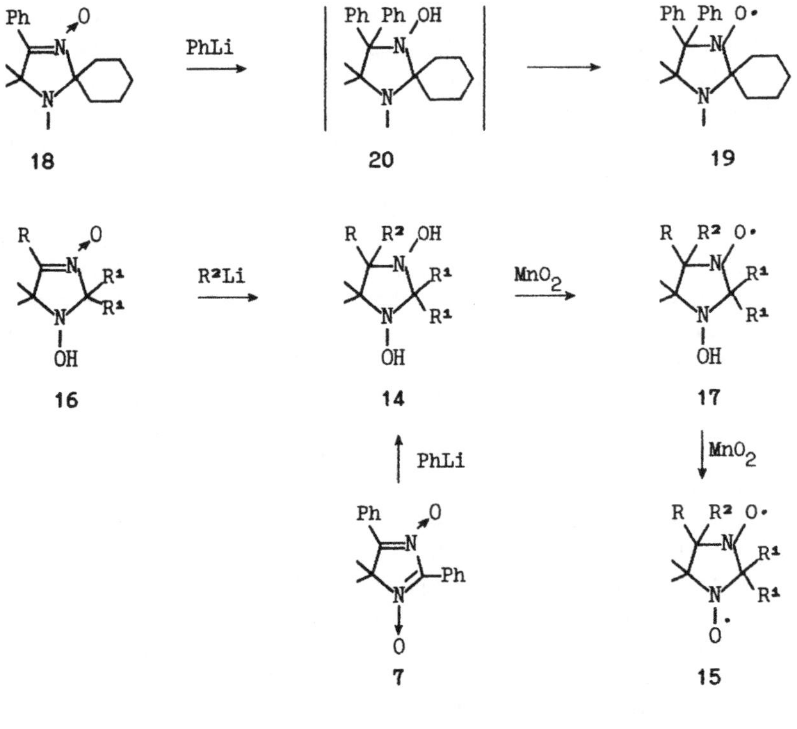

SCHEME 4.

Taking into account that one of the traditional methods for the generation of the nitroxyl group is the reaction of sterically hindered nitrones with organometallic compounds and subsequent oxidation,[8] the derivatives of 3-imidazoline-3-oxide **16** might be suggested to be potential precursors of various monocyclic imidazolidine biradicals. However, it is known that in the reactions of 3-imidazoline-3-oxides **16** with organomagnesium compounds, the imidazoline heterocycle undergoes cleavage to form acyclic α-hydroxylaminooximes (compare to Chapter 3, Scheme 19).[9] It has been suggested in Reference 10 that 1-hydroxy-3-imidazoline-3-oxides **16** should react similarly with organolithium compounds, and other 1-substituted derivatives have been proposed for use as dinitroxide precursors in the reactions with organometallic compounds instead of the 1-hydroxy derivatives **16**. Thus, the use of 1-methyl-substituted 3-imidazoline-3-oxides or the tetrahydropyranyl derivatives led to the dihydroxy derivatives of type **14**, but it was not possible to oxidize them into the corresponding biradicals **15**, and monoradicals **17** were obtained in small yields (compare to Chapter 3, Scheme 11).[10] Based on these data, the authors of Reference 10 have assumed that the necessary condition for stability of monocyclic biradicals is the presence of a spiro group in the 2-position of imidazolidine heterocycle (compare to Reference 7).

To verify this assumption, 3-imidazoline-3-oxide **18** was allowed to react with phenyllithium. Subsequent oxidation smoothly leads to the monoradical **19**. To obtain the corresponding biradical **15**, a sequence of transformations was performed including the acylation of the hydroxylamino derivative **20**, its oxidation, removing the acyl protection, and subsequent oxidation in mild conditions (compare to Reference 7). However, the heterocycle undergoes destruction at an early stage of oxidation of the N–CH$_3$ group to the nitroxyl group in H$_2$O$_2$/Na$_2$WO$_4$, and the only reaction product that was isolated was benzofenone.[12]

As opposed to reactions with organomagnesium reagents, the reactions with organolithium derivatives proceed without heterocycle cleavage, forming the products of addition at the nitrone group — dihydroxyimidazolidines **17**. The addition of butyllithium and especially phenyllithium occurs under mild conditions and gives high yields. However, the reaction of imidazoline **16** (R = C$_6$H$_5$, R^1 = CH$_3$) with methyllithium is rather slow and is incomplete even when the reagent is taken in a 10-fold excess. In the reaction of 4-alkyl-substituted 3-imidazoline-3-oxides **16** (R = alkyl) with phenyllithium, no addition products are formed, possibly, because of metallation at the alkylnitrone group (compare to References 4 and 11).

In the oxidation of the dihydroxy derivatives **14** with MnO$_2$ in ether the biradicals **15** (R^1 = CH$_3$, R = R^2 = C$_6$H$_5$; R = C$_6$H$_5$, R^2 = C$_4$H$_9$) were isolated. The biradicals **15** are bright red crystalline products stable in the solid state at 0°C over an unlimited period of time and unstable in solutions: at 20°C they virtually completely decompose after several hours. The biradical **15** (R^1 = R^2 = CH$_3$, R = C$_6$H$_5$) is somewhat less stable and may not be isolated in an analytically pure form. The biradical **15** with a spirocyclohexane group in the 2-position [R = R^2 = C$_6$H$_5$, 2R^1 = (CH$_2$)$_5$], which is a close analog of the defined monocyclic biradical (R = R^2 = CH$_3$),[7] was unexpectedly found to be unstable. An attempt to concentrate the solution or its storage for several minutes results in decomposition of the biradical with liberation of nitrogen oxides and formation of a complex mixture of products.[12]

Cleavage of 3-imidazoline-3-oxide heterocycle by organomagnesium reagents is attributed to possible existence of this compound as a mixture of two tautomeric forms, cyclic and acyclic, and addition of the organomagnesium reagent at the nitrone group of the acyclic form.[9] However, 4-phenyl-2,2,5,5-tetrasubstituted 3-imidazoline-3-oxides **16** react with organolithium compounds exclusively at the nitrone group of the cyclic form to form the products of addition at C-4. With hydrogen being one of the substituents in the 2-position of the imidazoline heterocycle, compounds **16** can actually exist in solutions as a mixture of two tautomers.[13,14] When compounds **16** which exist in solution as cyclic tautomers react with phenyllithium, the addition occurs at the nitrone group of heterocycle with intermediate formation of the dihydroxy derivatives of type **14**. In the case of compound **16** (R = 2-pyridyl), the water molecule is eliminated with intermediate formation of compound **13**. Further oxidation leads to the imino nitroxide **11**. In the oxidation of imidazolidines **14** (R = H), the water molecule is not eliminated and nitronyl nitroxides **8** are formed. Compound **8** (R = H) may not be isolated individually because of its low stability, but its further oxidation by potassium ferrocyanide in the presence of NaH leads to the stable nitronyl dinitroxide **21** (compare to Reference 15).

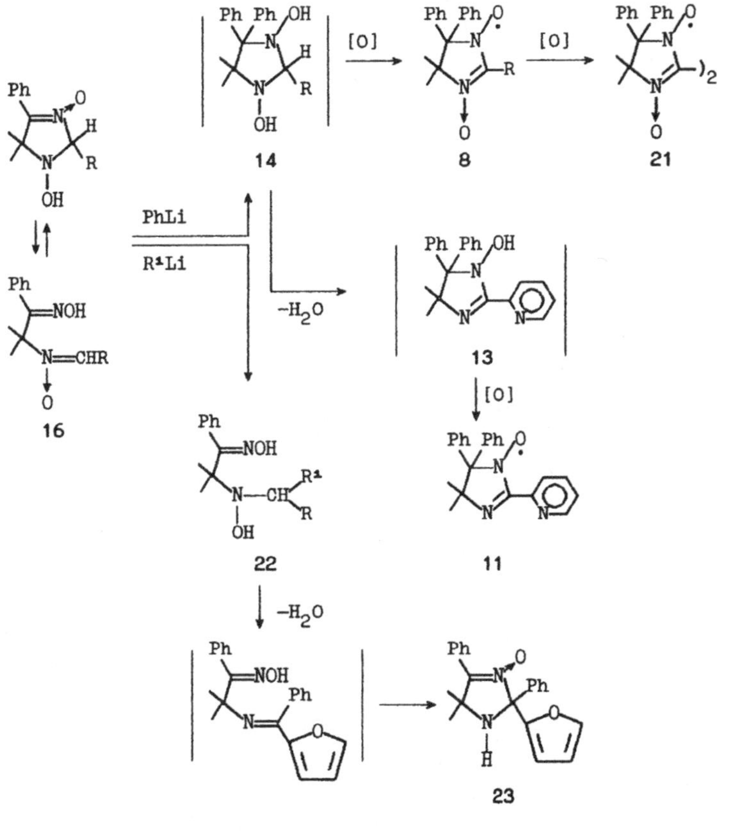

SCHEME 5.

Compounds **16** (R = C_6H_5, α-furyl) existing in solution predominately in an acyclic tautomeric form react with phenyllithium exactly in this form, forming hydroxylaminooximes **22** at the first stage. In the case of imidazoline **16** (R = α-furyl), resulting compound **22** is dehydrated; further cyclization leads to 3-imidazoline-3-oxide **23**.[19]

It was unexpectedly found that the product of oxidation of hydroxylaminooxime **22** by MnO_2 was not the 3-imidazoline-3-oxide derivative of type **16** but the stable acyclic nitroxide **24** containing hydrogen atom at the α-carbon atom of the nitroxyl group (it was formed in a quantitative yield). This compound was isolated individually and proved to be stable in the crystalline state at 0°C for an unlimited period of time. Probably, a sufficient condition of stability of such nitroxides is the presence of two phenyl groups bounded with a carbon atom bearing the hydrogen atom. Other stable nitroxides with an α-hydrogen atom have also been synthesized and isolated in a similar way. Moreover, the product of interaction of phenyl-*N*-*tert*-butylnitrone with phenyllithium, hydroxylamine **25**, also forms stable nitroxide **26** upon oxidation. This nitroxide is well known as a spin adduct and was even synthesized for use as a spin probe,[16,17] but was not isolated and characterized individually. It should be noted that the radicals **24** and **26** do not undergo disproportionation to form the corresponding hydroxylamine and nitrone and are not transformed into nitrones upon oxidation by MnO_2.[20]

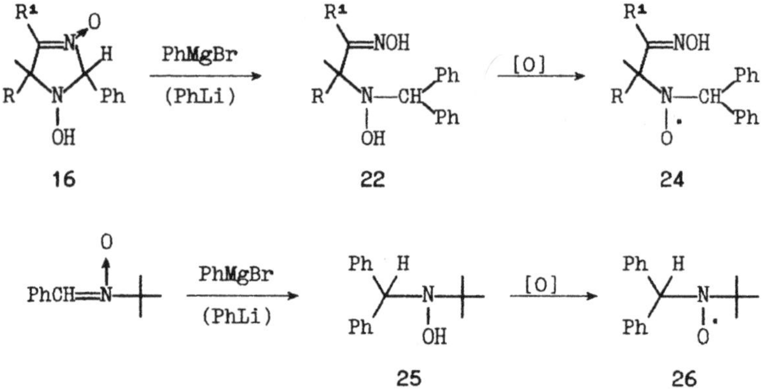

SCHEME 6.

An analogous approach has been used for the synthesis of pyrazine mono- and dinitroxides.[21] The comparatively low stability of monocyclic imidazoline biradicals might be attributed to the 1,3-position of nitroxyl groups in the heterocycle. Therefore, piperazine dinitroxides with a 1,4-position of nitroxyl groups might prove to be more stable and thus form a new class of monocyclic biradicals.

Pyrazines **27, 28** do not react with methylmagnesium iodide and phenylmagnesium bromide. Even with a 10-fold excess of methyllithium, the addition occurs only at one nitrone group of pyrazine **27**, and subsequent oxidation leads to the monoradical **29** (R = CH$_3$). When pyrazine **27** reacts with phenyllithium excess, smooth addition takes place at both nitrone groups to form dihydroxy derivative **30**. In the reaction of a 1.5-fold excess of phenyllithium with pyrazine **27** with subsequent oxidation, the monoradical **29** is formed (R = C$_6$H$_5$). In a similar way, excess butyllithium reacts with pyrazine **27** to yield a mixture of isomeric dihydroxypyrazines **29**.

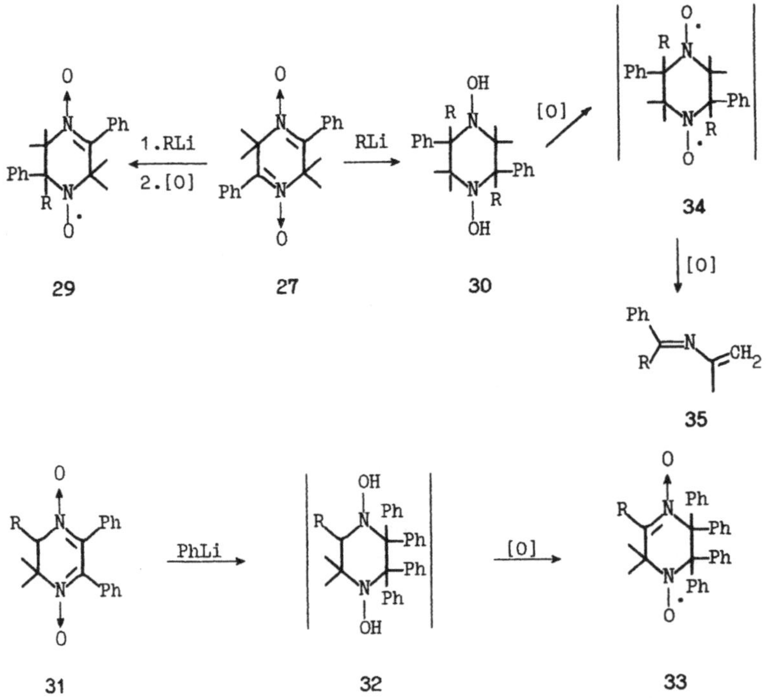

SCHEME 7.

The reaction of 2,3-dihydropyrazine-1,4-dioxides **31** with phenyllithium excess also proceeds at both nitrone groups. After oxidation of initially formed dihydroxy derivatives **32**, nitroxyl monoradicals **33** have been isolated.

In the oxidation of the dihydroxy derivatives **30** by MnO$_2$, the biradicals **34** were not isolated, possibly, because of their low stability; the reaction products were α,β-unsaturated imines **35**.

Contrary to pyrazines **27**, **30**, the pyrazine **28** molecule contains methylnitrone groups which does not form adducts in the reaction with phenyl- or butyllithium. This is, obviously, due to metallation reaction of the methylnitrone group (compare to Reference 4). After treatment of pyrazine **28** with phenyllithium and ethylbenzoate in sequence, the products of mono- **36** and diacylation **37** are formed.

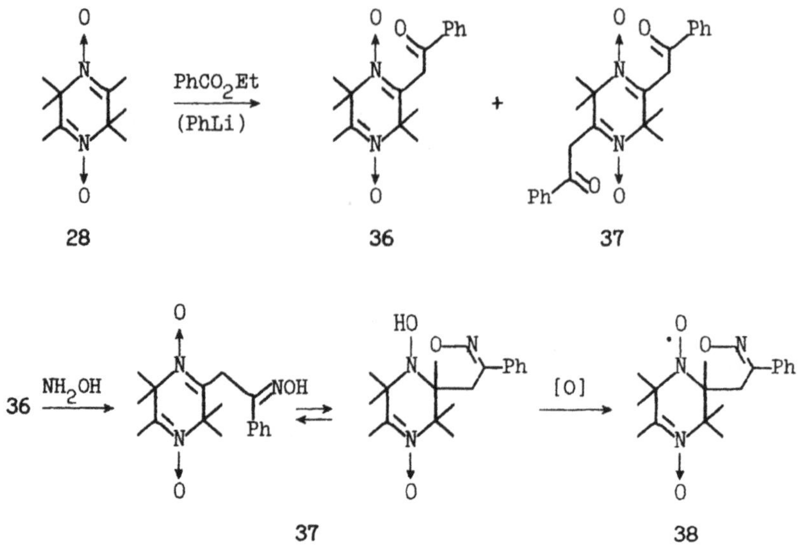

SCHEME 8.

Compounds **36** and **37** exist in solution as a mixture of enolized and nonenolized tautomeric forms (compare to Reference 18). Each of the enolized forms seems to be represented by enehydroxylaminoketone and enolnitrone forms. When pyrazine **36** reacts with hydroxylamine, the spirobicyclic compound **38** is formed which is readily oxidized by MnO_2 to form the stable nitroxide **39** (compare to Reference 11).[21]

REFERENCES

1. **Reznikov, V. A. and Volodarskii, L. B.,** Reaction of 2*H*(4*H*)-imidazole oxides with lithium organic compounds — a new route to stable nitroxides of 2(3)-imidazoline series, *Izv. Akad. Nauk Russ. Ser. Khim.*, in press.

2. **Larionov, S. V.,** Imidazoline nitroxides in coordination chemistry, in *Imidazoline Nitroxides*, Vol. 2, Volodarsky, L. B., Ed., Boca Raton, FL, CRC Press, 1988, chap. 3.

3. **Kobrin, V. S., Volodarskii, L. B., Tikhonova, L. A., and Putsykin, Yu. G.,** Isolation and properties of products of the covalent hydration of 4*H*-imidazoles, *Khim. Geterotskil. Soedin.*, 1087, 1973; *Chem. Abstr.*, 79, 137040P.

4. **Martin, V. V. and Volodarskii, L. B.,** Study of tautomerism and the chemical properties of β-oxonitrones, *Izv. Akad. Nauk S.S.S.R., Ser. Khim.*, 1336, 1980; *Chem. Abstr.*, 93, 204544k.

5. **Kobrin, V. S.,** Synthesis and Properties of 4*H*-Imidazol Derivatives, Ph.D. dissertation, Institution of Organic Chemistry, Novosibirsk, Russia, 1977.

6. **Laer, R. W.,** The chemistry of imines, *Chem. Rev.*, 63, 489, 1963.

7. **Keana, J. F. W., Norton, R. S., Morello, M., Van Engen, D., and Clardy, J.,** Mononitroxides and proximate dinitroxides derived by oxidation of 2,2,4,4,5,5-hexasubstituted imidazolidines. A new series of nitroxide and proximate dinitroxide spin labels, *J. Am. Chem. Soc.*, 100, 934, 1978.

8. **Keana, J. F. W.,** Synthesis and chemistry of nitroxide spin labels, in *Spin Labeling in Pharmacology*, Holtzman, J. L., Ed., Academic Press, Orlando, FL, 1984, chap. 1.

9. **Volodarskii, L. B., Martin, V. A., and Kobrin, V. S.,** Formation of oximes of *N*-alkylhydroxylamino ketones during the reaction of 1-hydroxy-3-imidazoline-3-oxides with Grignard reagent, *Zh. Org. Khim.*, 12, 2267, 1976; *Chem. Abstr.*, 86, 72520t.

10. **Martin, V. V., Volodarskii, L. B., and Vishnivetskaya, L. A.,** Interaction of sterically hindered imidazoline oxides which are precursors of stable nitroxyl radicals with organometallic reagents, *Izv. Sib. Otd. Akad. Nauk S.S.S.R., Ser. Khim, Nauk*, 4, 94, 1981; *Chem. Abstr.*, 96, 6648.

11. **Reznikov, V. A. and Volodarskii, L. B.,** Interaction of β-oxo nitrones of imidazoline and pyrrolyne with nucleophilic reagents, *Khim. Geterotsikl. Soedin.*, 912, 1991.

12. **Reznikov, V. A., Volodarskii, L. B., Spoyalov, A. P., and Dikanov, S. A.,** Monocyclic biradicals of imidazolidine series, *Izv. Akad. Nauk Russ., Ser. Khim.*, in press.

13. **Volodarskii, L. B., Kobrin, V. S., and Putsykin, Yu. G.,** Preparation and alkaline cleavage of 4*H*-imidazole *N*-oxides, *Khim. Geterotsikl. Soedin.*, 1241, 1972; *Chem. Abstr.*, 78, 4188a.

14. **Kirilyuk, I. A., Grigor'ev, I. A., and Volodarskii, L. B.,** Synthesis of 3-imidazolines and 3-imidazoline 3-oxides containing a hydrogen atom at C-2, *Izv. Sib. Otd. Akad. Nauk S.S.S.R., Ser. Khim. Nauk.*, 2, 99, 1989; *Chem. Abstr.*, 111, 232670g.

15. **Ullman, E. F., Call, L., and Osiecki, J. H.,** Stable free radicals. VIII. New imino, amidino and carbamoyl nitroxides, *J. Org. Chem.*, 35, 3623, 1970.

16. **Janzen, E. G. and Haire, D. L.,** Two decades of spin trapping, in *Advances in Free Radical Chemistry*, Vol. 1, JAI Press, Connecticut, 1990.

17. **Kotage, Y. and Janzen, E. G.,** Bimodal inclusion of nitroxide radicals by β-cyclodextrin in water as studied by electron spin resonance and electron nuclear double resonance, *J. Am. Chem. Soc.*, 111, 2066, 1989.

18. **Reznikov, V. A. and Volodarskii, L. B.,** NMR spectra of heterocyclic nitrones. VI. About tautomeric equilibrium of β-oxonitrones — derivatives of 3-imidazoline-3-oxide, *Khim. Geterotsikl, Soedin.*, 192, 1991.

19. **Reznikov, V. A. and Volodarskii, L. B.,** Reaction of sterically-hindered 1-hydroxy-3-imidazoline-3-oxides with phenyl lithium, *Izv. Akad. Nauk Russ. Ser. Khim.*, in press.

20. **Reznikov, V. A.,** to be published.

21. **Reznikov, V. A. and Volodarskii, L. B.,** Interaction of dihydropirazine-1,4-dioxides with lithium organic compounds. Synthesis of nitroxides derivatives of tetrahydropyrazine oxides, *Izv. Akad. Nauk Russ., Ser. Khim.*, in press.

INDEX

T - #0530 - 101024 - C0 - 234/156/13 - PB - 9781138562004 - Gloss Lamination